AF679780

Forests and Fires: A Paleoethnobotanical Assessment of Craft Production Sustainability on the Peruvian North Coast (950-1050 C.E.)

David John Goldstein

BAR International Series 2318
2011

Published in 2016 by
BAR Publishing, Oxford

BAR International Series 2318

Forests and Fires: A Paleoethnobotanical Assessment of Craft Production Sustainability on the Peruvian North Coast (950-1050 C.E.)

ISBN 978 1 4073 0901 9

© D J Goldstein and the Publisher 2011

The author's moral rights under the 1988 UK Copyright, Designs and Patents Act are hereby expressly asserted.

All rights reserved. No part of this work may be copied, reproduced, stored, sold, distributed, scanned, saved in any form of digital format or transmitted in any form digitally, without the written permission of the Publisher.

BAR Publishing is the trading name of British Archaeological Reports (Oxford) Ltd. British Archaeological Reports was first incorporated in 1974 to publish the BAR Series, International and British. In 1992 Hadrian Books Ltd became part of the BAR group. This volume was originally published by Archaeopress in conjunction with British Archaeological Reports (Oxford) Ltd / Hadrian Books Ltd, the Series principal publisher, in 2011. This present volume is published by BAR Publishing, 2016.

Printed in England

BAR
PUBLISHING

BAR titles are available from:

BAR Publishing
122 Banbury Rd, Oxford, OX2 7BP, UK
EMAIL info@barpublishing.com
PHONE +44 (0)1865 310431
FAX +44 (0)1865 316916
www.barpublishing.com

TABLE OF CONTENTS

List of Figures

List of Tables

ABSTRACT

During the Middle Sicán period (C.E. 950-1050) on the North Coast of Peru, artisans developed a sophisticated tradition of ceramic and metalworking production amidst dry coastal forests of the region. Organic fuel resources, specifically wood, clearly played a vital role in the manufacture of these objects; however, this component of production has been largely overlooked. Thus, a major gap in our understanding of the relationship between Sicán period production and the local landscape has developed. The Sicán Archaeological Project (SAP) suggests that the production of metal and ceramics during this period likely placed the local fuel resources under considerable stress. Yet, an evaluation of the archaeological data is essential to assess the degree of overexploitation, identifying the fuels used, their contexts for use, and their role in local ecology.

This study interprets how Middle Sicán artisans met their fuelwood requirements for production in light of easily endangered forest resources. An examination of the archaeological charcoal from Middle Sicán period kilns, hearths, and metal furnaces permits the reconstruction of fuel use and the ecological setting of production. This unique site demonstrates the concurrent production of metal and ceramics, as well as the presence of domestic activity. Using wood anatomy of fuels recovered from archaeological features, I identified the fuel materials of different use contexts. Modern comparative collections of wood and other organic materials were used to determine the plants present and interpret the ecology of the fuel remains. With these data, I reconstruct ancient artisans' fuel selection for cooking, firing ceramics, and working metal. At the same time, I examine fuel quality and generate a picture of the forest from which these materials were extracted.

The research presented here combines species determinations, an assessment of ecological and morphological variation of the representative species, and an understanding of forest composition and structure to maximize the interpretive potential for the Huaca Sialupe botanical remains. These analyses yield a thorough determination of fuels used and fuel quality; they also qualify their contexts for use and allow their inclusion in craft production models. This approach has been underrepresented in the region's archaeological research. This work is line with other recent research in ancient charcoal and fuel use that brings closer towards developing a global application of this methodology to integrate the study of organic resources, e.g. fuel wood, within broader research models of ancient and modern craft production.

CHAPTER 1
INTRODUCTION TO THE PROBLEM

Development of the Research

The social and natural dimensions of craft production have been widely ignored in archaeological investigations in the New World as a whole; likewise, cross-disciplinary examination of fuels (their identity, conception, management, etc…) has been woefully inadequate (Arnold 1985; Johannessen and Hastorf 1990; Lentz 1999; Lopinot and Woods 1993; Moorehart et al. 2003; Sillar 2000; Pearsall 1983; Winterhalder et al. 1974). Rarely are ecological concepts and the artisan's perception and intuition of fuel resources approached on a sophisticated, analytical level. As a result, fuel, the 'energetic basis of craft production,' is discussed in a manner that often separates local ecology from resource use. Resources are only available when they are perceived as viable by the artisan and when they are not overtaxed; archaeological discussions of craft production, however, often lack this critical link between ecological and cultural perceptions of the environment. The lack of sophistication in our analyses reduces our reconstructions of craft production to a bare skeleton and negates any sense of holism in anthropological investigations of artisans.

The past years of sustained research by the Sicán Archaeological Project (SAP) in the Lambayeque and La Leche valleys of north coastal Peru have combined archaeological fieldwork (Shimada 1981; Shimada 1995; Shimada 1997c) with experimental archaeology (Goldstein et al. 2007, Shimada 1994b, Shimada and Wagner 2001) and ethnoarchaeology (pottery production studies in the Morropé and the Chulucanas regions; e.g., Shimada 1997c; Cleland and Shimada 1998) to reconstruct the social, technological, and ecological contexts for metal and ceramic production of the Sicán period (C.E. 800-1375; Cleland and Shimada 1992; Shimada 1985; Shimada et al. 1999; Shimada and Merkel 1991; Shimada and Wagner 2001; Shimada and Wagner 2007; Wagner et al. 1998). The products, copper alloy metalwork, including paper-thin *tumbaga* (gold, silver, and copper alloy) sheet, and blackware bottles were both important vehicles for displaying and distributing the symbolic representations of the 'Sicán deity' (Cleland and Shimada 1998; Shimada and Wagner 2001). This symbol is central to Cleland and Shimada's (1998) 'aesthetic locus' within Sicán culture. These products stand in stark contrast to the standard or common material culture of the Middle Sicán that consisted of a metal tool kit fabricated from copper-arsenic alloys, and stamp impression, *paleteada*, or undecorated oxidized redware ceramics (Bezúr 2003; Shimada 1995; Cleland and Shimada 1998). The difference in the relative status assigned to these materials was derived from comparisons of royal upper elite tombs excavated at the site of Sicán in 1991-2, 1995-6, and 2006, to lower level elite and domestic settings at Sicán, Huaca de Pueblo Batán Grande in 1983 and Huaca Sialupe in 1999 and 2001. Huaca Sialupe was ostensibly one important site for the production of these artifacts that were the focus of the Sicán 'aesthetic locus' (Shimada 2000; Shimada and Montenegro 2002; Shimada and Wagner 2001).

Much of the SAP's research has focused on the social and symbolic dimensions of Middle Sicán ritual and status. At the center of this world was the image of the Sicán deity (Figure 1.1), a symbolic motif that appears at the end of the Middle Horizon Period on the North Coast of Peru (C.E. 800-900). This figure appears on blackware ceramics and tumbaga foil artifacts in the elite shaft tombs from the Middle Sicán period at the main site of Sicán. Cleland and Shimada (1998), based on Maquet (1979), argue that the deity was the standard bearer of religious/political identity, and one focal point of the 'aesthetic locus,' the technological and aesthetic apogee of the Middle Sicán. In particular, the highest expression of the deity's representation was on specially engineered gold-copper alloy sheet metal, fashioned into the regalia used in elite mortuary practice. In effect, these alloys were composed specifically to produce materials that were both malleable and stable enough to make costumes, e.g., masks, gloves, and clothing, and transform elite personages into a physical representation of the deity on Earth (Cleland and Shimada 1998). Alternately, continued work in the area by the SAP demonstrates that the image of the elite regalia became a symbolic representation of power and cosmology itself, and these symbols were then added to other technological media, e.g., ceramic blackwares and textiles (Cleland and Shimada 1998). Hence, the image of the Sicán deity permeates the material culture of the Sicán period. Yet, the different media on which the deity appeared was restricted to the highest technological and aesthetic achievements of the period.

As a result, the deity's representation is clearly tied to artifacts that are the cultural capital of Middle Sicán elites. Fortunately, sustained research in the area has located production facilities for these artifacts, one of which appears to have been the site of Huaca Sialupe, the topic of this dissertation. More recently, work has begun to fill in the data gap that exists surrounding the technological, utilitarian, and political economic dynamics of the Sicán metallurgical and ceramic production. The research program from 1999-2001 at Huaca Sialupe on the north coast of Peru provides new insight into the technology and organization of craft production that spanned an estimated 100 years during Middle Sicán period (C.E. 900-1100; Figure 1.1; Goldstein and Shimada i.p.; Goldstein et al. 2007; Klaus 2003; Shimada et al. 2004; Shimada and Wagner 2001; Shimada and Wagner 2007; Taylor 2002).

Our excavations (Shimada and Montenegro 2002; Shimada 2000) and experimental work (Shimada and Wagner 2001) focused on the production debris from the multi-craft workshop. This research has demonstrated that consistent supplies of organic fuels, consisting of charcoal, wood, animal dung, and other dried plant materials, largely provided the foundation for ceramic and metallurgy industries (Shimada 1998; Shimada et al. 1999; Shimada and Merkel 1991;

Figure 1.1 Sicán Deity as Represented in the Modern Day

Shimada and Wagner 2001). No studies at Huaca Sialupe, however, have examined fuel resource use as the key energetic input, the efficient use of which is critical to both industries Experimental research has suggested that fuel may have been a limiting resource to production due to the ecological constraints of dry tropical forest regeneration, the forests that are the dominant fuel source for the La Leche River Valley (Shimada and Merkel 1991). In step with this more holistic approach to understanding the organization of craft production and resource use by craft producers, my research investigates the social, economic, and ecological dynamics associated with the production of these 'symbolically charged' artifacts.

This dissertation analyzes the archaeobotanical remains from the site with particular attention paid to fuel resources to provide a basis for examining issues of labor organization and resource management. Specifically, I employ a paleoethnobotanical approach to understand fuel use by Middle Sicán metalsmiths and potters who shared the workshop space 1000 years ago. This approach adds to the extensive literature the SAP has generated that focuses on the inorganic resources for metal (Shimada 1994b; Shimada and Merkel 1991; Shimada et al. 1999) and ceramic production (Cleland and Shimada 1992; Shimada 1997c; Shimada and Wagner 2001; Wagner et al. 1998). A combination of ethnobotanical fieldwork, ethnohistorical and archaeological research, and the analysis of archaeological plant remains permit the reconstruction of fuel use and resource apportioning at Sialupe in the Middle Sicán. The multi-disciplinary nature of paleoethnobotanical research combines the sub-disciplines of anthropology with other natural and social sciences, allying it particularly well with the overall holistic approach of the SAP (Shimada and Wagner 2001; Shimada and Wagner 2007). From the platform of this dissertation I wish to address the following issues:

1. 1.From looking at the organic remains from fuel-use areas, what can we discern about how labor was organized around these two industries? This question can be addressed in two parts:

    - To what extent can we reconstruct a holistic understanding of organic resource use across the site?

    - To what extent did ceramic production and metalworking complement or compete with each other in regard to the critical fuel supply and use?

2. What can we discern from the botanical remains regarding the local ecology and did craft production in the Middle Sicán impact local forests?

It is from the frame of paleoethnobotany that I attempt to answer these questions and add data-backed reconstructions of the ecology, organization and technology of Middle Sicán craft production; in particular, to those crafting traditions that impacted or contributed to the transmission of cosmological and ideological information through the deployment of politico/religious symbolism on ceramic and metal media.

Research Issues: Crafting and Resource Perception

During its estimated 100-year existence, the workshop at Huaca Sialupe manufactured a variety of products in at least two media: ceramic and metal. In ceramic production, artisans relied on molds and small, manageable kilns to produce both oxidized and reduced fine wares. Metalsmiths in adjacent workspaces fabricated, among other activities, thin sheets of arsenical copper and gold alloys using simple but efficient furnaces. Metalsmiths cut, shaped and/or joined together the sheet metal to produce predominantly small utilitarian objects and ornaments. The reconstruction of the on-site activities is based on excavation of the site in 1999 and 2001, and the experimental reconstruction of the kilns and furnaces, described below, during July 2000.

This research draws on established notions of ancient craft production and of crafters as the interface between culture and nature demonstrate that interpreting fuel use is germane to understanding the social and ecological realities of ancient production. Scholars such as Ingold (2000) have proposed that technical skill and the manufacture of artifacts are best understood where artifacts are the direct results of human interface with their environments, providing an effective medium for reconstructing human and environment interactions. Ingold's (2000) ideas complement Lemonnier's *Elements for an Anthropology of Technology* (1992) concepts and suggest that there are cultural implications to resource selection and the technological composition of any artifact. Together, these approaches argue that reconstructing ancient technologies and interpreting the decisions made throughout the crafting process, enable the anthropologist to understand how artisans view natural resources and how decisions are brokered regarding resource allocation. The archaeobotanical record provides a unique dataset for approaching craft production at Sicán from this type of agent-oriented perspective, enabling the reconstruction of how ecological resources were apportioned at the site.

Reconstructing fuel resource use allows me to see where certain fuels were required for certain technological activities, and how more general fuel use intersected and with the main elements of production at the site. From this point of view I can interpret how resources are directed around a site. This perspective lends me insight into how the site was organized economically and socially. From the SAP's perspective, as certain manufactured goods are valued over others it may be that certain fuel resources were valued over others at the production site. Under this condition I ask if the same values of the goods carry over into how resources are valued? If they do, then overarching political and religious considerations directly influence craft production. If not, then craft production is more autonomous at Huaca Sialupe, presenting the opportunity to reconstruct the relationship between the craftspeople and their resources locally at the site. This idea has been put forward by many of my colleagues based on different data sources (Klaus 2003; Shimada and Wagner 2007; Taylor 2002). Having flexibility in resource selection and to generalize resources to serve their production needs would indicate that the overall domestic economic activities at Huaca Sialupe may have been weighted equally or more substantially relative to meeting production demands of high status goods, potentially under elite control.

In my research I see fuel remains as representatives of the resources perceived as 'meaningful' by ancient artisans. They are items that have been viewed, valued, selected, and brought to bear on a certain need (Graber 2001) This notion permits me to reconstruct meaning in the archaeobotanical assemblage in different possible ways. First, if they were exploiting a specific resource that was somehow prized or given primacy based on cultural assessments of fuel quality, then their relationship to the environment was largely mediated through the constraints of the technology that they primarily relied upon. If, however, they are exploiting a range of resources, even though there are better resources available, then they are conceiving of their environment in a way that's more mediated through other concerns, e.g., sustainability, combined production of artifacts and food, overlapping resource concerns. In the end, the conclusion of my research lets me reconstruct how resources were used, and perhaps how the artisans working at Huaca Sialupe in the Middle Sicán perceived them.

Fortunately, this approach dovetails well with Shimada and Wagner's (2001, 2007; Shimada 1985; Shimada 1994b) framework for holistic interpretations of regional craft production, the main research paradigm employed by the SAP. Their holistic method is a five-fold approach to integrating long term regional archaeological fieldwork with experimental, archaeometric, and ethnoarchaeological studies of craft production primarily through excavations of workshops. First, their goal is to spend long periods of time studying cultural developments regionally to develop a context within which material elements of a society can be more fully understood. Second is the location and sustained excavation at production sites to understand how artifacts are produced. Third, is the collaboration with materials researchers in archaeometry to develop field and laboratory methods that best elicit information about artifact production. Fourth, are experiments that illustrate the production of certain elements of material culture. Fifth is the consecutive testing of their products using the same archaeometric means to reconstruct products, technologies, and understand the material realities within the archaeological context of these goods. All of these steps seek to not only recognize important elements of an ancient society's material culture, but understand their context in terms of production, consumption, and reproduction of these goods (Shimada and Wagner 2007). The approach is similar to that also proposed by Costin (2000). The SAP, however, includes long-term research with intensive on-site and laboratory collaboration between materials specialists and anthropologists. In the instance of this study, paleoethnobotany and anatomical determination are used to quantify and qualify resources used in production, and these are described against ethnobotanical and ecological information acquired from the region of study. From a theoretical standpoint, I couple ecological and ethnobotanical understanding of the environment with the determination and interpretation of archaeobotanical data to reconstruct resource apportioning at the site, and posit how the local crafting economy and ecology operated in the Middle Sicán.

This research also follows the material and cultural historical approach of Matson (1966; 1989: 23-24), who argues that pyrotechnologies used in ceramic and metal production developed through a series of synergistic processes: metal furnaces and ceramic kilns developed parallel to one another technologically, with ceramic technology serving as a backbone for the development of subsequent pyrotechnologies. Due to similar concerns over fuel exploitation, these associated industries, metal and ceramic production, employed complementary fuel use strategies. At Huaca Sialupe we have exactly this situation where ceramic and metalwork are taking place in the same workshop. As a result, as Matson predicts, we can reconstruct, through looking directly at the fuel remains, how this particular system of resource sharing and apportioning functioned for Middle Sicán craftspersons.

In the Middle Sicán, we know that the certain ceramic and metal products were used in particular contexts, e.g., elite burials vs. domestic activity, and likely had differential value in the Middle Sicán (Bezúr 2003; Cleland and Shimada 1998; Shimada 1985, Shimada 1995). Accordingly, various SAP researchers have argued that resource apportioning between these industries should have a signature in the archaeological record that testifies specifically to these distinct social lives, and previous work in Sicán period contexts indicates that this may have been the case (Cleland and Shimada 1998, Goldstein and Shimada i.p., Goldstein et al. 2007, Shimada and Merkel 1991, Shimada 1994a,b). For instance, in Shimada's (1985) article looking at Middle Sicán resource procurement and consumption, he characterizes the mundane nature of arsenical-copper alloy tools and their wide distribution across status boundaries in burials. This situation contrasts heavily with the restricted use of *tumbaga* gold-silver-copper alloy artifacts recovered only from elite royal shaft tombs at the Sicán site core (Shimada 1995). These high quality items were surely visible status markers when these regalia were deployed ceremonially, and the materials themselves must have been ostensibly understood to carry the charge of status in very particular political and ceremonial situations.

Well-polished and black ceramic bottles bearing the image of the Sicán deity, the principal symbol of the Sicán religion, were a signature craft product for middle and high status burials of the culture (Shimada 1995). These items were clearly valued, exchanged, and even imitated in form and finish well outside of the Sicán territory (Shimada 1985; Shimada and Wagner 2001; Shimada and Wagner 2007). The large-scale production of true black pottery is one of the major Sicán legacies. The Middle Sicán artistic and technical standard bearers, however, were metal objects, particularly those of precious metals available only to the social elite (e.g., Cleland and Shimada 1998; Shimada 1994a,b). Shimada argues that black pottery, in essence, was a simplified version in reduced size, portable, and reproducible form of precious metal vessels in form, color and iconographic contents (Shimada 1995). These media and technologies in conjunction with the appearance of the Sicán deity reinforce the idea of the 'aesthetic locus' for Sicán society, and demonstrate the potential importance of the goods produced at Sialupe across a range of social strata.

The unique opportunity with the archaeobotanical data from Huaca Sialupe's production contexts is to use the plant-based ecological formation of these resources to connect the ancient environment to the cultural sphere of production. Matson's (1966) observations about the importance of fuel apportioning, combined with Ingold (2000) and Lemonnier's (1992) observations that artifacts are the products of culturally mediated value judgments about resources, demand that we look directly at the resources used in artifact production. As the majority of the SAP's previous work concentrated on the accurate reconstruction of craft technologies, looking directly at how fuels and non-fuel resources were consumed at Huaca Sialupe opens a new line of evidence for holistically reconstructing craft production by linking ecological and cultural resources with technological reconstruction.

Research Issues: Ecology

Even today only few discussions of human-environment interactions accept that humans are an important part of discussions of ecological reconstruction in Neotropical archaeological work (Crumley 1994; Balée and Erickson 2006). Smith (2007) has taken this perspective recently with regard to Australian indigenous peoples demonstrating that modifications to the landscape, e.g., canals and terraces, or remains of their extractive processes, e.g., mining, are part and parcel of human niche construction, just like our fellow ecological community members. This kind of perspective is lacking in much our archaeological literature. Some attempts in the lowland Neotropics have accepted that much of the diversity and soil development in the Amazon basin was due to these kinds of niche construction processes (Balée and Erickson 2006). Yet, descriptions of human action in the past as ongoing and continual aspects of ecological give-and-take remain rare. Given the higher social value metalwork held relative to pottery and the diverse role metals played in the Middle Sicán, some scholars have suggested that metal production received preferential material and labor support from the social elite (e.g., Cleland and Shimada 1998; Shimada 1994a,b). More specifically, the SAP team has argued that Sicán metalworkers had greater access than potters to supplies of high-quality hardwood fuels, a potentially limited resource in the region's ecologically specialized dry tropical forests (Proyecto Algarrobo 1997; Shimada and Wagner 2007). This hypothesis is contrary to Matson (1966) who argues that in any multi-crafting situation, which includes ceramic production, ceramics are always the backbone element of production. Through the analysis of charcoal from production contexts, this study revisits these claims from a paleoethnobotanical approach and reconsiders issues of multi-craft interaction at Huaca Sialupe.

The paleoethnobotanical approach necessitates that cultural, historical, and ecological research are correlated for the Middle Sicán (C.E. 900-1100). To bring these three fields of data into a comparative framework required that I collect ecological and ethnobotanical information about my survey area, sorted and analyzed archaeological material from solid and sealed archaeological contexts, and established a historical understanding of the research area through looking at published materials on the Middle Sicán and Andean archaeology and ethnography. To fulfill the first two portions of this agenda, Shimada, Newsom, and Goldstein in 2002 secured a National Science Foundation Dissertation Improvement Grant to examine the local ecology of the La Leche Valley (Figure 1.2). Field reconnaissance of the site region familiarized me with the local environment, plants, and climate, as well as enabled me to sample relevant materials to construct a modern comparative collection from which archaeological plant remains could be identified.

This portion of the research follows in the tradition of PEB studies that rely on ancient charcoal to aid in archaeological reconstructions of chronology, technology, and paleoenvironment (Fritts 1976; Newsom 1993; Schweingruber 1996; Stahle 1996; Smart and Hoffman 1988; Thompson 1994). I aimed, additionally, to generate a data set that offers a fine grained perspective of the regional flora used during the Middle Sicán that complements the climatological, geomorphological, and material record previously executed by the SAP and others (Craig and Shimada 1986; Delvaud 1984; Hocquenghem 1998; Shimada 1994a; Shimada et al. 1991).

At present, the local forests adjacent to the research area are non-extant: over 3,000 years of cultivation, shifting desert boundaries, and changing climate have effectively changed the established pre-human/Holocene ecology of disturbance and resilience in the area (Chapter 2). Presently, no non-human induced successional processes are visible, and we see no stands of primary, equilibrium, or 'virgin' forest in the area. Fortunately, the Peruvian government has recently established the area around the main precinct of the Middle Sicán period site of Sicán, closer to modern Batán Grande than Huaca Sialupe, as the National Historical Sanctuary, Bosque de Pomác. This area is under the supervision of two Peruvian governmental institutions, the National Institute of Culture (INC) and the National Institute for Natural Resources, Ministry of Agriculture (INRENA). While the sanctuary is about 22 Km from the actual site location of Huaca Sialupe, this portion of the final cataract of the La Leche river valley, is at approximately the same altitude as Sialupe, and offers, in contrast to the surrounds of Sialupe itself, areas adjacent to the park where the study of forest and farming ecological interactions can be suitably studied. It comprises 150,000 Km^2 of protected dry tropical forest ecosystem, potentially the largest in the area and in South America (Proyecto Algarrobo 1997). Actual work within the sanctuary was not permitted for the purposes of this research; however, areas adjacent to the protected area with tree cover and standing forests were used for modern reconnaissance.

At present, the dry coastal forests of the La Leche River Valley are seen as scarce and endangered and threatened environments in danger of constant encroachment by local human habitation (Cuba Salerno et al. 1998; Proyecto Algarrobo 1997; Gentry 1995). While the pre-colonial industries of ceramic and metal production have long since disintegrated, with only some small-scale ceramic production occurring in discreet areas, the modern forces of overgrazing, farmland clearance, and charcoal production from the ecologically important hardwood species (*Prosopis juliflora-Prosopis pallida;* Pascieznik et al. 2001) of the dry forests, compound the human extraction disturbance on the forest ecology of even the most protected areas (Baca 1958; Arroyo Balderon 1997).

With these factors in mind, I aim to describe the ecological forces and their observable implications in the modern era. I use these observations as a narrative with which to explore the potential for ancient craftspersons, operating on the scale of organization and production seen at Huaca Sialupe on their environment at the end of the first millennium of the Common Era.

Research Issues: Paleoethnobotanical Interpretation

From a historical perspective, while scholars have developed paleoethnobotany over the past thirty years in New World Archaeology, most archaeological projects in the region have largely avoided their application as a methodology for reconstructing activities, cultural preferences, and resource use. This selective disregard of both paleoethnobotanical methodologies and archaeobotanical datasets is particularly salient on the North Coast of Peru; although the stunning preservation of organic remains is well documented throughout the 120 years of archaeological fieldwork in Peru (Reiss and Stübel 1880-87), and a comprehensive tome of ancient plant remains and their significance (Towle 1961) was produced 46 years ago, few projects have included paleoethnobotanical studies into their research designs. This state of affairs is likely due to the overwhelming reality of the shear abundance of ancient plant materials in these sites. I speculate that fear of the costly and potentially slow process of recovering and determining the identities of these remains, as well as the need to develop expertise in plant biology, is the main culprit responsible for its absence among most archaeological projects in the region, past and present.

Few archaeological studies mention plant taxa and other organic materials used as fuel in Peruvian prehistory (Cleland and Shimada 1998; Shimada 1997c; Tschauner et al. 1994; Wagner et al. 1998), and the few studies of ceramic and metalworking technology development outside of the SAP's work have largely ignored fuel use and the fuels used (Raymond et al. 1998; Russell et al. 1998; Uceda and Armas 1998). Overall, the lack of either quantitative or qualitative information on specific fuels used in particular contexts limit severely our ability to reconstruct the role of fuel resources in both the economy and ecology of past societies. This dissertation presents an in depth attempt to recover, examine, and integrate this dataset into a holistic model of craft production through archaeological investigation in North Coastal Peru for the Middle Sicán.

The dataset for this dissertation consists of charcoal deposits from kiln, furnace, and hearth features excavated

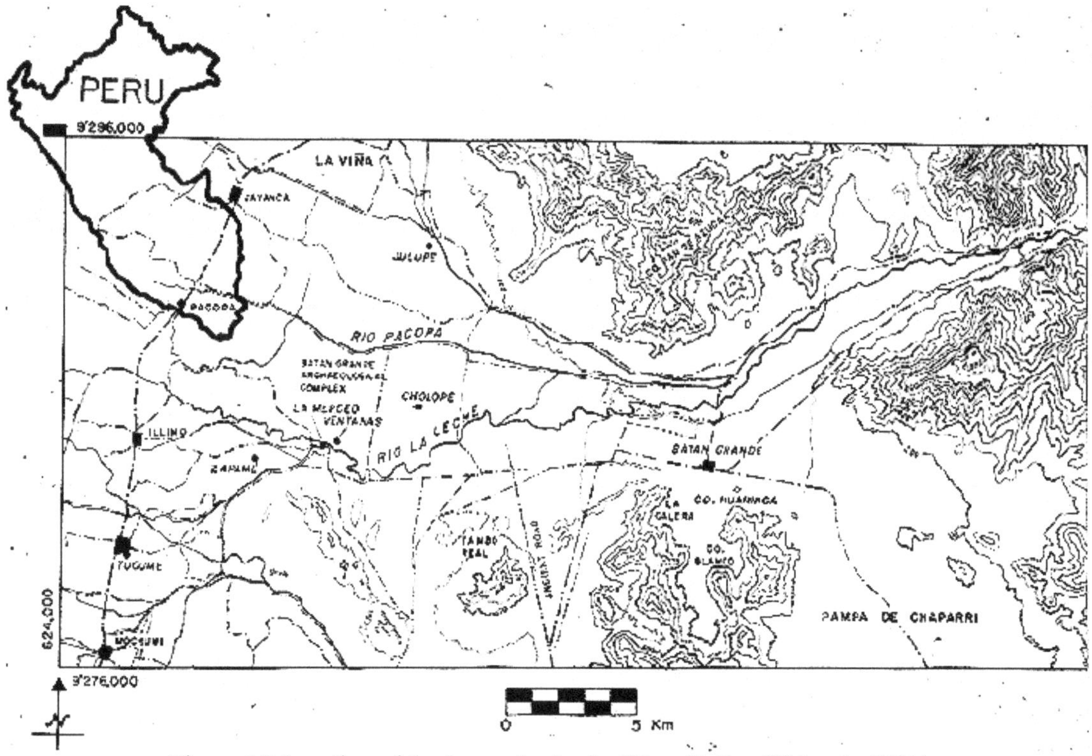

Figure 1.2 Location of the Lower La Leche River Valley (Shimada 1981)

in 2001 by the SAP at Huaca Sialupe. The goal is to reconstruct ancient fuel resource use by examining the charcoal data sets in light of the technological models and cultural histories already reconstructed for the organization of production (Shimada 1985; Shimada et al.1999; Shimada and Wagner 2001; Shimada and Wagner 2007). Charcoal samples were examined using reflected light microscopy to identify, where possible, the materials being used, the plant parts represented, and growth stress presence / absence in each piece of charcoal. From each context examined, I compare different features, e.g., hearths, kilns, metal furnaces, to interpret the specific materials ancient craft producers used to build fires to fulfill their social and economic needs in instances of craft production, fuel refinement, and provisioning. Identification of both the genera and plant part (branch, trunk, or root) highlights the preferences of the ancient craft persons, and can indicate strategies and intensity of individual resource apportioning. Locating these socially and ecologically related strategies of woody plant resource procurement, permits the identification employment of these forest management practices in the past, such as coppicing as expressed in tree growth morphology (Newsom 1993) or the use of certain plant materials for certain products (Miller 1985; Johannessen and Hastorf 1990).

I spent six months in the research area working with local people to identify and interpret the utility for plant resources around the National Historical Sanctuary, Bosque de Pomác in 2002. During this period, I collected a comprehensive set of comparative wood and plant material samples. These include wood from 69 species of woody vegetation, from three different ecological zones that typify the modern vegetation of the Lower La Leche River Valley, and plant herbarium vouchers for species identification and verification. The end result is a physical floristic record of the available plant resources in the area that can be compared to the ancient samples taken from Huaca Sialupe for the positive identification of the plant resources used in craft production.

The paleoethnobotanical approach is patterned after suggestions made by Newsom (1993) and by previous research accomplished in the southern Peruvian Highlands by Lennstrom and Hastorf (1995) and Johannessen and Hastorf (1990). These three cases suggest a physical reality that 'trees are good to think with' (Bloch 1998) when considering ancient human-environment interactions. As far as the paleoethnobotanical interpretation of wood charcoal is concerned, socio-cultural anthropology has only started to develop ethnographic research that outline the critical importance woody flora play in the historical identity and construction of society worldwide (Rival 1998; Jones and Cloke 2002).

Simultaneously, and somewhat isolated from the socio-cultural research, the role that trees play in structuring western society (Bloch 1998; Scott 1998) and the construction of western archaeological science (Nash 1999; Ballie 1990), has recently come to the forefront in the analysis of contemporary anthropological thought. In the case of Scott (1998) forest management under proto-capitalist, enlightenment-driven economic systems is viewed as not only a symbol of 19th century German economic organization, but is described as structural basis for understanding the evolution of modern German society. Both Nash (1999) and Ballie (1990) narrate the scientific contributions that both dendrochronology and forest ecology have made in archaeology and anthropology over the past 100 years, and that there has been, in turn, social impact as to how modern society interprets and integrates the ancient past into the present. In both instances, trees are no longer metaphorically significant in our human conception of nature and society, e.g., family tree, phylogenetic tree, but have become the research tool themselves. From this observation, there remains a disjuncture between the archaeological approach to trees through dendrochronology, dendroclimatology, and wood identification, and this developing ethnohistorical trend in socio-cultural anthropology. I demonstrate through this research how the modern state of the forest can be taken together with archaeological charcoal to produce integrated or 'holistic' (*sensu* Shimada 1994b and Shimada and Wagner 2007) reconstructions of ancient technology and a complex prehistory. This reconstruction recognizes the importance of woody flora in the development and maintenance of craft production. The products of which were the symbolic materials that gave substance to Middle Sicán identity, and were likely significant in how all Middle Sicán social relationships were manifest. By looking at how production needs were accommodated in the past we gain necessary insight into the historical developments that led to the modern biodiversity in the area.

Organization

On a chapter by chapter basis I identify how Middle Sicán artisans perceived and used their organic resources in craft production; moreover, I identify the important role fuel played not only in the organization of production, but also in the environment and society at the workshop level. To begin, Chapter 2 is an overview of dry forest dynamics as they relate to the situation on the North Coast of Peru in the 10th century based on my 2002 fieldwork. This includes data from my own ethnobotanical research in the Batán Grande area, the ethnohistorical and ecological evaluation of coastal resources in Northern Peru (Delvaud 1984; Rostworowski de Diez Canseco 1989; 1991; Shimada 1994a), and the modeling of these types of forests by Gentry (1995).

Chapter 3 presents a current summary of findings related to Middle Sicán history and society during its artisanal apogee at Huaca Sialupe, couched within the broader context of the regional history of archaeology in the area and the time frame of the Late Intermediate Period (C.E. 800-1375). I focus on the reconstruction of the manufacturing technology of the particular items made at the site. It outlines the SAP's conclusions, based on previous experimental and archaeological data, relating to the potential overuse and forest endangerment during the Sicán period. From this survey it is clear that the combination of archaeological contexts, methodological advancements, and theoretical components culminate in the investigation of Huaca Sialupe and uniquely reconstruct fuel use and ecological stability in the region. The chapter closes with a look at the organization of the site to develop a basic portrait of daily life at Sialupe, specifically its domestic component. It is here that the ethnoarchaeological studies of ancient metal furnace, kiln, and charcoal kiln constructions will be discussed relative to how different parts of the site were used. This discussion strengthens the assignment of 'characteristic use' qualities to these fire-use features as they are discussed in both the methods and analysis Chapters 5-7. This interpretation is critical if we are to understand how the production site of Sialupe may have articulated with the rest of the Sicán political economy.

Chapter 4 examines the theoretical background for the analysis of the Sialupe production contexts from the perspective of the present understanding of Middle Sicán political economy. This chapter also examines how resources were used and perceived by artisans at the site in light of their domestic or household needs, and tries address the overall human perception of resources and the incorporation of the natural into the cultural realm. I begin with the current understanding of Middle Sicán political economy from labor organization and trade network data. This overview of the Middle Sicán political economy, articulates with Chapter 3's discussion of the technologies employed at Sialupe, and attempts to locate the archaeological signature that political influence may have left at the site. What follows is an outline of traditional avenues for interpreting charcoal in Andean archaeology, e.g., dating, species determination, ecological information, and fire construction. Next, follow three sections that look at how I propose to tie charcoal remains to human actions associated with resource selection and resource management by ancient people. This avenue of thinking considers resource choices as they fit into technological models; the second part concentrates particularly on the problematic crafting, set within a framework of household decision-making. In this case, I pay particular attention to the difficulties of teasing apart household economic processes that have bearing on how craft production is performed, from actual multi-crafting activities being performed by crafters within the workshop. The chapter closes with a final word on perception of resources by crafters who are situated in a unique ecological and cultural position to decide how natural resources and ecology are shaped to conform to human needs.

Chapter 5 describes the archaeological and paleoethnobotanical analysis methods used for recovering the data presented in this dissertation. In particular, I outline how the comparative botanical collections were made, the observations about the ecology made in Chapter 2 were designed, and the specific techniques used for archaeobotanical recovery methods. In conclusion, Chapter 5 describes the manner that the taxa present in the assemblage were determined, and the relationship between plant growth and charcoal that permits ecological as well as cultural interpretations of the fuel remains at Sialupe.

Chapter 6 and 7 are descriptions of the analyses of the different data sets taken from fire-use features at Huaca Sialupe. Chapter 6 opens with a discussion of the how archaeological and paleoethnobotanical sampling was dealt with in the context of the comparative framework used in this dissertation between different parts of the site and the features. The specific discussion of the archaeobotanical data begins by looking at taxa ubiquities across the site to characterize the general subsistence and fuel resource base at Huaca Sialupe. Then the two larger excavation areas of Mound I, dominated by ceramic production features, and Mound II, the metalworking area, are compared. In each case the wood charcoal remains and the other organic macroremains recovered, e.g., seeds, tubers, faunal materials, are discussed. In particular, I describe the implications, predicated on the taxa present, of the ecology of the plants present and the organization of certain activities at the different ends of the site.

Chapter 7 is completely dedicated to the feature-to-feature comparison portion of the analysis. This section looks at ways of comparing the different resources recovered from different features in both the ceramic and metalworking areas to understand how resources were used in the different production activities. This section refers back the end of Chapter 3 assignments of different fire-use features, Mound activities, and the room typologies across the site for the Middle Sicán occupations. Here, I reconstruct the productive activities and their attendant pyrotechnologies at the site, treating the Middle Sicán period as a single occupation phase; each reconstructed process, ceramic, food, and metal production, aids in understanding the organization of the concerted efforts between metalsmiths and ceramists.

Finally, Chapter 8 presents the conclusions I derived from looking at the charcoal remains from Huaca Sialupe. I argue, based on the assemblages of archaeobotanical materials, that various alliances and continuities existed between all of the craft production activities at the site, and that, true to household economic models outlined in Chapter 4, distinct activities are superimposed and difficult to tease apart (Halperin 1994). There is a general sense, however, that ceramic production and food production at the site, while with a distinct connection of fuel complementarity (*sensu* Matson 1966), are distinguished from metal production, not only physically, but also in their fuel materials. In the final analysis it appears that the cornerstone of this separation of metal and ceramic work, at least in so far as fuel resources are concerned, was an interdependent system that required the conversion of wood fuel, the residues of reduced fired ceramics, into charcoal for metal working through the use of blackware kilns and possible on-site charcoal production.

While it is not clear if metal production was favored over ceramic production, it does appear that the two fit well together in terms of what resources were needed and how they were cleverly allocated. Even though certain metal and ceramic products, tumbaga sheet and blackware, were prized, at the site of Sialupe where these items were produced, their production was in concert and accord with the other fuel demands that the other activities at the site required. The concluding portion of the chapter focuses on the implications for future research in the region, and strategies for mitigating the limitations of the research presented here in the research design for future paleoethnobotany projects.

CHAPTER 2
ENVIRONMENTAL BACKGROUND

The site of Sialupe consists of five small low mounds located approximately 7 Km northwest of the city of Lambayeque and 15 Km inland from the Pacific Ocean on the north coast of Peru (Figure 2.1); it measures approximately 250m x 400m. Sialupe is located in an area of sparse settlement and irrigated farmland that today is home to non-industrial agriculture. The site was occupied from the early Middle Sicán period (C.E. 900) through the Inca (Chimú-Inca) period (C.E. 1375-1532). The workshop areas discussed here consist of two conjoined Mounds I and II, within the five mounds that comprise the site.

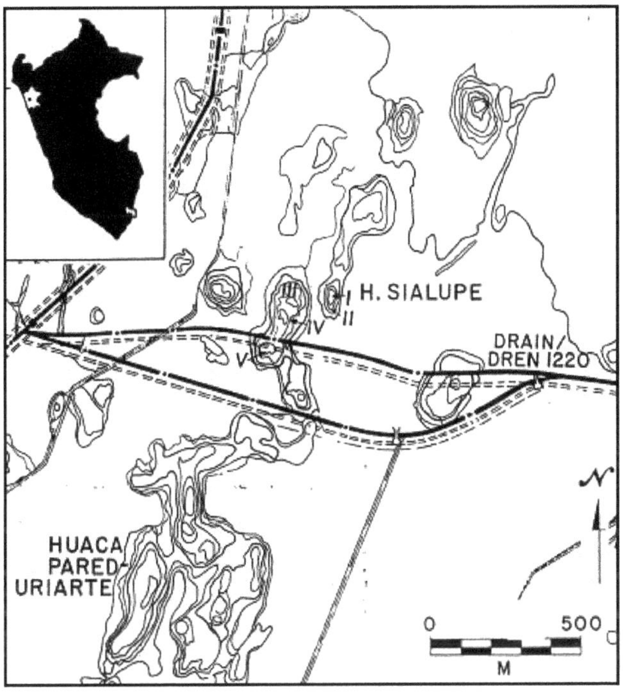

Figure 2.1 Location of Huaca Sialupe (modified from Shimada 2000)

This chapter presents the ecology of the area around the site and regionally as is relevant to the production levels of the Middle Sicán. This discussion focuses on the two ecological factors that appear to have circumscribed life and production, specifically fuel resources, at Huaca Sialupe. First, an El Niño/Southern Oscillation (ENSO) event punctuated the Middle Sicán production levels at the site, indicating the vulnerability and importance of orbitally forced climate change at the site in terms of human and environment interactions. Second, the discussion of ENSO events leads into how these climate events dominate the reproductive capacity of woody plants in the region and the delicacy of dry forest resources. The production activities at Huaca Sialupe relied heavily on renewable fuel resources, especially wood fuel, from these dry coastal forests of the Peruvian North Coast. After examining the ecological dynamics that constrained organic resource extraction during the Middle Sicán occupation, I look at the ceramic and metal production situated in their historical and cultural contexts for the Middle Sicán period.

Huaca Sialupe and its Climate

In the earliest phase of its research, the SAP attempted to reconstruct paleoenvironment in the La Leche Valley and adjacent coastal area (Shimada 1981; Shimada et al. 1982). The geomorphological work carried out by the SAP indicated that no agricultural activity had taken place in the immediate vicinity of the site of Sicán, the Middle Sicán capital, in what today is delimited by the National Historical Sanctuary, Bosque de Pomác, Batán Grande, Peru (Craig and Shimada 1986). This work also determined that the modern climate and environmental processes do not differ significantly from that of the Sicán period: work by Craig and Shimada (1986) indicates that periodic inundations, due to El Niño/Southern Oscillation (ENSO) rains, have been consistently episodic and greatly varied in their intensity throughout this geological epoch, the Holocene. The sediment deposits appear erratic and catastrophic, indicating a recurrent regime of droughts and floods for at least the past 1500 years (Craig and Shimada 1986). The work of Shimada et al. (1991), using ice cores from glaciers from the Southern Peruvian Andes, identified a series of geoclimatically forced severe drought events in the Common Era. This confirms Craig and Shimada's (1986) data for the broad regional trends of precipitation for the Lower La Leche River Valley research area. While rainfall varies between 20 mm to 79 mm of precipitation annually, the median rainfall for the period between 1937 and 1963 was 32.7mm of rainfall annually (Delvaud 1984:26). Most rainfall in the area falls above 2000m in the upper intermontane valleys, with most coastal precipitation falling in the coastal plain as either fog or rare intermittent storms (Delvaud 1984). Hence, the majority of the water reaching the coastal plain is through the extensions of the river delta system where the La Leche and the Lambayeque Rivers meet just before the Pacific Ocean.

In terms of the application of Holdridge's (1947; Tosi 1964) climate and vegetation model for the area the dominant vegetation supported by this rainfall regime, taking latitude and evapotranspiration into account (Delvaud 1984:26), is thorny woodland. Radiocarbon-dated lake diatom strata from Southern Ecuador corroborate the reliability of these sequences and the application of Holderidge's (1947) model (Shimada 1994a). Both the climate and geomorphological data demonstrate that today's dominant climatic regime, including ENSO events, was similar to that of the Sicán period in the La Leche Valley, which allows us to extrapolate modern phytogeography onto the past. Additionally, the presence in some areas, of rather extensive, long-lived and evolved dry tropical forests with thorny vegetation dominating the landscape further confirms the probable consistency of the climate regime through the Holocene.

We must keep in mind, however, that these observations regarding climate do not discount the impact of the past 500 years of landscape modifications by humans, e.g. overall decrease in plants/animal populations and community biodiversity (Craig and Shimada 1986; Shimada 1995). Even so, given the general similarity in climate regimes between the Middle Sicán and the present, we can utilize modern local flora and their dynamics as comparative frameworks for the identification and interpretation of archaeobotanical samples and their ecological interpretation.

ENSO Events

Southern Oscillation weather events have become more and more prominent in the modern discourse about weather and periodic or episodic climate events globally. In particular, this phenomenon, known also as El Niño, has been the focus in reconstructing climate change from the early Holocene through to the present. Recent investigation and climate modeling indicate that ENSO activity is due to global shifts in weather patterns, indicating a universal interconnectedness between the Earth's weather patterns, human activity, and global history. Moreover, it stands as a common link between human-environment interactions through the history of human habitation in South America.

The ENSO phenomenon has been covered in the natural and social science literature extensively, and an exhaustive discussion of that literature is of little use here. What is important to this study is that ENSO events occur; they play a dominant role in the local ecosystem and have the potential for an archaeological signature in terms of how the people of the Middle Sicán period reacted and incorporated these events into their worldview. El Niño/Southern Oscillation phenomenon broadly describe the cyclical global weather phenomenon that is characterized by (DeVries 1987; DeVries et al. 1997; Hocquenghem et al. 1992; Wells and Noller 1999; Rollins et al. 1986):

1. The weakening of the dominance of the upwelling south-north cold Humboldt ocean currents running along the western coast of Peru.
2. The depression/incursion of warm water currents originating in the equatorial inter tropical convergence zone into areas normally dominated by the Humboldt Current. It appears that this warm water body originates in the Indian Ocean system some nine months prior to its impact on South American weather systems. Over the past 80 years the sea surface temperatures have increased during ENSO events between 2-8°C with instances of 6-8°C being extremes punctuated by intermittent rises of 2°C
3. The warm sea temperature increases influence of Atlantic based low pressure weather systems on the western slopes of the Andes, permitting the shedding of precipitation along the western drainage areas. Overall increase in annual precipitation can be in excess of 100%.
4. The intrusion of warm waters brings with it lower levels of dissolved oxygen and salinity that results in a change in plankton levels and a shift in sea fauna from cold water to warm water species.

The impact of the Atlantic Low Pressure weather systems on the Pacific drainage zones of the Western Andes is profound. This leads to a vast increase in precipitation in areas that are desert, due to the 'normal' dominant influence of the Humboldt Current. Additionally, the river drainages swell and flood under the impact of the intense highland and coastal precipitation. These events create catastrophic flooding that results in crop destruction, habitat displacement, and potential for synergies with other precipitous disasters, (e.g. famine, plague, etc.) that affects the human and other plant and animal communities throughout the regional ecosystem. To a great extent, these events play different roles depending on elevation, incident slope, seismic stability, and human habitation. For instance, lowland floodplain settlements can simply flooded, while upland settlements in the steep intermontane valleys can be completely washed away or buried. The importance of this cycle has not been underestimated in the modern and historical record, and is argued to have been a driving force in cultural and ecological change throughout the course of Andean Prehistory (Holmgren et al. 2001; Mabres Torello et al. 1993).

More directly to the point, the impacts of ENSO events and their cyclical occurrence on the Peruvian North Coast, and in particular the modern department of Lambayeque, where the site of Huaca Sialupe is situated, are well known (Craig and Shimada 1986; Hochquenghem et al. 1992; Kaulicke 1993; Mabres Torello et al 1993; Macharé and Ortlieb1993). Both modern and historical record keeping indicates the occurrence of ENSO events associated with river avulsion, flooding, and general destruction on scales of time that can be measured in nearly 20-year increments (Craig and Shimada 1986; Kaulicke 1993; Mabres Torello et al. 1993; Shimada et al. 1991). This periodicity, however, is irregular with 20-year flood stage events, often occurring within 3-10 years of each other (Shimada 1994a; Wells and Noller 1999). Many Andean archaeological scholars have expressed the important impact of this system by focusing on the inundation events themselves (Keefer et al. 1998; Kolata et al. 2000; Moseley 1978; Moseley 1992; Moseley and Feldman 1981; Ortloff and Moseley 1983; Ortloff and Kolata 1993); other scholars have focused on the importance of the pre/post event climate that is dominated by successively drier to serious drought conditions (Shimada 1997a, Erickson 1999).

Shimada (1994), in his review of the ENSO based literature and his assessment of potential impact on the development of complex societies on the North Coast of Peru, recognizes that it is the synergy between these forces of extreme drying and wetting, drought and floods, that provided a dynamic stress to cultural activity in the area. In response, people developed strategies for moving water away from settlements in times of inundation. For instance, Schaedel (1992) reports that the pre-Hispanic wall protected the city of Chan Chan, and most likely post-flood labor motivation removed sand and flood alluvium from settlements, canal systems, etc (Shimada 1994a). In addition to post-hoc tactics, social and technological strategies for resistance to droughts, e.g., extensive canal systems conjoining river valleys, and curvilinear irrigation systems, are what sustained the development and continuity of large settlements and population demographics on the North Coast. Additionally, supplementing agriculture with marine resource exploitation and camelid herding surely added to community resilience both pre and post ENSO events (Shimada 1994a). In either case, flood events or extreme drought, we see that the pre-Hispanic populations were aware of the regional climate pattern. They were capable of devising strategies anticipating drought and facilitating relief from inundations, and incorporated this dynamic into their worldview (Kaulicke 1993; Narvaez Vargas 2001).

As these measures can be modeled archaeologically, our reconstructions of the pre-Hispanic period also need to include provisions that pre-modern peoples were, while knowledgeable of the phenomenon, similarly unable, as we are today, to predict the severity of ENSO related disasters. The most dramatic evidence of this comes from the site of Moche during the Moche IV-V period transitions, approximately C.E. 550-600 (Shimada 1994a). Here catastrophic flooding followed by extreme aridity apparently contributed heavily to the site becoming undesirable for human habitation. At the same time pre-modern peoples were able, evident in their sophisticated agricultural and technical solutions to living in arid deserts served by the Andean watershed, to anticipate that ENSO events would eventually occur and that there would be renewal once the water subsided.

Implications for Human Reaction

The human implication for recurrent ENSO activity, for me, is best considered in contrast to the Temperate Zone model of seasonality. In the Temperate Zone agricultural scheduling, provisioning, and settlement strategies are centered on the acknowledged spring, summer, winter, and fall schedule. As has been long acknowledged in tropical agriculture and ecology, Temperate Zone climate and ecology models do not work well for modeling Tropical Systems that tend to be dominated by a single wet and dry seasonality (Brown 1970; Upton 1996; Whitmore 1998). This dual period seasonality is often perturbed by global climate systems, e.g., ENSO events, and/or regional shifts in climate zone activity, e.g., depression or shifting of the Intertropical Convergence Zone. Discussions with local farmers in the La Leche River Valley indicate that Andean Coastal agriculture in the region anticipates this traditionally held wet and dry cycle. It also includes an intuitive acceptance of the periodic possibility of an ENSO episode, and the potential for 100% fluctuations in annual rainfall in the Lower La Leche River Valley (Delvaud 1964:24-30). As a result, the seasonality model here is not a simple annual cycle, but is expanded to anticipate the periodic small ENSO events over the course of decades, and also to anticipate that these can be catastrophic (Mabres Torello et al. 1993; Kaulicke 1993). Whereas rainfall is seldom seen along the coast, rather with nightly fog accounting for the majority of coastal precipitation, the greatest effects of increased rainfall are felt in the upper elevation areas of Lambayeque. In the region of Sialupe the lasting impacts of ENSO events, from a precipitation point of view, are most likely in swelled river deltas and flooding. These effects include the potential for destruction as well as fertility and stimulating forest growth that these phenomena include. In Lambayeque, then, it is best to consider seasonality in terms of its effect on the non-agricultural environment, not as an annual cycle, but as a 20 year or more cycle that, though driven by the annual cycles of wet and dry tropical climate, is overshadowed or dominated by long term cycles of change related to ENSO events. These catastrophic events are anticipated, but neither their occurrence nor intensity is readily predicted.

Modern agriculturalists in the research area clearly recognize this longer cycle of seasonality and have incorporated agricultural strategies that take advantage of the cyclical nature of ENSO events. During non-ENSO event years, modern, non-mechanized agriculture is dominated by extensive irrigation that are canal fed field systems. Indeed, at times, both gasoline generators with electric pumps, combined with cement lined feed canals are used. The field system itself uses one of two different methods at present for agricultural production: furrowed fields with high walls for retaining water that are flooded between 2-3 times a growth period, and fields with meandering irrigation throughout that are filled with water periodically during the growing season. Most farmers also, notably, have landholdings in and around the flood stage riverbanks along the La Leche and Lambayeque rivers, as well as the larger canals that overrun their banks at flood stage. These fields, naturally, are not in production every year: they depend on fertilization and irrigation as floodwaters recede after an ENSO event. Often these plots have tree crops on them that are tended even in non-ENSO years. The plots can be farmed, in addition to irrigated holdings, for up to a year and a half after the flood stage has been reached. In fact, some smallholders extract, given labor availability, up to three growing seasons off these plots that are often as large or larger than irrigated systems. These can be used as additional income sources, or ways to reduce dependence on federally regulated and taxed irrigation water. The use of ENSO event fields illustrates that these field systems are incorporated into the annual wet/dry agricultural system intuited by the local agriculturalists of the La Leche River Valley.

The presence of flexible farming strategies indicates that the indigenous people of the region incorporate the reality of ENSO events into their worldview in a way that highlights the lasting potential of these events, not just focusing on single event catastrophes as reference points. Most importantly, the event itself, often lasting months, comes to incorporate long term impacts that these events have on the local environment. In other words, the vision of how the agricultural system works, from an indigenous point of view, may, for every 20-year cycle, incorporate some five or more years in a 20-year cycle. These people can then anticipate the use of these other systems, e.g., flood plain agriculture, extensive upland orchard crops, in place of or in tandem with their irrigated fields. The fact that these fields exist and are used indicates that the local conception of ecology includes ENSO level 'destruction.' Yet it also allows for ENSO opportunities to be anticipated on an intuitive and practiced level by the people living in the area. If they can plan for ENSO related events in their agricultural cycle, the residents of these forests are more than likely aware of the impact of ENSO events on local ecology, including forest regeneration and resource cycling. This point is critical to understanding the potential for how people living in the La Leche River Valley, past and present, view non-agricultural ecology: in particular forest regeneration, and sets up arguments for modeling the use of forest resources in antiquity.

Dry Tropical Forests of the La Leche Valley

The site of Sialupe is located just southeast of the modern pottery-making town of Morropé, in the Department of Lambayeque, at the southern boundary of the Sechura Desert. Dry tropical forest dominates the few areas of the desert's southern edge minimally disturbed by human activity (Tosi 1960; Proyecto Algarrobo 1997). At present the area is suffering from drought and desertification from the encroachment of the Sechura Desert from the North. This is partly due to the long-term effects of the Neogene uplift that created the Sechura Basin, indicating that there has been a trend towards aridity in the region. This process has been exacerbated by pronounced silting processes and deforestation that are predominant in the modern era (Shimada and Wagner 2007). The effects of desertification are thought to have been more limited in the past.

The dry tropical forest is the dominant ecosystem throughout the region, as this type of vegetation and rainfall regime is encountered in undisturbed and protected areas throughout Lambayeque and the other North Peruvian Coastal Provinces or Departamentos. The regional woody flora consists of xerophytic trees, low shrubs, and cacti. The tree species are adapted to an extremely dry environment, mean precipitation <40mm/year (Delvaud 1984:26), and typically the biodiversity of woody plants is manifest of two main families Capparidaceae (caper family), e.g., *Capparis* spp. (four species), and Fabaceae (bean family), particularly *Prosopis* sp. (three species) and Acacia spp. (two species) (Proyecto Algarrobo 1997). These three tree genera comprise the dominant mature stature trees in the modern forest in the region.

Presently, the cessation of modern charcoal production and agricultural clearing are at the heart of a regional campaign to stop deforestation in the area and preserve the remaining areas of dry coastal forest (Risco 1998; Cuba Salerno et al. 1998). Dry coastal forests are extremely vulnerable to deforestation by ecosystem disruption due to slow annual growth of dominant tree species, uncertain periodic water and nutrient recharge (Maas 1995; Cuba Salerno et al. 1998). As the primary dominant tree species in mature forests, *Prosopis* sp. tends to be long lived, 200-400+ years, and some are in their adolescence up to at least 60+ years of life (Ocampo 1991). In most instances, however, there are very few extant trees over 40 years of age (Ocampo 1991). Due to the modern problems of forest fragmentation in the area, it is unclear what other potential woody species may or may not be equally as long-lived. Forest resources in the surrounding area are quickly disappearing as they are cut to create agricultural fields, supply fuel, and create building materials (Risco 1998). Forest regeneration, e.g., seed germination, is highly dependent on the presence of water, especially inundation due to ENSO events, to be successful.

Dry tropical forests are highly specialized ecosystems that rely on a limited number of tree species to maintain soil structure and provide shelter to other plants and animals critical to forest regeneration (Bullock 1995; Beaumont 1989; Cuba Salerno et al. 1998). In some cases these may indeed be true keystone species, few in number offering highly specialized ecosystem services, and in others their sheer number contributes to overall ecosystem health, e.g., leguminous trees offering shade and nitrogen fixation benefits (see distinction made by Búrquez and de los Angeles Q. 1994:10-24). Trees in the dry tropical forest, however, are critical for conserving the main factor that limits growth in these environments, the access to and retention of water. Trees in the dry tropical forest provide shaded areas that cool the land surface and prevent the evaporation of ground water. Root and trunk systems keep water tied up in terrestrial nutrient supply systems so that they are not lost to the atmosphere. Lastly, most plants are adapted to xerophytic environments with waxy or small leaves, deep taproots, diurnally dormant stomatic systems, so that they do not dry out in periods of extreme drought. This forest is, as an ecosystem, very specialized, demonstrated by its myriad responses to a hot and constant tropical sun in an area of limited to non-extant rainfall (Gentry 1995).

The specialized nature of these forests is best expressed through growth form diversity (Medina 1995). In the case of Neotropical rain forests, biological diversity is the most notable element, boasting one of the highest diversities of living organisms, plants and animals per biome on Earth (Whitmore 1998). For the rain forest, however, in terms of the growth form that woody plants occur, e.g., shrubs, trees, lianas, vines, they are limited. In comparison, the dry tropical forest, while comparatively lower in overall biodiversity, tends to include a higher diversity of growth forms among the plant species present, e.g., cacti, trees, palms, etc. In this case, ecologists argue that almost every plant growth form present on Earth can be encountered in dry tropical forests as an indicator of their very specialized nature and potentially of their age (Medina 1995). Each individual species fills not only a niche, but its life form is distinctly adapted to the niche that it fills (Medina 1995). With that said, it is critical to understand the potential for each individual species to contribute to the local ecological community as a whole. In the case of this research, the

Tree Species	*Acacia* spp., Corresponds to local *A. macracantha*	*Capparis* spp., Corresponds to local *C. angulata* or *C. ovalifolia*	*Cercidium* spp., Corresponds to local *C. praecox*	*Prosopis* spp., Corresponds to three species of *Prosopis* spp. available locally	*Quercus alba* (for comparison with temperate woods)	*Eucalyptus globulous* (for comparison with modern reforestation materials)
Caloric Value (Dry Wood) KJ/Kg	19054	20101	18782	20737	17340	20719

Table 2.1 Comparison of KJ/Kg Values of Select Wood Genera, Highlighting Woods Local to the Research Area, *Quercus* and *Eucalyptus* are added for Comparison (Harker et al. 1982)

focus is on species humans collect for fuel use, i.e. fuel wood.

The common way for measuring the health of a temperate forest is to generate site-index plots of the stature and diameter at breast height (DBH) of dominant tree species in the area (Schwartz 1997). When charted these data present a visual representation of the stature (height), relative age (DBH), and structure of the forest under study, permitting an estimation of overall forest health. Figure 2.2 presents an attempt, in an idealized form, to examine the health of the modern forest that I surveyed in 2002 based on the vital statistics of the dominant forest species.

Granted, not all trees take the same growth form, nor do they grow at the same rate. What Figure 2.2 permits, however, is an estimation of the overall health of the forest by considering the height of the trees present relative to the species and ecological niche that each species occupies. A larger quantity of taller trees tends to demonstrate an older forest, whereas only larger trees with no or few young trees indicates that something is retarding forest regeneration (Fralish 1997). Certainly, this potential mature stature is determined by both what trees grow in a given area and the regional climate regime. Yet, the population of the taller trees should be of the species that is most representative of the full stature or 'healthy' mature forest. In the case of Figure 2.2 the overall trend is best described as one of a forest in regeneration, most of the *Prosopis* sp. trees, the most representative species of Lambayeque's forests, are conspicuously short, with a few large trees in the area. The few areas of supposedly mature stature forest measured by Proyecto Algarrobo (1997) in the National Historical Sanctuary, Bosque de Pomác, is between 15 and 20 meters high. In the case of the site index created for this project it seems that few of the species that can attain this height, e.g., *Prosopis* spp., have been able to do so. This trend indicates that ecological, here largely human, forces are preventing the dominant forest species from achieving a full canopy height. This situation, in turn, may be preventing the establishment of a healthy community of local dry forest flora, and probably fauna. Further study is required in the area to be certain.

Given the history of the area, a recently (40 year interval, considering the life span of the trees (220+ years)) enforced area of conservation within an area of extreme disturbance, e.g., agricultural clearance and fuel predation, this is not a surprise. My site index, however, is artificial: it conflates several areas within the Upper La Leche Valley (Chapter 4) where we could locate forest cover. Given the constraints of not being able to work within the boundaries of the National Historical Sanctuary, Bosque de Pomác, this site index is the best estimate of what we have for the forest stature outside of the park.

I present three main observations regarding the site index graph in Figure 2.2. The first is that *Prosopis* spp. is the dominant higher stature tree among the species intensively examined; among those, a folk distinction between female and male trees indicates that female trees are most prevalent among those high stature trees (see discussion below; Table 2.2). The next largest tree by size grade is *Acacia macracantha* that does not appear, based on my observations, to achieve a stature greater than 12 m in height and 30 cm in DBH. Although it is thought regionally to be a tree slightly lower in stature, 9 m (Zevallos P. 1986: 322). Lastly, the majority of woody species are present in a mid-range height of seven to two meters in height and 25 to five centimeters in DBH. From the perspective of a human being, approximate average height in Peru 1.65 m, this is a dense forest, with a high ceiling. Most vegetation is branching at about eye level to the human body (Figure 2.3). The collected ecological data also demonstrated that the forest understory is dominated by underbrush and small bushes in my model modern forest, which I attribute, based on my observations of the dry forest ecology in the study area, to adventitious growth in disturbance areas, around forest clearance or natural tree fall. If we take this as a panorama of the modern forest, then what are the implications for fuel resource selection and what are the visibly important economic species (beyond fuel) to people dwelling within and around these areas? This is explored below.

Dominant Tree Species in the La Leche Valley

Previously published archaeological and experimental research in the area identifies only two species of primary wood fuel from select contexts, *Prosopis pallida* and *Capparis angulata* (Shimada et al. 1998; Shimada and Merkel 1991; Shimada and Wagner 2001; Wagner et al. 1998). These studies identify *Prosopis* sp., locally *algarrobo*, as the primary high quality fuel resource available in the area; *Capparis angulata*, locally *sapote*, although comparable in its energy release to *Prosopis* spp. (Table 2.1), produces a charcoal that is rapidly consumed in the fire and is considered secondary (Shimada 1981; Shimada 1985;

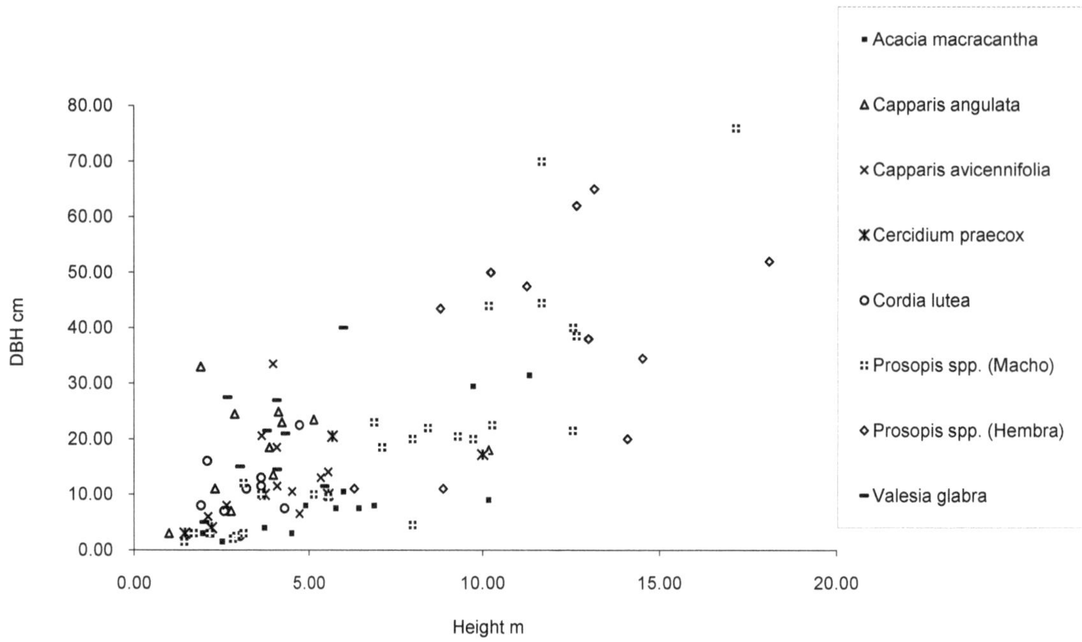

Figure 2.2 Site Index of Arboreal Species Visited in 2002 Reconnaissance

Shimada and Wagner 2001). Both were heavily used and traded in the Colonial period as wood fuel, and were used to produce goods, such as furniture or fences (Ramírez 1996).

Rostworowski de Diez Canseco (1981), however, cites 16th Century documents that mention 15 other economically useful trees for construction and fuel in the region, indicating that current conceptions of ancient fuel resources are incomplete. For instance, other economic activities surely competed for fuel resources, e.g., construction, but fall outside of the analysis and study that I am pursuing here. In the region of Huaca Sialupe, *Prosopis* spp. and *Capparis* spp. represent only seven of the dominant woody plant species (Table 5.3; Cuba Salerno et al. 1998; Proyecto Algarrobo 1997). Because *Prosopis* spp. (Fabaceae) meets the physical demands of simultaneous industries of sustained ceramic firing, metal smelting and metalsmithing, the SAP has proposed that it was the cornerstone of Middle Sicán production (Shimada 1985; Shimada and Merkel 1991; Shimada and Wagner 2001). If true, then from what we know about both dry tropical forests and the ecology of this species, the implications for forest destruction are profound.

The three *Prosopis* sp. species (*P. pallida, P. juliflora, P. affinis*) provide more than just high-grade wood and charcoal fuel, this genus is the dominant member of the mature dry forest ecosystem and has many modern economic uses (Delvaud 1984; Proyecto Algarrobo 1997). As such, these are one of the hallmark species of life form adaptability to these environments. Present research, however, worldwide continues to have difficulty classifying these species (Pasiecznik et al. 2001). Here we use *Prosopis pallida* as the most general species, and *Prosopis* spp. as a generic term that encompasses the three species. *Prosopis* spp. both sustains soil fertility as a nitrogen fixer and helps maintain soil surface water through its lateral root system (Ocampo 1991; Proyecto Algarrobo 1997). Additionally, leaf litter and lateral root systems enhance soil surface nitrogen (Nordt et al. 2004). The species plays a significant role in preventing soil erosion and desertification through both the shade and soil cohesion offered by its branch and root systems, respectively (Luna-Suarez et al. 1998). Additionally, trees in this genus offer high protein animal fodder in the form of their bean-like fruits, that are also used to create a sweet syrup *algarrobina*, and fermented beverage, *chicha de algarrobo*, for human consumption (Risco 1998). Local populations use the resin from the trunk as an adhesive and as a local remedy to fill dental caries (Risco 1998). These trees require at least 100 years to grow to mature

Figure 2.3 A View of the Dry Tropical Forest Near Batán Grande, Lambayeque

Figure 2.4 *Prosopis* spp. The *Milenario* of La Zaranda, Lambayeque

stature (Ocampo 1991). It is not merely the dominant feature of what informants see as a 'forest (*bosque*)' but it is the trademark species of the area known for its beauty, longevity, durability, fodder, shade, drought resistant and fuel properties.

Prosopis spp. is clearly the most 'famous' of the trees in the area among denizens. In almost every small settlement (n=15) visited in our forest survey one or several algarrobos stand in the center of the settlement, next to the football pitch, school, and seat of community government. In nearly every farm visited algarrobo trees are selectively left next to the standing structures or in the animal corrals to serve as shade and fodder. If they are not present, then these trees are often being tended nearby and guarded in hopes that they will provide shade, animal fodder, and potentially an architectural feature or support in the coming decades. Their utility is also due to their ability to be tended or pruned easily when certain hard wood branches are needed for door lintels, axe handles, turkey roasting, etc (Elfadl and Lukkannen 2003; Pasiecznik 2002). Elfadl and Lukkannen (2003) have demonstrated that the annual cutting of the stem at the base repeatedly, effectively coppicing, or lateral branches off the stem, pollarding, is not only possible with Prosopis sp., but that if tended over the long term leads to a predictable source of wood fuel and construction materials. Finally, in several communities, like that of La Zaranda, both within and outside of the boundaries of the National Historical Sanctuary, Bosque de Pomác, some trees are 'legendary,' such as the *milenario* (Figure 2.4), a tree potentially more than 300 years old. These trees, as with this one, are often seen as forest 'guardian spirits' that operate within a conflation of local myth and history regarding the permanence and threat to the forest. This topic will be discussed in more detail below.

The genera *Capparis* and *Acacia* are likewise important, recognized by almost all humans in all age groups as salient to locals. *Capparis* spp. is also a relatively slow growing genus that has two main species present in the research area. *Capparis ovalifolia* (syn. *avicennifolia*), locally *vichayo*, mainly appears in areas of low or no forest cover, around settlements and in agricultural field boundaries. It is a plant that fruits twice a year and is a source of animal fodder and human famine food. Additionally, it is easily pruned and when maintained, new and predictable shooting serves as a source of renewable fuel. *Capparis angulata* (syn. *scabrida*) (*s(z)apote*) is a potential keystone species of dominant *Prosopis* spp. forests. They are rare in forest stands, but they are always present. It appears mostly in and on the boundaries of forested areas, can be managed for fuel wood, is mostly valued for its shade and as kindling for fires. As mentioned above, this species is often the second species used for charcoal production in these forests of North Coastal Peru. Both species within the genera *Capparis* are also often used for tool handles, or in one case, the workbench itself (Figure 2.5), and secondary construction materials. The tree is, however, susceptible to *Scylotid* beetle infestation when still wet. Once completely dry, however, it appears to be durable and resistant to infestation. *Capparis* sp., although second to *Prosopis* spp. in charcoal production, is still a significant economic resource in the dry forest.

Of secondary importance and mention after algarrobo among locals is *Acacia macracantha*, locally known as *faique*. The tree is a quick colonizer of fallow land and is often used as mammal fodder. The tree produces fruit in its first year of life, and often fruits twice a year, at the end of the wet season, as well as just before the Andean winter. Some Fabaceae species, including *Prosopis* sp. have been observed flowering twice in one year in this area and may be an adaptive response to more temperate climate regimes during its

Figure 2.5 *Capparis angulata* used as a workbench, a piece of *C. angulata* is being formed into an axe-handle.

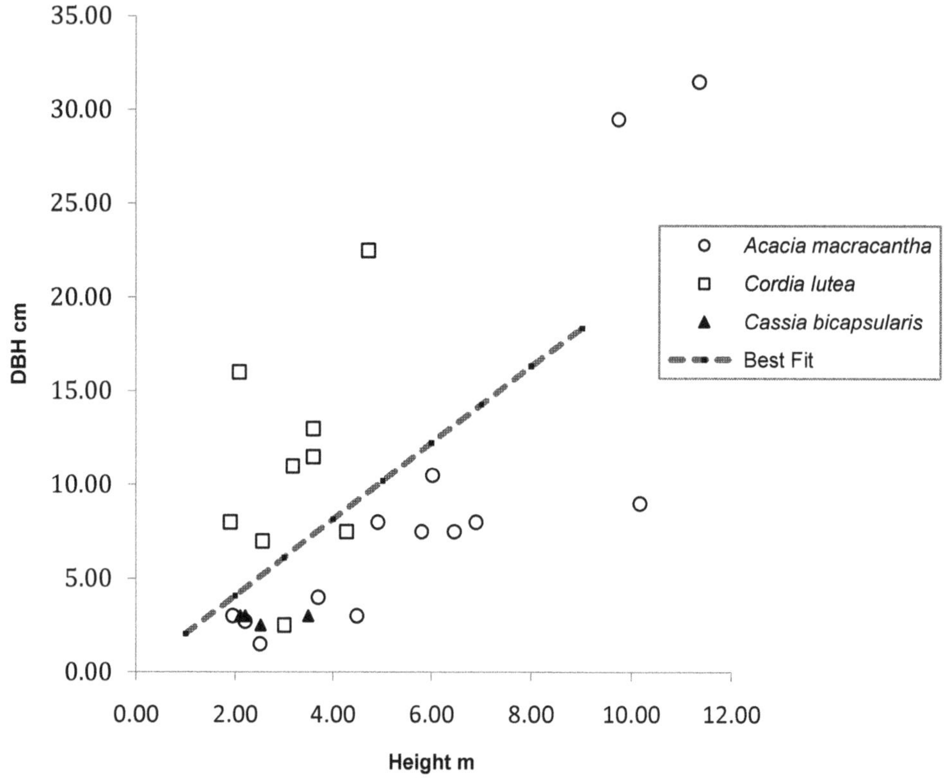

Figure 2.6 *Acacia macracantha* DBH Relative to Two Shrubby Taxa

evolution. Additionally we noted that *Acacia macracantha* plays a significant role in forest regeneration in the area, discussed below. During the 2000-2 field seasons, I made a number of observations regarding the tree ecology of the area. In the late Andean Spring of 2000 the area received an abnormal amount of rainfall, e.g., more than 79 mm (Delvaud 1984) the high end of the annual rainfall scale, due to ENSO activity. This phenomenon happened to coincide with a slump in productivity in the sugar company, formerly cooperative and hacienda, in Batán Grande. The result was large areas of fallow cane fields, being left to regenerate after the rains. In the 2001 and 2002 seasons I was able to see the regeneration of the forest species *Acacia* and *Prosopis* in the local dry tropical forests in the fallow cane fields around Batán Grande. My inspection indicated that *Acacia macracantha* was the dominant sun-tolerant species, growing nearly two meters in height after the first year. Given that the area had previously been used to fodder cattle, and that *Acacia macracantha* pods are a common animal fodder in the area, it is thought that *Acacia* represents a pioneer or early successional species. *Prosopis* sp. and *Capparis* sp. species seedlings appeared, but growth was extremely retarded until shade cover was established, establishing that the saplings of these two genera are shade tolerant. The retarded growth indicates that these are mid-successional or primary species, dependent on established/establishing stands of Acacia to thrive. These observations are critical to understanding the potential for forest regeneration in the area now and in the past.

During the 2002 season, discussions with local farmers and residents about the lack of *Acacia macracantha* extant in the few stands of full stature dry forest in the area raised the issue that these trees are plagued by insect larvae, *cosu blanco*, lit trans. white grub; they compromise the root system, which over time destroy the stability of the tree, making it vulnerable to termite infestation and death. Locals argue that the larvae are from a white moth, which I never observed, and are more likely to be beetle larvae (Creffield 1996) as is discussed in Chapters 6 and 7. Informants explained that the entire process of tree destruction takes between 10-15 years for the trees to die. Indeed, survey of the few stands of older *Acacia macracantha* trees in the survey area indicated that of the 14 trees examined, the average height was 5.8 m with a diameter at breast height of 12.75 cm, (avg. b.h.=63 cm), suggesting that their growth potential may not be fully realized (Figure 2.6). *Acacia macracantha* trees over 8 m high were only found as solitary trees with open globe-like branch growth indicative of their growth in open areas away from competing trees. Consequently, these trees are rarely seen in the high-stature mature forests of the area, confirmed by the tree surveys in the National Historical Sanctuary, Bosque de Pomác (Proyecto Algarrobo 1997).

The growth pattern of *Acacia macracantha* diverges from early successional forest species in that it appears to die away as the population reaches maturity without a clear cluster of age grades present in the population. This trend is visible when compared with *Cordia lutea* in Figure 2.6 or *Prosopis* spp. in Figure 2.2. Even so, *A. macracantha* plays a crucial role in the regeneration cycle of a forest: during those first 10-20 years of growth, they provide shade for shade tolerant saplings that are vulnerable to exposure until they become more sun tolerant as they mature. In particular, this successional relationship was observed in both young *Prosopis* spp. and *Capparis angulata* populations. Although well documented for trees in temperate forests and in studies of disturbance in Neotropical rainforest ecological systems, these inter-genera relationships are rarely observed and documented in Peruvian dry tropical forests, due to their extremely fragmented nature and the massive amount of disturbance caused by agricultural production. Instead they are often modeled from other forest types; however some efforts have developed literature in this discipline of dry tropical forest research (Bullock et al. 1995). In the modern forest system, charcoal makers too frequently cut the dominant tree species, *Prosopis* spp. and *Capparis* spp., as well as the *Acacia macracantha*, before they reach the mature age of somewhere around 30 years (Risco 1998). Essentially, both the pioneer and successional species in these instances are wiped out simultaneously to support a charcoal industry and clear land for agriculture.

The practice of intensive cutting has clearly had a significantly deleterious effect on the regenerative ability of high stature forest in the area. Given this intensive harvesting, people have not actually witnessed forest regeneration and succession here in the modern era; nor was it documented in the National Historical Sanctuary, Bosque de Pomác during its reforestation. Hence, forest ecologists and local farmers do not have experience with the relationship between Acacia growth and *Prosopis* regeneration, which may be a large contributing factor to overharvesting in the contemporary setting. The present propensity of charcoal makers and agriculturalists to cut these trees beyond regenerative capacity has affected significantly forest restoration throughout the colonial and modern era (Delvaud 1984; Risco 1998). *Prosopis* spp. charcoal production has become a critical deforestation problem over the past 200 years and has forced the regulation of forest products in the coastal areas of Peru (Ferreyra 1987). At present, rural forest residents often produce *Prosopis* spp. charcoal illegally for additional household income to be sold in local and regional markets, and/or domestic use for fuel (Risco 1998).

Impacts of ENSO Events on the Dry Tropical Forest

In this climatic system, water is the critical factor for agricultural production to regional dry tropical forest regeneration. When water abounds, the majority of the plants in the area are adapted to rapid growth; tree seeds in particular germinate. ENSO events, or even chance pluvial episodes, provide opportunities for forest regeneration; the rhythm of the dry coastal forests of northern Peru is dependent on these rain episodes (Holmgren et al. 2001). For my research, it is critical to note that these opportunities do not occur on yearly cycles. Hence, what we associate as the annual period of rebirth and regeneration in the temperate zone, a true spring, is a phenomenon that, for the major tree species in the area, occurs only seldom in a 20-year cycle in this environment. The ensuing period of drying is, from the perspective of the forest, the gravy period of slow, progressive growth.

The importance of these dynamic systems in ecological regeneration of dry tropical forests has only recently entered the literature. Holmgren and Scheffer (2001) have started to catalogue these kinds of ENSO/regeneration processes in a number of ecosystems, like those of the North Coast of Peru. I observed that adult trees of genera *Acacia*, *Prosopis* and *Capparis* are adapted to deal with ENSO events as episodic growth spurts. Young Acacia sp. immediately begins its accelerated growth in the dry hot tropical desert sun after an episodic rainfall. One of the few tree species that will both flower and fruit in its first year, *Acacia* sp. provides, almost immediately, a food source for livestock in the first season post inundation. Older *Acacia macracantha* with an often-infested lateral root system, on the other hand, responds to inundation of the clay/sandy soils with river rise and pluvial downpour by simply falling over. Blow downs and fallen/leaning *Acacia* sp. trees are a visibly common occurrence in agricultural field boundaries and river bottoms. *Prosopis* spp. and *Capparis* spp. also experience growth spurts following an inundation; they do not, however, flower until their fourth or fifth year of life. Mature *Prosopis* spp. and *Capparis* spp. trees have deep taproots, as well as lateral roots, that are resilient adaptations to weakening soils and rising groundwater, allowing them to survive occasional flooding and benefit from the abundant water.

Of the tree genera studied, those in the Fabaceae, *Acacia macracantha* and *Prosopis* spp., are both heavy fruiters, increasing their above-ground biomass by two or three times (Ocampo 1991; Risco 1998). In an ENSO event occurring around December or January it is not a question of if these trees will fall over, it is a question of when and how hard they will fall. While it is clear that some *Prosopis* spp. have deep taproots they do not prevent keeling even in the most extreme instances. Yet, while they do seem to be vulnerable to keeling over when laden with heavy fruits, keeling does not seem to kill the tree. This situation, however, is not the case for *Acacia* that both keels and dies. On the other hand, the *Acacia* is adapted to this environment in three distinct ways: first it is a fast grower with sun-tolerant saplings, and second it is a heavy fruiter with two fruiting periods per year. In both instances, these adaptations demonstrate *A. macracantha* as a pioneer and early successional species that permits quick and resilient repopulation of the species. Lastly, *Acacia* seeds are extremely resilient, passing through the gut of a ruminant animal, either a camelid, pig, or cow, and still remaining viable in the soil until the next ENSO event, sometimes a several year wait. Its rooting system is mostly adapted towards collecting surface moisture, however, and does not appear to be adapted to the weakening ground stability and potential for diluvial force. *Prosopis* spp. and *Capparis angulata* are adapted in similar ways to their environment, yet flower only once per year, normally; however, both include the early development of taproots during the principal days of germination and throughout their development in their shade tolerant youth. These characteristics serve these trees well during the onslaught of the ENSO events particular to the region.

One final and important observation that I make regarding dry tropical forest regeneration and the unique nature of the La Leche River Valley forests is regarding forest fires. Throughout the world we see that fire plays a dominant force in forest regeneration in dry areas (Block and Richter 2000; Maas 1995). One merely needs to look at the examples of the North American west to understand the importance of fire and to see seasonally dry forests as pyrogenic landscapes (Pyne 1998). Wherever trees grow and water is a limiting factor forest fires appear to be an important element of regional ecology. These are pyrogenic ecosystems. In our instance, of heavy deforestation in this historic period, forest fires in the La Leche Valley are not clearly documented in the manner that forest fires in adjacent areas of dry tropical forest are (Block and Richter 2000). Looking for documentation of forest fires in my research area may, in fact, indicate that during the Colonial or modern period there were few forest fires due to deforestation from logging, charcoal production, fuel procurement, and grazing of Old World animal species. My intuition, given my research and a potentially limited population pressure for the Middle Sicán (est. pop. 250,000, Shimada personal communication 2007), is that there was much more forest on the ground during the Middle Sicán. So it is interesting to note that, given my assumption for more forest on the landscape in the Middle Sicán, to date, we have seen no evidence of forest fires in the Colonial or pre-Hispanic period. I will return to this idea in the conclusion, Chapter 8.

Ethnotaxonomy of Algarrobos (*Prosopis *spp.)

The case of the famous milenario tree (Figure 2.4), and others like it (Figure 2.7), is an important one to consider, culturally. The architecture, in terms of growth form, of these trees is distinct, as is their place in local conceptions of forest ecology. In all cases where these ancient and revered trees are found in forested areas, away from human settlement, their branching and main bole is best described as spreading or network, almost spider like. For people living and working in the area, this form of growth is distinct to them, and frequently I had conversations initiated by my crew regarding the odd, rambling branching and horizontal trunk growth of these trees. The tree is described anthropomorphically as creeping (*arrastrandose*), crawling (*gateandose*), with its legs (*sus piernas*), arms (*sus brazas*), and on its knees (*en sus rodillas*) moving across the surface of the ground. This growth phenomenon occurs in the local typology of algarrobos (Table 2.2). The ethnotaxonomy for plants in general, is based on a principle of male (*macho*) and female (*hembra*). Most generally, tall, robust, and thorny trees are *macho*, and those without thorns, with lighter color leaves and more slender boles are *hembra*. The rambling and horizontal growth of these particular *Prosopis* spp. trees is characteristic of the female or *hembra*. These distinctions are not based in any way on sexual dimorphism, as the flowers are perfect (bisexual), and are more likely to be salient based on species or variety distinctions at the systematic level, see below. Most curious for me, is that the female trees are clearly the oldest trees, the most revered trees, and the most talked about trees in the forest, hence their being robust, historically, became of interest to me.

This quality provoked a discussion among my closest informants leading to developing a folk-typology of these trees. Local people classify the 'legendary trees' as female/*hembra algarrobo*. They do not have spines, and are clearly more resilient to ecological stress than the macho variety. One of the distinguishing characteristics of these trees salient to my informants is directly related to growth. These are the tap root systems that support the female trees, making this class of algarrobos both resistant to drought in their ability to reach deep groundwater systems, and to blow down/fall over due to inundation and storm systems. Ecologically, this explains their longevity and apparent abundance. Culturally these distinctions are important as male and female serve different purposes medicinally, in terms of beekeeping benefits, and for practices associated with traditional healing practice.

Figure 2.7 Keeled Mature *Prosopis pallida* with New Branch Growth Corresponding to 1998 post-ENSO Growth

My informants' observations demonstrate that female trees are the best adapted to long-term catastrophic disturbance, a potential for why they maintain a legendary status. The spider-like or network branching appears to be due to bole collapse during heavy rains/waterlogged surface soils, which often occurs during ENSO events and heavy fruiting periods of the algarrobo trees in the Andean summer. Once growth restarts, often months after the ENSO event itself, the taproot and the new orientation of the tree towards

Trait	Male Algarrobo (*Macho*)	Female Algarrobo (*Hembra*)	Kettle/Naive Algarrobo (*Pava*)	*Prosopis pallida* var. *Pallida*	*Prosopis pallida* var. *Armata*	*Prosopis juliflora*	Male *Acacia macracantha* Faique (*Macho*)	Female *Acacia macracantha* Faique (*Hembra*)
Architecture	Divergent and Fastigate Branching Branching starts close to ground. Like a broom.	Branching is more deliquescent. Globular Crown	Branching is deliquescent. But the overall effect is more Mimosoid, layered branches, giving a pagoda-like appearance. Like a pyramid	Branching is more deliquescent. Globular Crown	Divergent and Fastigate Branching starts close to ground. Like a broom.	Branching is deliquescent. But the overall effect is more Mimosoid, layered branches, giving a pagoda-like appearance. Like a pyramid		Tall, slender with divergent to orthotropic branching
Leaves	Wider with terminal leaves more abundant and bushy, pinnae are close together.	Smaller more delicate, more space between pinnae.		Smaller more delicate, more space between pinnae.	Wider with terminal leaves more abundant and bushy, pinnae are close together.		see algarrobo	see algarrobo
Leaf Color	Dark Green	Light Green		Light Green	Dark Green		see algarrobo	see algarrobo
Leaf Groups	Clumped	Smaller groups		Smaller groups	Clumped		see algarrobo	see algarrobo
Pinnae	Thicker	Thinner		Thinner	Thicker		see algarrobo	see algarrobo
Flower							very fragrant in the evenings	very fragrant in the evenings
Fruit	Yellow	Yellow-Purple	Purple	Yellow/Purple	Yellow	Purple	black	black
Fruit Quantity	same	same	same	same	same	same	more	less
Fruit Size	Wider	Longer/Slender		Longer/Slender	Wider		smaller (2-3 sections)q	larger (more than 2 sections)
Seeds	same, seed lasts longer than that of lucuma, each seed its own chamber in the fruit	same	same	same	same, seed lasts longer than that of lucuma, each seed its own chamber in the fruit	same		
Bark	Thick and Shaggy	Fine, Planar, Smoorth		Fine, Planar, Smoorth	Thick and Shaggy		Shaggier, Bole is dark, runs with sap	Smooth
Bark Hairs (Nectaries)	Yes	No	?	No	Yes	?	Yes	?
Wood Color	Black/Dark	Lighter like cinnamon		Lighter like cinnamon	Thick Dense Heartwood		Dark Cinnamon/ Purple	light cinnamon/ light yellow
Extent of Heartwood	Thick Dense Heartwood	Less Heartwood		Less Heartwoood	Thick Dense Heartwood		Dark Inner Heartwood	
Resin	yes, black	no/almost none		no/almost none	yes, black		yes, black	
Spines	Yes	no/almost none		no/almost none	Yes		larger	smaller
Spines (Size)	Big and Fat	Scanty, thin and small		Scanty, thin and small	Big and Fat			
Roots	More buttressed, without taproot.	more or less buttressed with tap root		more or less buttressed with tap root	More buttressed, without taproot.		Less susceptible to coso larvae.	more susceptible to parasitic moth larvae
Life Span	Falls in Rain, Grows like a Man (slow and strong).	Tall:Grows like a man	Grows like a man	Tall:Grows like a man	Falls in Rain		Strong:Grows like a woman, tall and thin, grows fast)	Shorter (10 years with the moth larva 'coso'):Grows like a woman

Table 2.2 Ethnotaxonomy of *Prosopis* spp. and *Acacia macracantha* with Ecological Information

sunlight, leads to these trees in high stature mature dry tropical forest in the region. In addition to the *macho/hembra* distinction, I also learned that most people can distinguish these two and a third different varieties of algarrobo. For the most part, as I indicate in Table 2.2, I believe that the two varieties of *P. pallida* var. *armata* and var. *pallida* that are accepted Linnaean taxonomic classifications are mirrored as salient categories understood by local residents of the Lower La Leche (Pastor Rozas 1993). Based on my interest in the identifying these trees through their anatomy this topic became of further interest to me.

The third variety of algarrobo is the hen-turkey/*pava algarrobo*. While there was only one in our research area to look at, it seems that the growth habit, leaf structure, and branching are all distinct, and much like what is recorded for *P. juliflora* (Pascieznik et al. 2001). In this instance, it seems that the ethnotaxonomy described in Table 2.2 roughly corresponds to Pascieznik et al.'s (2001) recognized classifications for these two species of *Prosopis*, *P. pallida* (*macho/hembra*) and *P. juliflora* (*pava*) that they document as interbreeding and ecologically indistinct species in the region. As I cannot approach this level of distinction in my own research, I simply state here that these species both exist in area, are recognized to the distinction of variety and species by local populations, and only present my data, as it relates to these species as *Prosopis* spp. in general.

The reason for including this ethnotaxonomy as a part of my interpretation the modern ecological system is to indicate that the local residents have particularly developed and savvy understanding of the living components of their regional forests. These nuances are elusive even to botanical field taxonomists. Yet the descriptive nature of their taxonomies indicate that cutting and carrying of resources is understood, at least for some of the residents of area, at very profound level of ecological and botanical knowledge. While I could continue to develop further ethnotaxonomies of all plants in the area, I focused here on *Prosopis* spp. due to its prominence in the local forest ecology, modern conservation efforts, cultural practice, and lastly as an exceptional high-heat long-lived fuel resource. Clearly, collecting a whole set of ethnotaxonomies would lead to a very well developed companion to any archaeological data set in the area, but in this case it allowed me to illustrate the taxonomic nuances of *Prosopis* spp., and explain some of the problems associated with its taxonomic distinction.

Where this section of the research becomes critically important is that it locates forest resources, and trees in particular as historical and culturally important aspects of the landscape. Understanding the role that landscape features play in society, like long lived trees, is one way to begin to decipher and reclaim histories of land use in the region where these datasets are otherwise elusive. Basso (1996), Bloch (1998), and Santos-Granero (1998), have all demonstrated that landscapes function as mnemonics for historical narrative and identity formation. In the case of the trees of the La Leche River Valley, paying attention to their ethnotaxonomy, and ethnobotanical salience begins to elucidate a whole set of land use practices, history construction, and identity valorization. These trajectories open up the potential to locate historical realities that are not yet recorded in the area, and may lend a deeper understanding of ancient practice and its continuity with the Lambayecano past. This particular approach has been particularly useful in trying to understand the anthropogenic effects on Lowland Amazonia, and in reclaiming an understanding of landscape change and use over the past 5,000 years where only scanty environmental proxy data exists (Balée and Erickson 2006). In the end, work dealing with the myths (Narvaez Vargas 2001) and colonial ethnography (Ramirez 1996) in the area, coupled with more intensive ethnobotanical fieldwork, may fill the many gaps that exist in our present understanding of landscape history in the La Leche Valley.

Forest Stature Implications for Huaca Sialupe

In sum, the kinds of trees present in the archaeological assemblage have the potential to suggest the existence of some of the dynamics that are visible within the modern ecological systems. For instance, the ratios of *Acacia* to *Prosopis* may indicate the successional status of the forest. The presence of late successional trees in the wood assemblage may indicate the presence of mature stature forest in the area during the Middle Sicán. With the above profile of both forest and climate dynamics, we have for the first time a rough picture of what pre-Hispanic forests looked like in the area; more importantly, we have established a baseline from which to interpret the fuel resource situation Middle Sicán artisans may have encountered on a daily or periodic basis.

To properly model or interpret the archaeological charcoal remains from the production contexts at Huaca Sialupe, we look to the modern forest to understand the dynamics of climate, seasonality, and species-species interactions. The different relationships, as observed, are far from complete, but serve to illustrate the basic dynamic among the dominant species sought as economic resources in the modern forest context. The interactions described above explain both the lack of *Acacia macracantha* in mature forest, as well as the growth architecture (Table 2.2) of most *Prosopis* spp. trees. At the same time, they explain how people intuit tree growth, forest health, and the impact of episodic ENSO disturbance in these forests. Granted this interaction is presented here within the modern context; regardless, pre-Hispanic indigenous peoples of this area were most likely similarly aware of their own versions of ENSO events and the impact on their lives and resource base.

Above, I developed a picture of the modern forest dynamic within which I believe the site of Huaca Sialupe was situated. With these factors in mind, we can turn to the situation that circumscribed the ancient workshop's realities. These included the artisans' needs themselves to procure the resources necessary to produce the fires for production at the site. To understand the potential for projecting this picture of the modern forest into the past, we now turn to a discussion of the cultural history of the area, during the 10th century to locate other factors likely affecting resource use and allocation at Huaca Sialupe.

CHAPTER 3
SITE ORIENTATION AND HISTORICAL BACKGROUND

Perhaps the most significant symbol of Middle Sicán material culture is the depiction of the Sicán lord or deity. Today it remains a key hallmark of Peruvian culture: along with *Macchu Picchu*, the famous *Tumi de Oro* of Peru (Figure 3.1) picturing the Sicán deity captured the imagination of the world during the 1960's, as it was caught surreptitiously leaving the country. It subsequently became a symbol of the archaeological riches of the Andean past, as well as an emblem of indigenous Peru (Figure 1.1). The design and the god figure, known from this ceremonial *tumi* or knife, composed of an alloy of gold and silver is also a benchmark for Sicán period mythic history, status and power, and technology. Making it a key component of Cleland and Shimada's (1998) discussion of the 'aesthetic locus,' of Middle Sicán society. Unfortunately, as a looted or *huaqueado* object, it has no provenience. But, after twenty-nine years of research by the SAP, we can reconstruct its historical and cultural context, including its emblematic or political meaning as well as the technology and economy that produced it. Following Shimada's example (1994) of using the *Tumi* as a metaphor for understanding the social and cultural developments of the Middle Sicán, we will develop a material and technological understanding of the Sicán period in this section.

History and Technology of the Middle Sicán

The *Tumi de Oro*, a cast artifact of gold and silver, is a type of artifact that the metalworks portion of Huaca Sialupe could have produced. The artifact itself presents us with two key variables that were important to Sicán polity identity and offer insight into the reconstruction of Sicán period political structure. The first is the image of the Sicán deity; the second is the use of gold-alloyed with arsenical copper in the production of a high status and high quality artifact. Artifacts like the *Tumi* and the equivalently emblematic and important gold alloy masks, which also use the Sicán deity as a symbol, are found throughout the world in art collections and museums. Yet, despite their wide distribution and distinction, their archaeological contexts were heretofore little understood before the efforts of the SAP began investigating archaeological contexts in the La Leche River Valley in 1978 (Shimada 1981). This work culminated with the excavation and recovery of two elite tombs in the principal ceremonial area of Sicán period activity located near Batán Grande, Lambayeque, which is now protected as the National Historical Sanctuary, Bosque de Pomác (Shimada 1995). Excavations and investigations continue in this area.

These excavations have enabled us to provide a context for these high quality artifacts, *tumbaga*/gold tumi knives and *tumbaga* masks; but more importantly the association of these artifacts with sepulcher-like elite tombs, monumental architecture, and other luxury trade items presents a window into Sicán period political economy. In all cases, these developments have a direct bearing on how we interpret the remains at Huaca Sialupe where goods, similar to those recovered in the Sicán site core, were produced during the same time period. The Sialupe contexts, in comparison to those of the ceremonial center at Sicán, present a radically different view of daily life and economy from the recovered tomb contexts. As a result, it is critical that these two distinct, but complementary sets of archaeological contexts, Sicán elite burials, the depositional or consumption loci of these high status artifacts, and Huaca Sialupe, a potential manufacturing center for these artifacts, be examined in tandem to reconstruct the historical context of the Huaca Sialupe production activities.

Figure 3.1 Sicán Deity on *Tumi de Oro* (modified from de Lavalle 1989)

In 1991-2 and 1995-6 Shimada and his project (1995; Shimada et al. 2007) excavated two deep elite Middle Sicán Period shaft tombs in the middle of the National Historical Sanctuary, Bosque de Pomác at the site of Sicán, below the Huaca Loro (Figure 1.2). In both instances the tombs were replete with iconographic representations of the Sicán deity both on blackware ceramics and gold sheet metal objects. Also, a wide variety of imported materials were present, e.g., feathers, shell, and gemstone ornaments. The chosen material for displaying the symbol was painted textiles, blackware ceramics, ceramics covered in sheet metal, tumbaga sheet metal masks and ornament, and cast tumbaga and copper alloy objects. In some cases, the ceramics, depicting the deity, were covered over in tumbaga sheet, again indicating the idea of the Sicán deity as the 'aesthetic locus' (Cleland and Shimada 1998 Shimada 1995). Where to be treated in the utmost, it must be coated, covered, or depicted in *tumbaga*, all other representations being secondary (Shimada and Wagner 2007). In comparison with other graves excavated by the SAP during the past 20 years of excavation in and around the main precinct of Sicán, it has become clear the highest degree of treatment of the Sicán deity, *tumbaga*, these items and their symbolic representations do not occur in non-elite burial contexts (Shimada 1995). Instead copper-arsenic metal tools, ostensibly without the Sicán deity, do occur with a generalized frequency throughout the Middle Sicán populations indicating that these metal goods were available without the same amount of elite currency or symbolic charge (Cleland and Shimada 1998: Shimada 1994b). The political and organizational aspects of these materials are discussed in better detail in Chapter 4.

From the SAP's work we know a great deal about how ceramics and metalwork were made, and have some key ideas of how these goods circulated. Middle Sicán ceramics can be broken down into two categories for the purposes of understanding material culture and technology. *Paleteada* wares, produced by a distinctive paddle and anvil technique, were the more ordinary decorated wares of the period (Cleland and Shimada 1992). These, together with large ollas and larger storage/fermenting vessels, were made fired in oxidation kilns yielding ceramic redwares (Cleland and Shimada 1998; Shimada 2000). There are also the famous reduction fired blackware ceramics formed in molds with the standard Sicán deity motifs that were the currency used in elite and upper status burials (Shimada and Wagner 2001; Shimada and Wagner 2007; Wagner et al. 2001). The deep black metallic sheen was an achievement of technological complexity that was apparently attempted from the Formative Period through to the Middle Sicán, when that was finally achieved (Shimada and Wagner 2007). Shimada refers to this as an aesthetic locus, borrowing a term from Maquet (1979), indicating that sheen and metallic qualities of artifacts were premiums for Sicán elites (Shimada and Wagner 2007). The development of blackware technology led to Sicán products having a rapid and wide distribution throughout the Andes (Lyon 1991:32).

Additionally, there are two different kinds of metalwork visible for the Middle Sicán, ordinary copper-arsenic alloy metal tools that apparently were a common denominator for Middle Sicán La Leche River Valley residents, e.g., chisels, needles, blades and spoons, (Bezúr 2003; Shimada 1984; Merkel et al. 1994), and the high status gold, copper, and/or silver sheet metal materials seen in elite Sicán burials (Bezúr 2003; Shimada et al. 1999; Shimada and Wagner 2007). Moreover, for the Middle Sicán we know that there was textile production culture of finely colored cotton and wool materials (Shimada 1995; de Lavalle 1989; Shimada and Montenegro 2002). From these finds and archaeological contexts, we see that the iconography that became so symbolic for modern Peru (Figure 1.1), as visible on the *Tumi de Oro*, permeated Middle Sicán culture, yet in combination with certain material culture is restricted in its distribution. Additionally, we see that its presence in the archaeological record is concurrent with other trends in material culture and developments in technology and trade.

Craft Production and its Organization at Sialupe

During its approximately 100-year existence, the Middle Sicán workshop at Huaca Sialupe manufactured a variety of products in at least two media: mold-made, oxidized and reduced fine ceramic wares, and formed copper arsenic and gold alloy sheet metal. The SAP's interpretation of the site relies on a team of materials scientists, archaeologists, botanists, and artisans. Based on the remains at the site, we have derived the following understanding of production (Shimada and Wagner 2007):

1. Ceramic and metalworking were concurrent activities at Sialupe.
2. Metal objects produced at the site included both high status materials, e.g. gold alloy sheet, and utilitarian objects, e.g. Cu-As (Copper-Arsenic bronze) needles and spoons (Bezúr 2003).
3. Similarly, ceramic production consisted of high status burial goods, e.g., molded blackware spouted vessels, and utilitarian wares, e.g., with a wide variety of deep and shallow bowls (Shimada and Wagner 2007; Taylor 2002).

The social correlates of these material remains lead us to believe that artisans at the site were producing materials for high status ceremonies, as well as for local non-elite consumption; we have recovered evidence of both high and low status produced goods from the Middle Sicán ritual center at Sicán in distributions that demonstrate only the restricted consumption by non-elites of tumbaga alloy materials that display the Sicán deity (Cleland and Shimada 1998; Shimada 1995; Shimada et al. 1999; Shimada and Wagner 2001; Shimada and Wagner 2007). At the same time, Sicán artisans may have produced goods for a socially and economically diverse local Middle Sicán population and geographically removed foreign consumers, evidenced by the presence of Sicán copper arsenic tools made from sheet metal and *naipes* (axe monies) in burials and along the central Andean Coast from the Jequetepeque River in the south to Central Ecuador in the North (Shimada 1985). From these data, it is possible that the products of these artisans were integrated in both the local and external economies known to Middle Sicán society.

Huaca Sialupe Ceramic Production

At Huaca Sialupe, the remains of both oxidation and reduction kilns with distinct morphologies were found at the downwind, north end of the Mound I portion of the workshop (Shimada and Wagner 2001; 2007). This location is logical for removing the heat, smoke, and debris generated by kilns and furnaces. Oxidation and reduction kiln features have distinct morphologies (Shimada and Wagner 2001; 2007). Reduction firing was done in elliptical/teardrop-shaped kilns with the large firing chamber oriented downwind (Figure 3.2; Shimada and Wagner 2001; Shimada et al. 2003); oxidation firing employed a similar ovate shaped, but larger kiln. In all cases, our archaeometric data suggest that local clay sources were responsible for the majority of ceramics produced at the site (Wagner et al. 2001).

Field and laboratory firing experiments (Wagner et al. 2001) as well as a comprehensive archaeometric analysis of samples of excavated and experimentally fired ceramics have demonstrated that the widely valued glossy Middle Sicán black pottery owes its distinctive appearance to a number of factors. The vessel to be fired was first thoroughly burnished in a leather-hard state to

Figure 3.2 Reduction Kiln

create a charged surface, onto which a smooth layer of amorphous carbon from organic compounds (such as dried llama dung) formed during firing under strongly reducing conditions at temperatures of between 800 and 900°C (Shimada and Wagner 2001, 2007; Shimada et al. 2003; Wagner et al. 2001). Thin-section microscopy and scanning electron microscopy showed that carbon penetrated a few millimeters into the body of the ceramics, contributing to the uniformly and truly black coloration. The metallic sheen found on many Middle Sicán black vessels is related to the formation of graphite crystals on the surface through prolonged exposure of the carbon to temperatures of at least 900°C (Wagner et al. 2001). Shimada and Wagner (2007) have recently posited that it is this metallic quality of the sheen that was critical for Sicán elites when selecting their vessels as some incompletely fired vessels were found in elite tombs covered in *tumbaga* sheet. The main products of blackware production were single and double spout mold made bottles. The double spout bottles, in particular, appear to have been the currency of elite mortuary practice.

Oxidation firing employed a similarly ovate kiln structure that was slightly larger in size (Figure 3.3). These were built of adobe bricks set in mud mortar and placed in pits circa 125 × 60 cm wide and 35 cm deep. The preserved portions and scattered structural remains suggest that the kilns used incurving walls to create semi-closed firing chambers. All but one of the kilns were oriented north–south, taking advantage of the prevailing winds for drafts and smoke removal. Associated with these kilns is a series of elongated beds of ash, charred twigs, and pulverized charcoal found nearby that we interpret as the remains of vessel preheating areas used before firing. Because of the overall good state of preservation of these features we can reconstruct a sizable portion of both oxidation and reduction firing processes.

In both cases, oxidized and reduction wares, a number of tools were used for producing the ceramics. Most importantly were piece molds that formed a large part of the ceramic data set recovered from Sialupe. Shimada and Wagner (2007; Taylor 2002), indicate that mold remains demonstrated the need to create some form of predictable product. On the other hand, however, there are several instances where molds, and the matrices from which they were formed, included forms or shapes that combined the Sicán deity with other motifs. This tendency indicates that, contrary to the way molds are usually thought of in reconstructions of ancient ceramic technology (Arnold 1999), Sialupe artisans were actually experimenting with form and symbol using molds. Other tools in the Sialupe tool-kit include turning plates, burnishing stones, matrices, molds, formed appliqué, prepared clay bodies, and unfired bottle or plate bases. Most interestingly was the recovery of several inscribed/marked molds indicating that certain makers/artisans marked or claimed them. While the number was insufficient to complete a clear systematic study, it did offer that at least two different groups of artisans were using the site to produce similar, or the same, products. Another indication was that some supervisory control of production was present, either by artisans or patrons, at the site at least in terms of the production of particular bottle forms (Shimada and Wagner 2007). Based on the experimental work carried out in 2000 the SAP has been able to contextualize at least some of these production remains in terms of the overall ceramic production process.

Experimental Ceramic Production

Complementing the archaeological insights are those derived from the SAP's firing experiments. We conducted a series of firing experiments at the southern edge of Huaca Sialupe in July 2000.. The goal was to gain comparative data for understanding the material and behavioral correlates of Middle Sicán ceramic firing and metalworking and the physical performance of the associated kilns and furnaces. Such experiments afford a unique opportunity to observe the decision-making processes and behaviors involved in successful firing, which is commonly regarded as the most challenging stage of pottery making (Shimada and Wagner 2007). The 2000 experiments built on experience and knowledge gained from similarly conceived research conducted in 1992 using a 3,000-year-old kiln and its modern replica (Shimada et al. 1998; Shimada et al. 2003; Wagner et al. 1998).

Figure 3.3 Oxidation Kiln

José Sosa, a professional ceramist, prepared a full-scale replica kiln and pottery used in the 2000 experiments (Figure 3.4). Sosa is from Chulucanas, Piura, a well-known pottery-making town, who has a good working knowledge of pre-Hispanic ceramic technology. The replicas approximated the measurements, materials, and techniques as determined by our prior excavations and attendant archaeometric analysis of relevant samples (Shimada and Wagner 2001; Shimada et al. 2003; Wagner et al. 2001). Fuels for the experimental firings consisted of dry, fallen hardwood branches (1–4 cm in diameter) collected from the forest floor. *Prosopis* spp. was used at a ratio of 5:1 by weight to *Capparis angulata*. Both species of wood exhibited bark beetle burrowing on the outermost wood surfaces of the collected materials. Our ceramist used leaves and small twigs as kindling, and occasionally to boost combustion in the later stages of firing (Figure 3.5). He also charged the kiln with crushed, dry llama dung (brought in from Callejón de Huáylas, La Libertad) and immediately and thoroughly sealed it with sand and mud to create a reducing atmosphere.

The reducing conditions in the kilns not only produced the carbon/graphite layer on the surface of the ceramics, but also reduced the hardwood fuels to charcoal. This charcoal remained in the kiln in significant amounts after firing was complete (ca. 6 kg hardwood charcoal). We left the kiln with its load of some dozen vessels to cool naturally for at least two hours (to a temperature of ca. 100°C) and, once, overnight. Sosa continuously monitored the kiln for the entirety of each firing. Independent of Sosa's own observations, we placed four nickel-cadmium thermocouples, connected to digital thermometers, in various locations within the kiln to monitor its performance. Our experiments produced glossy black pottery that closely approximates that produced 1,000 years ago at the Huaca Sialupe workshop. The kilns also effectively reduced the majority of the substantial branches used for fuel to charcoal, and not to ash. They also yielded vessels with incomplete reduction and mottled appearance, just as we have documented among wasters recovered from Huaca Sialupe. Subsequent archaeometric work has confirmed that we were successful in recreating, physically as well as chemically, Middle Sicán blackware technology through our experimentation (Shimada et al. 2003; Wagner et al. 2001).

Evidence from the workshop indicates that a wide range of relatively small, decorated vessels were mold-made at the site. Some vessels were embellished with appliqué or stamped decoration (Taylor 2002). Important to note is that across the site we have recovered a variety of mold remains and entire mold fragments (Shimada and Montenegro 2002; Shimada and Wagner 2007). These indicate that the primary iconographic canon of the Middle Sicán, the Sicán deity, was fully employed at the site. Mold-made ceramic technology ensured the accurate and systematic deployment of the appropriate iconography. Additionally, discreet marks, potentially indicating artisanal ownership of individual molds, or at least authorship, appear on many of the molds recovered (Taylor 2002).

To date we have determined that at least two different groups of artisans were working to create mold made ceramics that deployed the deity's image (Shimada

Figure 3.4 Experimental Kiln

and Wagner 2007). In every case (200 of circa. 2000 mold fragments), it is clear that molds were probably made on site, as the clay signature is similar to that of the local clay sources. The appearance of the iconography, its apparent standardization through mold technology, and the marks on the exterior mold surfaces indicate that production of these ceramic types, both oxidation and reduced wares, with or without the presence of the Sicán deity, was organized and not a haphazard conglomerate of artisans sharing space. Clearly they were capable, and likely did, produce high status goods that were appropriate for deployment to elites. Shimada and Wagner (2007), as I will indicate in Chapters 4 and 8,

Figure 3.5 Fuel Prepared for Firing

believe that these artisans were largely working autonomously under the supervision of a master craftsperson that was directly responsible to deliver goods to elites.

Huaca Sialupe Metal Production

We also documented a cluster of metal working furnaces at the south end of the site, each consisting of at least six furnaces. Furnaces were constructed from large, inverted ceramic urns firmly planted into the floor, known locally as *botijas* or *porrones*. These are oxidation-fired ceramics and could easily have been produced on site, especially if large open pit kilns were used as described by Cleland and Shimada (1998). However, no remains of Middle Sicán period open pit kilns were recovered at the site. An undated open pit kiln was recovered from the south side of Mound III, but only minimally excavated and no relative dates were recovered (Shimada 2000). The excavated metal furnaces contained beds of charred fuels (ca. 8 L) and each had a round 8 cm draft hole cut into the body near floor level, facing the direction of the prevailing wind; like the ceramic kilns, craftspersons positioned the furnaces to maximize the natural draft. Additionally, each furnace included two short walls flanking the draft hole to funnel wind towards the fuel inside the furnace (Figure 3.6). The slag, spills and scrap metal recovered in association with these furnaces indicate that the furnaces were used for melting ingots, as well as for the alloying and annealing of hammered sheet metal (Shimada and Wagner 2001; 2007). The SAP's reconstruction of Middle Sicán copper alloy technology indicates that copper alloy ingots were formed near to mining operations in the La Leche River Valley. These ingots were transported to and made available to artisans at a site like Huaca Sialupe where they would turn the ingots into artifacts (Merkel and Shimada 1991; Shimada and Wagner 2001; 2007).

Figure 3.6 Metal Furnace (Feature 8-'01)

Experimental Firing of a Metal Furnace

Similar to the SAP's investigation into ceramic production, in 2000 we reconstructed one of these metal furnaces as an experiment to understand how they functioned. We constructed our replica metalworking furnace with locally-made adobe bricks, clays, and a *porrón* (large ceramic storage urn) that was made in the nearby town of Morrope closely matched in size, shape, materials, and technique to the Middle Sicán originals previously documented at Huaca Sialupe (Figure 3.7). Unfortunately, as there have been no indigenous copper-arsenic metal workers making objects in Northern Peru for over 500 years, there was no expert available to help us run the furnace. Instead the experiment was to see how the furnace worked relative to the natural drafts, the fuel charge, and its morphology. We charged the furnace with *Prosopis* spp. charcoal, taken from the fuel remaining from the kiln experiments, to the height just below the bottom of the circular draft hole that faced the prevailing wind. Our experiment showed that, with only natural wind fanning the charcoal bed, the furnace is capable of achieving an internal temperature between 1,000 and 1,100°C within 20 minutes from the time of ignition. Egressing air at the top of the furnace also reached 800°C.

Moreover, it was evident that we could easily control the furnace temperature to suit the variety of metalworking tasks, such as melting ingots, alloying metals, and annealing metals, found in association with the original furnaces. We accomplished this level of control by adjusting the draft entering the furnace with a small potsherd and altering the internal height of the charcoal fuel load. The experiment effectively demonstrated that the Middle Sicán furnaces made with inverted urns were highly efficient and flexible devices well suited for a variety of metalworking tasks. Additional details of the experimental setups and the results of an archaeometric analysis of fired vessels and

Figure 3.7 Experimental Furnace

the kiln lining are described elsewhere (Shimada and Wagner 2001; 2007; Shimada et al. 2003; Wagner et al. 2001).

The SAP research suggests, based on our experiments in 2000 and the experiments carried out at Huaca del Pueblo de Batán Grande, that Sicán period metalsmiths utilized wood charcoal as the fuel of choice, as it is a physically stable fuel at high temperatures, around 1200°C (Merkel et al. 1994, Shimada and Merkel 1991, Shimada and Wagner 2001, Shimada et al. 1999). This observation led to the construction of an experimental small-scale charcoal furnace during my 2002 field research. Observations by Shimada (1985) and recent work by Cayo (2003) demonstrate that charcoal can be produced in quantity at the household or small scale for use in metal production (Cayo 2003) or for general household use and resale in the case of my 2002 work. In the field, I observed that one person easily could collect 500 Kg of *Prosopis* spp. and construct a small charcoal kiln to produce around 250 Kg of hardwood charcoal. The labor investment to set the kiln and collect the wood requires a single day for one person; the firing takes four days. The main limiting factor in the modern case is the need for a steel axe and/or chainsaw if large branch or trunkwood is desired.

The labor investment and productivity of these small-scale furnaces depend on the quantity of wood used, but this experiment demonstrates the ease with which charcoal can be prepared. The wood is arranged in a set of parallel lines on an area of ground cleared of loose soil and debris. Once arranged, a layer of *poña*, or kindling, collected from the surrounding area is laid on top of the logs. *Poña* consists of leaves, refuse, small twigs, and animal dung gathered from below the trees in the area surrounding the kiln (discussed further in Chapter 5). The entire pile is covered with compact peds of soil dug from the perimeter of the piled logs and kindling (Figure 3.8). Hardpan clods are preferred, and loose dirt is saved for the last, uppermost covering of the kiln above the clods. Overall, the processing of 500 Kg of wood into charcoal results in a pile 1.5 m high with a surface area that covers 4 × 6 m. A small area is left open at the downwind side where a small initial fire is started, called the *caja* (Figure 3.9). Once the *caja* catches fire, it too is covered, and the fire is left for several days to burn through the entire pile. During the second day, small vents are made at the end of the kiln opposite to the *caja* by inserting long straight sticks about 1 cm in diameter to let the smoke and gases escape. When the firing is complete, the charcoal makers carefully excavate the pile; the carbonized branches are removed intact, often in single pieces.

From my observations it is not surprising that the branches of *Prosopis* sp., or any of the woods used at Sialupe, retained their overall structure during and after firing, and survived buried some 1000 years in the soils of Northern Peru. Even after eight continuous days of firing in a charcoal kiln; most woods from the dry tropical forest maintains their overall form, structure and appearance of a tree branch. Reduction kilns, see Chapter 3, function along the same principles as a charcoal furnace and would have produced similar results with similarly carbonized wood structures. As a result of this experiment, it became clear that charcoal making could have easily been done on a small scale, left a relatively limited archaeological signature, and could have been incorporated into metalworking production in close proximity to actual workshops.

Figure 3.8 Charcoal Kiln Laid

Metal production, however, is clearly a two-part process. First ingots must be prepared from excavated ores, and then these materials are worked to fabricate objects of ornamentation and utility (Shimada 1994b). Please note that quality fuel charcoal is necessary throughout the process and charcoal production probably went hand and hand with both parts of Middle Sicán metal industry. From our excavations we see no indication for the production of copper alloy ingots at Huaca Sialupe.

For the most part, this observation makes sense. Archaeological and experimental work at Huaca del Pueblo de Batán Grande, some 40km into the La Leche River Valley and the nearby site Cerro Cementarios/Cerro Huaringa-- both closely situated to the Sicán main site-- were the likely sources for ingots worked in the Middle Sicán. Work by Shimada et al. (1982), Merkel and Shimada (1991), and Shimada (1994) have reconstructed the overall production of arsenical copper alloy ingots in the La Leche River Valley. Along the sides of the upper and lower valleys are a number of

Figure 3.9 Covered Kiln/Stoking the *caja*

sites where copper ores were extracted throughout antiquity and into the modern day (Shimada 1994b). Merkel et al. (1994) describe that ores found in these areas adequately characterize the copper alloy ingot processing remains and artifacts found at Huaca del Pueblo de Batán Grande. Shimada (1994b) and Merkel et al. (1993) have argued that the kinds of workshops evident in the Sicán period indicate that small-scale, local production of copper alloy ingots probably fueled the demand for copper alloy in the valley. It appears that Middle Sicán ingot production took place in close proximity to ore sources around the La Leche Valley. Additionally, these mines and processing centers, like Cerro Cementarios/Cerro Huaringa, were in close proximity to both the mines and the valley floodplain, the site of the dry tropical forests that were the main fuel source for production. As there are no ore sources within sight of Huaca Sialupe, it is likely that ingots were brought to the site to be worked.

We do see, however, small metal molds, crucible fragments, small metal tools, etc, all at Huaca Sialupe indicating that some metal foundry work, aside from sheet production, took place at the site (Shimada and Wagner 2007). As with ceramic production at the site, the metal technology necessitated communication and a strong economic relationship with areas in the La Leche River Valley to procure copper alloy ingots, including those with precious metal contents. Taken together, the depiction of deity iconography on ceramics and the processing of precious metals and copper alloy at the site also required a degree of political interaction with elite sectors of society that elsewhere, especially in elite funerary contexts, appear to be the recipients of some of the artifacts made at Huaca Sialupe, in particular gold sheetwork and double spout blackware bottles.

The final part of this chapter is an overall description of the site of Huaca Sialupe in relationship to its metalwork and ceramic production features. Additionally, however, I highlight features besides those described above that indicate other activities were taking place at the site, and that craft production, in terms of Middle Sicán elite goods, was not the only form of production that we see.

Huaca Sialupe Site Orientation

As stated before, the site of Sialupe consists of five small low mounds located approximately 7 Km northwest of the city of Lambayeque and 15 Km inland from the Pacific Ocean on the north coast of Peru (Figure 2.1). Sialupe is located in an area of sparse settlement and irrigated farmland that today is home to non-industrial agriculture. The site dimensions are approximately 250 m X 400 m; it was occupied from the early Middle Sicán to the Early Colonial period (C.E. 900-1532). The workshop areas discussed in depth here consist of two conjoined Mounds, I and II, within the five mounds that comprise the site. The preserved remains of the workshop cover an estimated 1750 m^2 (ca. 70 m N-S x 28 m E-W), approximately 40% (ca. 600 m^2) of which was excavated during two eight week excavation seasons during June-August 1999 and 2001 (Shimada and Wagner 2007). The overall site boundary was established by the natural fall off of artifact densities recovered from the mounds during the 1996 and 1999 surveys (Shimada 2000).

The material relevant to this dissertation pertains to the 2002 excavations of the Middle Sicán occupation levels, during which we see the intense use of the site as a metal and ceramic production center. While production materials were scattered across much of the five-mound complex, the areas excavated exhibited, by far, the highest concentrations of fuel and manufacturing debris on the surface. Artifact production took place, to varying degrees, across much of the site; archaeological evidence, however, indicates that production centered most intensely and on the largest scale at Mounds I and II. From the surface collection of these Mounds in 1996, we recovered mold fragments, broken blackware vessels and sherds, all indicating ceramic production.

We have divided the 100 year Middle Sicán occupation levels into three different phases (A, B, C) centered on an abandonment of the site sometime around the Late to Middle Sicán transition (C.E. 1050), associated with a catastrophic rainfall event around C.E. 1050 (Shimada and Montenegro 2002; Shimada and Wagner 2007). Phase A begins sometime around 950 C.E. with the establishment of the site in the southern sector of Mound I where the main building that housed the ceramic workshop was located. Additionally, it seems that during Phase A, both ceramic and metal production was located in the Mound I complex initially. The overall preservation of these levels was poor, and we did not recover fully the contexts associated with this period. The Middle Sicán B floors form the majority of the contexts that were recovered pertaining to the Mound I and II workshop contexts. At this point in time metalwork was removed to the Mound II portion of the site exclusively. Middle Sicán C phase pertains to the post ENSO event reorga nization of the site where a ceremonial complex was constructed over the main workshop contexts at Mound I. Our discussion of fuel use at the site uses a more general time scale and covers the activities at the site, generally, as the Middle Sicán, focusing on the Middle Sicán B features up to the ENSO event of 1050 C.E., roughly 950-1050 C.E.

Mounds III-V (30 m due West of Mounds I and II) revealed contemporaneous Middle Sicán ceramic evidence of domestic habitation. Surface collections taken from these portions of the site indicate that they were in use concurrently with the production contexts of Mounds I and II. The refuse and surface finds during site survey in 1996 and 1999 associated with Mounds III-V do not demonstrate the same intensity of production remains, e.g., slag and mold fragments, as we find on the surface or in the excavated workshop contexts. These parts of the site are currently occupied by modern houses and subsistence farming compounds: Mounds III-IV occupied by the Sra. Tuñoque Vidaurre family and Mound V by Sr. Choso Santistiban (Shimada 2000). We need, however to further explore through excavation the relationship between the production contexts of Sialupe's Mounds I and II and the contexts of Mounds III-V at Sialupe to understand their interrelationship.

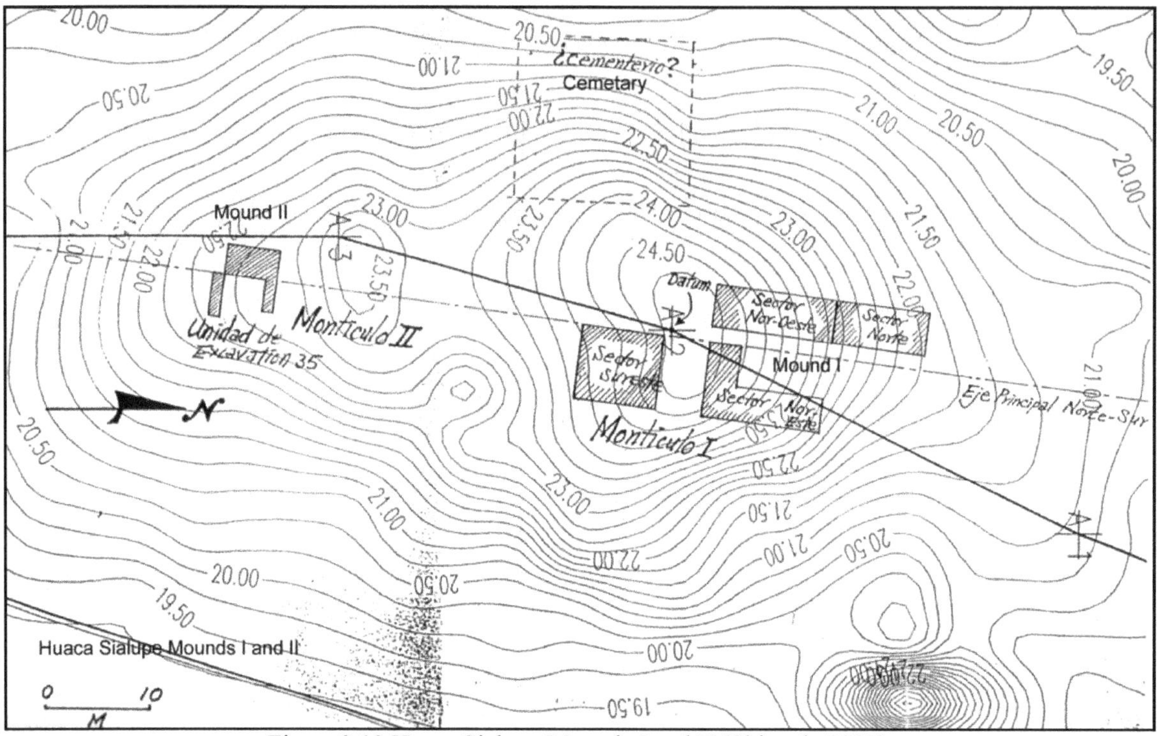

Figure 3.10 Huaca Sialupe Mounds I and II (Shimada 2000)

The site of Sialupe is one of several smaller sites in the area that represent settlements located around the larger Sicán-Chimú ceremonial-residential complex at Huaca Pared-Uriarte (Figure 2.1; Shimada and Montenegro 2002). Preliminary data from the surface collection indicate that the artisans at Sialupe were not engaged with supplying the bulk of the ceramic wares to this site, despite its proximity to the workshop 300 m to the South (C. Rospigliosi 2007). This indicates that the artisans at Huaca Sialupe were engaged instead in supplying the ceramic and metalwork needs of the regional economy of the La Leche delta and drainages, and were not specialists directly subsumed by the local ceremonial-residential center. We assume, overall, that a culturally salient site hierarchy existed in the Middle Sicán (Shimada 2000), and it is likely that Mounds III-V represent residences associated with the production units at Sialupe. We estimate that Mounds III-V represent dwellings for local craftspersons working at the Mound I and II workshops.

From the ceramic and metalwork production remains at the site we know that these were the central activities at the site; they were, however, embedded within a community of social and economic activities. Excavation of Mounds I and II demonstrate evidence of intensive craft production, ceramics and metallurgy, respectively (Figure 3.10). At a small cemetery adjacent to Mound I, we recovered human remains of 12 individuals, including children, from the Middle Sicán deposits, indicating a probable year round occupation of the site (Klaus 2003; Shimada and Montenegro 2002). These burials are severely fractionated probably due to the pre-Hispanic practice of returning to ancestor burials for unknown cultural reasons (Klaus 2003). The location of the Middle Sicán burial contexts close to the ceramic workshop, the presence of children in the burial population, and burial revisitation all indicate an intimate connection between local households and the workshop. As we look at the data from Mounds I and II at Huaca Sialupe it is easy to focus on production of status artifacts alone, yet clearly other attendant social, including domestic, or household processes, were taking place at the five-mound complex. These processes undoubtedly had an effect on how resources such as labor and fuel were both apportioned at the site. This idea will be addressed in the review of the datasets in Chapters 6 and 7 as well as in the theoretical framework outlined in Chapter 4.

Orientation: Mounds I and II

Based on floor plans and architectural remains we believe that Middle Sicán dwellings were similar in style and function to those in the area today. While we have not excavated the Middle Sicán households surrounding the workshop areas at Huaca Sialupe, excavations at other Middle Sicán sites (Shimada 1995) indicate that buildings were commonly constructed of *quincha*, mud plaster over woven reed and cane planar supports, buttressed between posts, set into low basal mud brick walls. Occasionally, these rooms were roofed with similar material and had floors of compacted earth (Correa Álamo 2000). The main workshop areas at Sialupe in Mounds I and II demonstrate this kind of construction. These construction technologies rely on locally available resources and are still employed in the area today (Figure 3.11; Huertas 1999: 133-142; Schaedel 1988: 72-101; Correa Álamo 2000: 73-79).

For the Middle Sicán levels at the site, activities primarily associated with metalworking took place at the south end of the site in areas designated Mound II; ceramic production was carried out at Mound I. While Mound II appears to have been largely open space, uncovered and open to the elements, Mound I consists of

Figure 3.11 Exterior House Wall Demonstrating *quincha* Construction

areas that were likely to have been covered. For the site tour, I begin with a description of the Mound II area and follow with a description of the more complex production areas of Mound I.

Orientation: Mound II

The area excavated at Mound II consists of a 15 m x 15 m square area with internal divisions and evidence of metal production (Figure 3.12). Running along the northern side of the area are the remnants of the exterior adobe block wall, the northernmost boundary of the Mound II production unit. Along the eastern third of the excavation area are a series of rectilinear rooms, with portions of adobe and/or *quincha* wall bases serving as boundaries. The eroded quality of the floors, lack of supports for roofing elements, and four bounding walls indicate that this space was most likely an open area. Remnants of *quincha* and adobe linear arrangements on the floor indicate that perishable walls may have been sometimes erected to divide space as needed. For the purposes of this analysis the majority of the metal working space is referred to as in the 'Open Area;' the small area bounded by at least three full walls is referred to as 'Room 3.'

A number of features are present in Mound II, which appear to be related to metal working production. Within the Open Area, we recovered the certain remains of four inverted porron type metalworking furnaces of the type described above (Figure 3.6; Features 59, 58, 70, and 8-'01). Most of these features contain small bits of charcoal, small branches, some uncarbonized remains, and a variety of domestic organic garbage. The large pieces of wood charcoal exhibit freshly broken surfaces. The broken urn feature in the center, north of the Open Area, is assumed to be another one of these *porrón* metal furnaces that was destroyed at some point prior to interment. There is also a feature that appears to have been a storage or production area for charcoal (Chapter 7; Feature 9-'01:Hearth 1 and Feature 9-'01:Hearth 2m Feature 12-'01), as well as cut-off bases of old *porrón*-style metal furnaces that contain charcoal (Features 19-'01, 10-'01, 20-'01, 21-'01, 14-'01). All of these features demonstrate exposure to heat, if not high temperature fires. Additionally, parts of the floors of both Room 3 and the Open Area of Mound II display evidence of extensive heating events, perhaps as expedient areas of hot coal use or removal from the furnaces. Finally, as will be discussed in Chapters 6 and 7, a large cache of compact and desiccated camelid dung was recovered to the northwest sector of the Mound II Open Area. Camelid dung is an important fuel in the modern Andean highlands (Sillar 2000; Winterhalder et al. 1974), and camelids were more present on the coast in the pre-Hispanic period (Shimada and Shimada 1985). The indication is that dung, stored for fuel, was being hoarded at this part of the site for use as fuel in the metalworks.

Shimada and Wagner (2007) in their summary of the activities that took place at Mound II indicate that metalworking was centered on two discreet clusters of furnaces (Features 8-'01 and 12-'01 and Features 58, 59, and 70). Additionally, the activities that took place in Room 3 were somehow unique as though segregated from the Open Area furnace clusters. The use of this room and its features will be explored further in Chapter 7. Bezúr (2003) indicates that copper arsenic residues were present throughout the Mound II Open Area complex indicating the forging of these metals. Additional information from Neutron Activation Analysis from Mound II (Shimada and Wagner 2001) indicates that up to 400 parts per billion (ppb) of gold residual trace elements were present in ash taken from at least one of the furnaces at Mound II. The background gold concentration in the site's soils is 1 to 2 ppb. The presence of gold in high concentrations indicates that it was probably being worked and forged at the site. At least one small piece of gold sheet was recovered from the charcoal residues during the 1999 excavations at Mound I (Shimada 2000), and pieces of copper alloy scoria, prills, etc., recovered from this sector of the site clearly mark it as a locus for metalwork. That said, however, it is considerably 'cleaner' in terms of its overall appearance from the Middle Sicán floors at Mound I where the ceramic production and firing took place.

Orientation: Mound I

Mound I, a 17 x 24 m excavated area (Figure 3.10; Figure 3.13), is a more complicated area to describe as it contains both a complex of rooms and a series of probably unroofed firing areas adjacent to the room complex. The South end of the Mound I excavations was dominated in the Middle Sicán by a roofed multi-roomed structure. Well-constructed adobe and/or *quincha* walls lined the exterior of the structure. The main room sets that indicate the contiguous use of space are Rooms 10, 13, and 29, Rooms 24, 25, 22, and 21, and Rooms 23 and 35. All of these room sets had direct access to the workshop firing areas adjacent to the building, the patio are labeled here as Rooms 38 and 39. The articulation between Rooms 30, 31, and 32 to the rest of the structure are unclear, but it is likely that there was step-over access

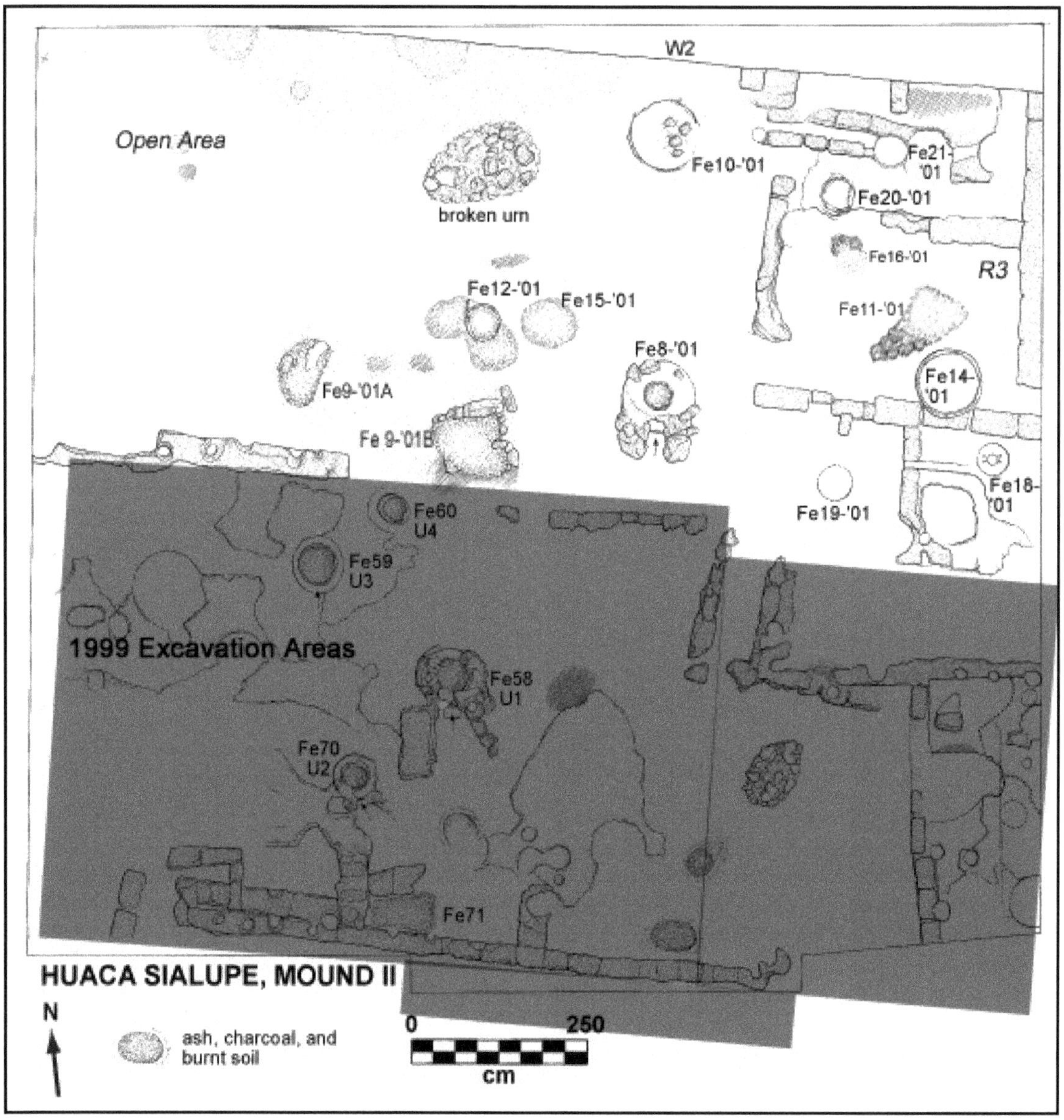

Figure 3.12 Features at Mound II

into these spaces from one another, either through Room 25 or Room 23.

The indoor spaces are mostly devoid of any fire-use features. Most of the features recovered from the rooms are either related to structural elements in the building, e.g., Features 64-66 that were column supports, or related directly to the activities associated with the building craft production, e.g., Feature 60, a cache of ceramic molds, Feature 67, a bowl set in the ground associated with a bone awl. Feature 69 represents a part of the roof/wall that fell in on Room 22, and indicates the use of *quincha*, more specifically made from *Gynerium* sp., a tall grass (bamboo relative), covered in mud and lashed with bark and leaves.

Most importantly, we recovered a variety of remains within the Mound I structure indicating where different parts of the ceramic making process took place. We recovered from Room 25 unfired mold made vessel fragments from molds recovered from the site, a potential ceramic turning plate, and several mold pairs and mold halves, many with makers marks (incised lines and other symbols), and some potential slip material, levigated kaolin. In Rooms 23 and 35 we recovered unfired bowl fragments and four, large prepared clay lumps, Feature 68. Their recovery, combined with polishing stones, mold and ceramic fragments and bone tools, indicate that most rooms on the south end of Mound I were related to the forming and finishing of ceramics (Shimada and Wagner 2007).

Relative to the issue of who was producing the ceramics at the site is the makeup of the ceramic mold dataset and the maker's marks. Marks recovered from the cache of molds from Room 23 did not match those found in Rooms 35. Both sets correspond to the stirrup spout and Sicán deity blackware vessels. Shimada and Wagner (2007) report that identical mold forms carried discreet mold vessels by room in the case of those recovered from Rooms 23, 35, and 25. Shimada and Wagner (2007) offer two potential interpretations. Either marks were designated for the artisans to distinguish between

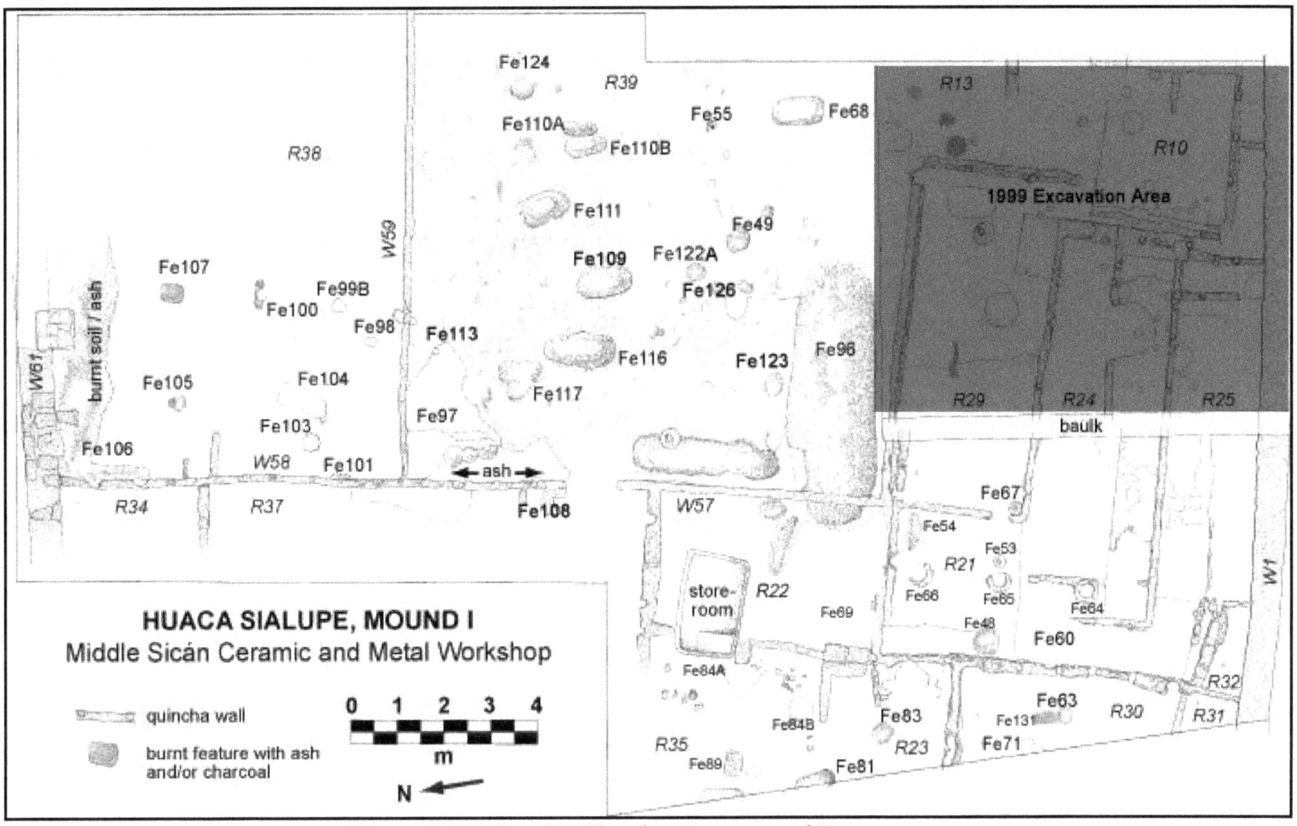

Figure 3.13 Mound I: Showing Features and Rooms

products produced by sponsor or by each artisan's ownership of the mold. While marks appear on the finished products as well, very few have been recovered and there are no conclusive matches.

The clear indicator from these markings is that space, either directed through sponsorship or by producers, was being shared by at least two groups of artisans working at the site (Shimada and Wagner 2007). Yet, the dataset, overall, is small in comparison to the total number of mold fragments and the SAP has only found two specific vessel matrices, representing the actual kind of piece made, that correspond to marked mold fragments. It may be the case, that the site was used, overall by a group of potters sharing most resources, save for those marked specially for certain individual potters or patrons. The further implications of marking, and the relationship to brick stamps will be addressed in the political economy section of Chapter 6.

Orientation: Rooms 38 and 39

The larger portion of the area excavated in the 2001 season contains the areas labeled Room 38 and Room 39. These two rooms were open areas surrounded by thin *quincha* stockade walls set on top of adobe blocks around the perimeter. This area corresponds to stockade or patio like activity enclosures attached to house structures seen in vernacular architecture of the colonial and historic periods (Schaedel 1988). Room 39 was the primary locus for reduction firing, described above; this process took place in the evenly spaced kilns set into the floor of the large, walled, but unroofed, area (Features 68, 109, 110A, 110B, 111, 116, 117: Figure 3.14). These kilns were elliptical or teardrop-shaped and their large firing chambers were oriented downwind. Their slightly concave, sloping, clay-lined bottoms were recessed about 15 cm below floor level and measured approximately 100 x 45 cm wide, with some degree of variation between them.

The excavated kilns were thoroughly blackened, including one with over 50 fine blackware sherds mixed among compacted charcoal powder. The associated floor of Room 39 was similarly blackened with charcoal. Fragments of collapsed linings indicate that the kilns originally had incurving sidewalls to create a dome-like firing chamber with a height about 30 cm above the floor. Ceramic kilns included relatively little charcoal; they generally appeared to have been cleaned out after use. The charcoal that they did contain, relative to their apparent volume, tended to be minimal and was composed of small twigs and branches. These features are also present in Rooms 35 (Feature 81) and Room 30 (Feature 131). Additionally, we see in southwest corner of the room Feature 96, an elongated shallow deposit of charcoal that was likely used as a preheating fire.

Room 38 (Figure 3.15) while of an almost equivalent size and shape to Room 39 appears to be devoid of ceramic kilns entirely. Instead we see a series of fire-use features with distinct signatures: five hearths (Features 98, 99B, 100, 105, 107), one broken interred porrón, as seen in Mound II (Feature 103), and a more generalized long ash and charcoal feature in a series of round lenses in the northwest corner of the room (Feature 106). Although kiln and metalworking furnaces are significant features at the site, hearths constitute the third and largest set of fire-use features that predominate in Rooms 38 and 39 at Mound I.

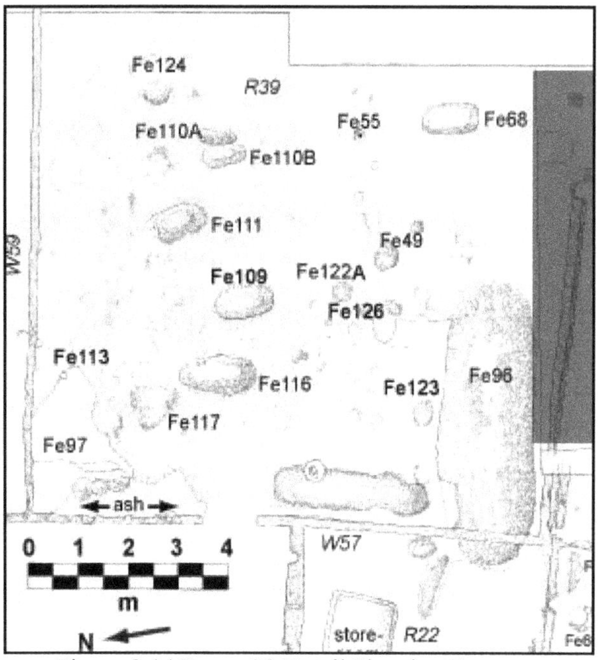

Figure 3.14 Room 39 Detail Showing Features

We distinguish hearths from the other fire-use features at Sialupe based on the following characteristics. Hearths occur in two different ways at the site, individual hearths (e.g., Features 105 and 104) and communal hearths (e.g., Features 106 and 97(Room 39)). Individual hearths are small, ca. 15 cm in diameter, and rarely larger than 30 cm. Seldom do these features have formed superstructures: they are typically shallow pits or depressions. Occasionally, however, hearths incorporated broken adobe block fragments, the rims or walls from broken large cooking or storage vessels or reused metal furnaces as firewalls or hearth boundaries (Feature 103). All of the hearth features demonstrate reuse as the ashy deposits occur in superimposed lenses, and occasionally reformed depressions or fire pits.

Communal hearths are areas of aggregated or agglutinated individual lens-like hearths (Figure 3.16). The communal hearth designation will be explained further in Chapter 7. We recovered two of these features in two aggregated areas associated with the Middle Sicán at Sialupe. The location and morphology of these features leads us to believe that they were ancillary to industrial production and pertain to the domestic economy of the site. The aggregated hearth features are what we believe to relate directly to suprahousehold food production. We have noted elsewhere that the wood charcoal and extent of burning in the hearths is qualitatively different from that seen in the furnaces and kilns (Goldstein et al. 2007). Instead, hearth charcoal represents branch wood from a wide range of sizes (2-5 cm diameter), representing the smallest twigs to remnants of large branch or trunk wood. All hearth features are also comparably 'dirty' relative to either the kilns or the furnace features: the hearths are ashy and sandy, often with ash extending between the individual hearth feature loci.

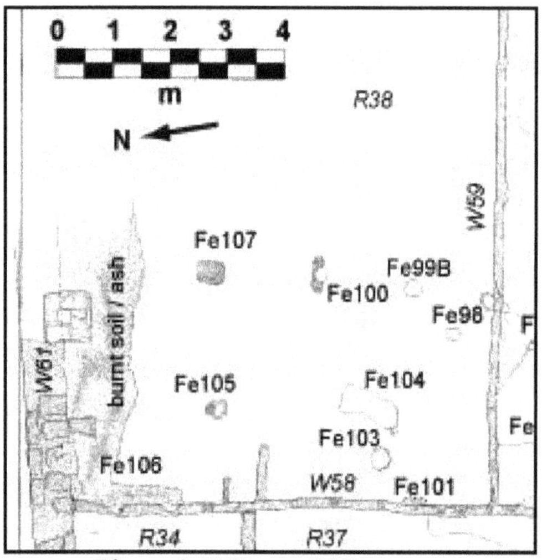

Figure 3.15 Room 38 Features

Overall what we see within the features of the Huaca Sialupe Mound I and II workshops are indicators of production activities. From a paleoethnobotanical point of view, my research agenda, this puts me in a unique position to take known or well understood features and qualify them by assigning them to specific tasks, like metal work furnace or reduction kiln. From there I can determine the contents of their fuel remnants and the compare them to other features. As with looking at mold marks or contiguous rooms, identifying differences or similarities in fuel remains permits us to associate or interpret fuel use activities and their associations throughout the site.

Figure 3.16 Communal Hearth Feature 97, Room 39

Demise of the Middle Sicán at Sialupe

The Middle Sicán occupation begins a long tradition of abandoning and reuse of the Mound I and II complex over time. The recognition of a dynamic environment is evident in the development of the site. The Middle Sicán phase of occupation is punctuated, likely around C.E. 1050, by an ENSO event that effectively demolished the site. Evidence of dissolved and dried adobe brick and mortar, fallen and rotted roofing material, pluvial strata resulting from pooling and drying rainwater, are all evidence of an ENSO event with heavy precipitation forcing abandonment of the Huaca Sialupe workshop areas. Just prior to this event, at the end of the Middle Sicán, ceramic and metal production at the site was most likely at its apogee (Shimada and Montenegro 2002). The enhancement of craft production site infrastructure, characterized by the construction of adobe walls and the reorientation of the site axis, occurs just prior to this period of abandonment. Disintegrated *quincha* or wattle and daub walls and laminated surface soil deposits associated with pluvial formation processes provide telltale signs of this flooding event. Most interesting was is the presence of stumps of young growth trees (2-4 years of growth based on stump diameters compared with site index calculations made in 2002) with profound *in-situ* root structures that were cut off at the base rather than removed all together. These serve as indicators that the site was abandoned, forest successional processes were able to begin in the post-ENSO environment, and then people returned in the Late/Middle Sicán transition to reinhabit the site. The subsequent phase of construction included the construction of a large altar in the middle of the Mound I patio and firing area.

The cycles of drought and pluvial onslaught clearly affected organization at the site over time. Outside of a dramatic ENSO event, perhaps a one-in-a- hundred-year event, and during what were probably protracted or increasingly drought conditions artisans at Huaca Sialupe built up their production infrastructure over the Middle Sicán. This surge indicated intensified demand as well as production of the workshop's signature goods. The end of this period of relative prosperity followed with the dissolution of the production facility as known during the Middle Sicán period. At the same time, a similar dissolution and reorientation between the Middle Sicán and Late Sicán also occurs with the conflagration of the main site core of Sicán in the Lower La Leche River Valley.

History, Culture, Production and Organization in the Middle Sicán

Returning to the hallmark metal objects of the Sicán polity, the *Tumi de Oro*, gold funerary masks, and blackware bottles, we see that there was a connection between the raw materials coming into the site of Huaca Sialupe, the attendant technology, and the potential for some of these important and socially charged goods to be leaving the site. This is especially the case as we see only the broken remnants and trace elements of the most precious of these materials. Without question, Huaca Sialupe was participating to some degree in the overall elite level economy. At the same time, many local products and non-elite goods were produced there. The relative abundance of plate and bowl fragments, as well as arsenical copper alloy tools and metal debris indicates that more modest goods were produced as well. This leaves us with an impression of a two-tiered articulation with the region and the locality as the norm for Sialupe artisans in the Middle Sicán. The examination of the inorganic components of production and trade, however, have brought only somewhat closer to understanding how local production integrated with the broader political economy, as well as with the daily lives of the craftspeople.

Where that leaves me is in a position to examine the organic components of production at the site. I analyze the fuel remains as signatures of the essential components of the production activities occurring at Huaca Sialupe. Their presence is a reflection of decision making processes made by people living in and around the site to incorporate parts of their living environment, wood as fuel, to address their production concerns. The patterns that they have offer clues into what production and consumption needs might be guiding these process. Chapter 4 presents the theoretical approach to interpreting these remains against the background of the Middle Sicán political economy. Chapter 5 describes the methods used in the analysis and development of the comparative collections. In Chapters 6 and 7, I return to the archaeobotanical data from the site to assess the relationship between local artisans and the political economy, as well as the relationship between the two different production sectors at the site. This evidence demonstrates that fuel use is generalized across the site, which has broad implications for the role of Huaca Sialupe artisans in the political and domestic economies of the Middle Sicán.

CHAPTER 4
A CULTURAL ENVIRONMENTAL FRAMEWORK FOR INTERPRETING CHARCOAL

Today wood fuel remains the most utilized natural resource that humans use to generate energy (FAO/UN 1983). At Huaca Sialupe, wood fuel was ostensibly being used to produce items associated with the transmission of the Sicán deity, the 'aesthetic locus' of the Sicán period, most visibly used in upper echelons funerary practice. Wood fuel, however, also represented the integration of an acquired knowledge of local ecology, resource management, and task scheduling with the production of these valuable objects. The craftspeople that maintained the workshops at Huaca Sialupe selected fuel materials to perform their tasks taking into account their own needs and efficiency. The main resource pool for the artisans' fuel resources and the engine that drove the workshops' energy needs were the local woods in and around Huaca Sialupe. A holistic understanding of the role of charcoal at Huaca Sialupe requires us to consider how the political economy intersected with local perceptions of the landscape and domestic economy, a three-fold approach.

First, I examine the role of the elite economy and its potential influence on the daily lives of artisans: were they concerned with using their resources to meet the expectations of political elites? Next, I consider different ways that charcoal assemblages can be interpreted in the archaeological record. Then I look at the technological-conceptual framework developed by artisans that influenced the particular manner that fuel was used and distributed at the site. Finally, I examine how artisans integrated their actions of making fuel choices within the broader framework of the site's economic activity. What I attempt to develop below, is an approach to understanding fuel use at Sialupe that marries multiple theoretical perspectives in anthropology and archaeology, including theory of political economy, domestic economy, technological choice and notions of perception of nature and the acquisition of natural resources into the sphere of human cultural activity. Yet, it just begins to pull these ideas together, and at the very least provides a basis for interpreting the data from Huaca Sialupe.

Reconstructing Sicán Political Economy

Marx invariably gave us the operative idea of political economy as we use it in the social sciences and economics today. Marx's concept of Political Economy focuses on the concept of labor estrangement from production through control of resources by established control of private property in capitalism (Marx 1844 in Tucker 1978:79). Marx's illustration of capitalism's political economy focused on the power dynamic between disenfranchised labor and the forces of production controlled by bourgeois capital and natural resource owned by the upper class. Within archaeology, it is often impossible to set up a proper Marxian dialectic due to fragmentary data sets and the reality of non-capitalist economic and social organizations.

As a result, I employ Earle's (1997) understanding of political economy, following from Service (1962) and Sahlins (1972), being more simplified and potentially more flexible archaeologically. Earle's conception is based on the maintenance of power and status in society. He argues that power is composed of a triad of military, economic and ideological control (Earle 1997:4). Earle's theoretical approach to resource control provides an interpretive framework for understanding craft production at Huaca Sialupe. Earle suggests that resources are divided into two frames: one is staple finance, the materials necessary for subsistence, e.g., agricultural infrastructure and products, and the other is wealth finance, materials that are rare or valuable, or those used for deploying ideology (Earle 1997:70). A political economy where elite or upper class individuals hold a monopoly on materials used to display status relies heavily on wealth finance. The control of these items reinforces the power structure, leaving non-elites in awe and unable to access or deploy these materials; social inequality is maintained. While Earle's is an extremely oversimplified way of considering the role that crafted products have in societies, it highlights that the products themselves, through their creation and meaning have implicit meaning. Mary Helms (1993), however, tempers Earle's enthusiasm for unilineal control. She indicates that artifacts are best understood paradigmatically, where it is only the meaning of objects and their materials in certain contexts that serves to reinforce power relationships in a social system (Helms 1993:69-75). Taken together, if artifacts were being deployed by elites and they play a fundamental role in the development of Cleland and Shimada's (1998) Sicán deity as 'aesthetic locus,' then their contexts for conception, production, and use have meaning. As a result, these meanings serve across a wide spectrum of situations both multivalently and mutilvocally to reinforce social inequality.

Of particular interest to me is that Earle creates a relationship between the way that products are controlled and made and the landscape that is generated. For Earle, any political economy, regardless of how its organization leaves a physical mark on the landscape, generates its own 'political landscape' (Earle 1997:147). In Earle's study (1997) of the chiefdoms of the Hawaiian Islands, he identifies in the characteristic features of political economy in the organization of agricultural and ritual space. First, irrigation ditches and non-indigenous 'canoe' plants indicate staple finance control (Earle 1997:84). Second, there are the *heiau*, the sitting places of ancestral chiefs and their attendants that are both sites of religious and political pilgrimage and loci of spiritual and historical remembrance (Earle 1997:175).

In the end, changes to the landscape due to the control, extraction, or deployment of resources makes physical what Earle calls a 'materialization of ideology' (Earle 1997:151). These are permanent marks cut into the landscape or commonly held beliefs about the local ecology that are artifacts of a functioning or past political economy; I add to this characterization and suggest that the effects of political economy are also visible in a region's local ecology. I believe that almost any archaeological site, in a comparative framework with other contemporary sites, presents the potential to be interpreted as a materialization of ancient political economy. In the case of Huaca Sialupe, we know that some of the artifacts were finding their way into elite burials, like those at the site of Sicán, and that precious metals were being brought to Sialupe, as part of the extensive trade routes ostensibly controlled by Sicán elites. From a purely economic standpoint, both of these extractive and additive potentials at Sialupe played a specific role in how fuel was being used at the site, and thus how artisans at the site extracted fuel resources from their available forests.

While the political economy model offers a framework for interpreting the economic important of the currency of Cleland and Shimada's (1998:137) 'aesthetic locus,' it does not serve to explain the articulation with domestic activities and the potential for their having been influenced by the overall political economy. Wolf, however, describes that for pre-capitalist economic formations one productive way of considering production is through a kin-based form of production (Wolf 1997:88). This case is especially relevant and important, as kin ties are one central aspect of culture that is destroyed through capitalism. While it is impossible to reconstruct kinship systems for the Middle Sicán, we have some indications from the site archaeologically that demonstrate that some form of kin structure influenced the manner that the site functioned on a day-to-day basis. First, as demonstrated in Chapter 3 the Sialupe workshop contexts are surrounded on three sides by small mounds that the SAP believe to represent house mounds. Second, we have a pattern of burials, repeatedly reentered directly adjacent to the workshop (Klaus 2003; Figure 3.10). Third, the proximity of the cemetery to production centers seems important for the Middle Sicán, as it is visible at other production loci for the period (Shimada 1995). Other activities that influenced the lives and development of the site, such as farming, food consumption, and herding, undoubtedly left their mark at the site (Chapters 6 and 7). Overall, this indicates that either one or a coalition of domestic groups is likely the focus of how production resources and labor were organized at Huaca Sialupe.

Wolf's (1997:89) kin-ordered mode of production recognizes that kinship is specifically elusive in the historical and, I add here, archaeological record, yet mainly co-residential groups structure domestic production. His argument is that in non-feudal, and non-capitalist modes of production, regardless of the political forces directed from elites, the domestic economy is responsible for all forms of production and reproduction. Hence, subsistence, Earle's staple finance, and craft items, some of what constitutes Earle's wealth finance, are fundamental aspects of the domestic economy or household/co-residential production. Wolf's (1997:268-274) capitalism is formed through the co-opting of these co-residential production units. Additionally, Wolf's use of the kin-ordered concept is designed for looking at cultural evolution (Wolf 1997: 94-95). For my purposes Wolf's kin-ordered production from a local, small-scale, or site-based perspective, indicates that resource use and management can be modeled at the level of a house or group of houses in a pre-industrial society. For Wolf, these patterns for resource use and management at this level are the beginning of a series of economic connections that lead from co-residential unit decisions to power related decisions of elite use or deployment of goods manufactured at the local/domestic/co-residential level.

Recently, Costin and Wright (1998) compiled a series of papers demonstrating that cross-culturally, regardless of differing political economies, co-residence and resource management at the household or domestic unit level always forms the core economic unit of craft production in pre-industrial societies. Wolf (1997:91) argues that aside from determining and reinforcing kin and affine relations, a domestic economy based on co-residence produces and reinforces norms of labor apportioning and natural resource extraction and use. Most critically, following from Polanyi (1944), kin-ordered production is the structural base and foundation for resource sharing, labor included. Hence, this base, from the simplest unit of production, influences the structure of the entire political economy. Where the productive activities of co-residential groups are the economic productive and reproductive locus of society in terms of the food produced for a society, the manufacture and trade in its wealth items, and the social reproduction of its values and norms. The repetition of these norms in daily practice, from a resource extraction point of view, impacts the local ecology and generates the materialization of power in the local landscape.

In the case of the Middle Sicán, some evidence points to an Earle (1997) styled wealth finance political economy. A large ceremonial center at the site of Sicán in the La Leche River Valley primarily consists of monumentally mausoleum-like pyramidal structures. Associated with these monuments are deep elite shaft tombs made for single individuals, who were interred with attendant sumptuous goods, occasionally including human sacrifice. These elite tombs contain not only large quantities of long distance trade goods, but also locally produced high quality ceramic and gold and silver alloy artifacts that display a central symbolic entity, the Sicán deity, Cleland and Shimada's 'aesthetic locus' of the Sicán period.

The tons of metal and ceramic materials recovered from elite Sicán tombs reflect the manifest power that elite individuals possessed, based on their ability to mobilize labor. Drawing on Earle's concept of the materialization of ideology in the landscape and Wolf's stress on the importance of co-residential production on the political economy of pre-industrial societies, I argue that the implications of a wealth finance system most likely left other indicators in the landscape and at the domestic economy or household level that may demonstrate how this political economy functioned; apart from the consumption of wealth items, an analysis of their production can provide additional insight into how the political economy was felt more broadly across the population (Shimada and Wagner 2007). That overall impact of potential demands placed on the Sialupe workshop might have ecological consequences as far as available fuel/forest resources were concerned. Hence, by reconstructing fuel use, and examining the materials used, we gain insight into the ecological ramifications of production at the site, and potentially can reconstruct how decisions were made regarding resources.

Current models of Middle Sicán political economy have largely developed from the perspective of elite consumption of wealth items; little research, however, has examined the production component of the political economy. Huaca Sialupe, a site of ceramic and metal production thought to be responsible for many of the sumptuous goods recovered from elite tombs at Sicán, enables us to evaluate this model of political economy from the other end of the economic transaction: the lives of the producing craftspeople. In interpreting features of ecology and material culture outside of the ceremonial center of Sicán, we have the opportunity to identify the degree of influence, if any, Sicán elites had on the functional aspects of daily economy at Huaca Sialupe. First, I will consider the existing model of Sicán political economy (Shimada 1995); from there, I examine how local artisans articulated with elite consumption, based on excavated material from Huaca Sialupe.

Middle Sicán Political Economy

The structures that house elite tombs, often the resting place of long distance trade items and sumptuous goods displaying the Sicán deity, have provided the starting point for understanding the economic relationships between elites and their supporters; ethnohistoric records documenting tribute and taxation have provided additional support to this present model. These two lines of evidence, elite tomb architecture and ethnohistoric tribute records, suggest that elite economic power focused more on labor control and tribute extraction, rather than control of local production activities and resources. Although Sicán elites strictly controlled access to exotic trade items, such as gold ore, the manufacture of luxury goods, like tumbaga, from these raw materials may have been left to local craftspeople. At this level, it appears that the effects of co-residential groups on production, ala Wolf (1997), had more influence than any efforts on the behalf of elites to control a wealth finance polity, *ala* Earle (1997).

Ceremonial Center Construction

Architectural construction methods have been critical to the interpretation of political systems throughout Andean archaeological research. In particular, archaeologists have utilized these methods to define the expansion of political control in the Middle Horizon (Isbell 1991), the taxation and tribute system employed during the Moche period (Moseley 1975), the exclusivity and power of the Chimú state (Moseley 1990), and the intellectual prowess and strength of the Inka Empire (Morris 1982). Studies of the structures at the ceremonial site core of Sicán are no exception. The heart of the main site consists of several enormous block pyramidal constructions, locally called *huacas*, constructed from adobe, a mud brick that remains the main traditional construction material regionally even today. Locally, *huacas* are primarily considered to be mausoleums to the dead, or ritual/ceremonial structures, which is largely the case at the main site of Sicán; some structures, however, demonstrate domestic or other use/functions, e.g., Huaca Arena (Shimada 1995).

The main components of architectural construction in the Sicán site core are constructed through what Shimada (1990) calls, borrowing from Kroeber (1930), 'chamber and fill' construction. Chamber and fill is executed through the creation of square adobe block chambers, roughly 5 x 5 m, that were filled with debris, often garbage. These chambers were then 'roofed' with large cut timber beams; preliminary analysis indicates that some of these beams are *Prosopis* sp., the likely dominant forest resource in the area during the Middle Sicán, Chapter 2. These were, in some cases, then covered with pieces of potsherds and plastered over with crude wet adobe. Further stories or levels of chambers were then constructed on top until the building achieved the desired height.

Adobe blocks during the Sicán period maintain the earlier tradition seen during the Moche periods of stamped or maker's marked adobe blocks. In this case the wet blocks, when initially formed, were impressed with either a symbol, e.g. an 'x,' circles, or an actual hand or foot impression, sometimes children's feet. The production and marking of adobe blocks is not particular to ceremonial architecture, as several of the adobe blocks used in Middle Sicán construction at Huaca Sialupe also exhibit these marks. Additionally, the trend in brick marking extends from early in the Common Era through the Colonial period. Distinct from the previous archaeological period Moche blocks, however, Sicán period blocks are slightly flatter and rectangular. This trend towards more tabular, rectangular blocks continued towards more elongate, loaf-like blocks by the Late Sicán and Chimú periods on the North Coast and in the La Leche River Valleys (Shimada 1990).

Cavallaro and Shimada (1988) suggest that the distinct construction techniques employed at Sicán represent a change in labor and political organization from Moche. During the Middle Sicán the way that marked blocks were used in construction was considerably different from earlier periods of monumental construction on the North Coast. Moseley's (1975) analysis of Moche period construction at the ceremonial site of Moche demonstrates that adobe blocks with maker's marks were divided into discrete groups, which he interprets as a direct labor tax system. The group responsible for the production of the bricks was likely responsible for their delivery and a specific portion of the building construction. The archaeological and architectural signatures of this practice are tall stacks of bricks in section-like portions of construction, where the blocks from each section demonstrate similar, if not identical, brick types. Moreover, the pyramid design of solid blocks also implies a taxation method of organization: no clear architectural or structural reasons exist for constructing monumental architecture with solid materials when other techniques, e.g., chamber and fill, could be used more efficiently (Cavallaro and Shimada 1988; Shimada 1997b).

Cavallaro and Shimada (1988) argue that the Sicán system is markedly different. First, unlike Moche construction, the preferred method of monumental construction was that of chamber and fill, indicating a more rapid deployment of architectural labor and construction. Second, this pyramid design required that individual chambers be constructed in concert to ensure an overall stable product. These lines of evidence demonstrate a more skilled approach to construction, which they interpret as a more intensive direction of labor. This method contrasts markedly to the slowly and more uniformly stacked layers of bricks showing similar stamps at Moche.

Finally, Shimada (1997) notes a more or less random mixture of brick stamps distributed throughout the construction. This pattern indicates that although bricks were perhaps marked to track effort inputs of different contributing groups, once delivered to the work site, the distribution of the bricks was directed by another labor pool. The method of pyramid construction divided labor into two components: the production of bricks by several local communities and the construction of the pyramid structure under the supervision of a low-level supervisor. Cavallaro and Shimada (1988) choose to represent their pattern of brick stamps within the constructions at Sicán as a 'sponsor's model.' Here different sponsors request the production of a certain quantity of bricks from a producer or workshop, these were then brought to the site and added to a common brick pool where an overseer determined where they were to be used in construction, yielding the pattern of randomly dispersed brick stamps throughout a monumental building at Sicán (Cavallaro and Shimada 1988:94).

This labor organization identifies at least a three-tiered hierarchy of political economy in Middle Sicán society: elites, low-level administrators, and producers from local villages. Important to stress here is that somewhere between elites and non-elites were people given the responsibility to oversee construction of elite monuments. Certainly the construction of pyramids was strictly controlled, and clearly choice was given to mid-level overseers to complete projects based on their available materials. What is less clear is the role that the co-residential production units played in the production of bricks. Certainly brick producers were interested in distinguishing their products with brick stamps, what exactly those stamps represent is unclear. Additionally, once those bricks were delivered to the construction site, it appears that supervisors viewed them at least as construction resources, with a de-emphasis on preserving their net value as products of a corporate entity.

Referring back to mention of the maker's marks on ceramic molds found at Huaca Sialupe, a similar pattern may be present there (Chapter 3; Shimada and Wagner 2007). Again, the recovered of molds for the same vessels with different marks may indicate that sponsors were involved with ceramic production. On the other hand, it may simply be that multiple producers were using the same workshop. Given the idea that certain rooms appear to be contiguous, while others are not, the idea of several people using the workshop may be important. Regardless when taken together, the ceramic molds and brick stamp information indicate two points. First, Middle Sicán elites were not concerned with controlling local production, rather only the delivery or sponsoring of the making of certain products. Second, some form of product control, either as sponsors or master artisans, indicates the likelihood for at least a three tiered set of social distinctions, outside of gender or age grade distinctions, in Middle Sicán society; commoner, sponsor/master crafter, elite.

An Ethnohistoric North Coast Political Economy

Ethnohistoric sources from the colonial period support a similar model of labor and resource control. Patricia Netherly (1984) and Susan Ramírez (1996) have both argued that a segmentary political system, where local elites were beholden to regional leaders, held sway in the allocation of resources during the 16th century on the North Coast of Peru. Just prior to and after conquest by the Spanish, the region was run by a series of *señoríos*, chiefly organized regional polities, whose holdings, including retainers and lands, were divided up into *parcialidades* (Netherly 1984; Ramírez 1996). These *señiorios* were effectively circumscribed territories, where a chief or elite political personage had control over its productive resources, and often had several residences within the territory. As a result, political headmen, moved around the territory during the year taking up residence at one of several palaces with their entourage during the year. Netherly (1984) indicates that water resources were controlled, and thus irrigated farmland, based on a system of *parcialidades*. Often adjacent or even landholdings of one *señoríos* surrounded resources of another, creating an archipelago-like landholding system. It is clear that lands would be irrigated by *parcialidad* per *señoríos*, indicating that the resource, water, was controlled. What happened to the crops produced on the *parcialidad*, however, in terms of their consumption and redistribution through the political economy, is not clear from Netherly's work (1984). Netherly's dissertation (1977:248) indicates that in the few instances where we see masters artisans or production supervisors of *parcialidades*, they are put in charge of negotiating the delivery of products from the residents of *parcialidad* with either the state or the authorities. This form of supervision presented by Netherly, could explain or be implicit in the pattern of brick stamps seen in Sicán monuments.

Ramirez (1996) elaborates other parts of this system. The principal characteristic of political organization, as far as the control of certain kinds of production were concerned, is that retainers and their products were extracted for redistribution and consumption by the heads of these *señoríos*. These heads then redistributed material to lesser elites who eventually redistributed it down the social ladder (Ramirez 1996). In discussions with Susan Ramírez and from her writings, it is not clear from colonial taxation records or from other historic documents that actual extractable resources were directly controlled by local elites under this situation (Susan Ramírez personal communication; Ramírez 1996). Instead, labor and the results of labor, the extracted items themselves, offered as tribute or obligatory tax goods, were the focus of political control of the economy. On the one hand, Netherly (1984) argues that water, a natural resource, was controlled and meted out via the political economy. On the other hand Ramirez (1996) argues that political economy functioned based on control of labor, and in some cases specialized labor, e.g., craftspersons or herders; the goods that were owed as the product of those labors in obligation to *seniorio* heads, and not the raw or local resources necessary for production, were the subjects of control. Again, the model presented by Ramírez is consistent with that of Netherly, and together describe an early colonial system of resource extraction that is mediated by mid-level officials where the emphasis is on the delivery of products, leaving resource management and labor control to structures more like Wolf's kin-ordered co-residential production units (1997). In terms of Earle's (1997) ideas of political economy, it would appear that the materiality of ideology describes a 'fractured' landscape, where elites appear to control both wealth and staple finance using intermediaries as an indirect means for extraction.

Both Netherly (1984) and Ramírez (1996), however, have strongly cautioned the use of this material to make direct historical inferences. The Chimú, Inka, and Spanish have all since overlaid their own political economies in the Sicán region during the 500 years between the Middle Sicán and the Colonial Period. Ramírez (1996) in particular is wary of making inferences regarding the Middle Sicán based on her reading of the pre-colonial political economy as the Spanish, Inka, and Chimú were all extremely hegemonic extractive situations that focused on movements of goods by controlling labor. Finally, this legacy of extraction in the La Leche River Valley system ended with the Spanish *reducciones*, *encomiendas*, and eventual implementation of the *hacienda* system that, while based on land control, was largely labor taxation/coercion situation.

While the La Leche River Valley clearly had seen by the 16th Century a variety of different forms of economic change, Netherly (1984) and Ramírez (1996) both provide some ideas of how resources were potentially apportioned in the region. Ideally, their work, coupled with the model Shimada and Cavallaro (1988) have developed from Sicán architectural construction, provide a preliminary framework for interpreting the archaeobotanical data from Huaca Sialupe in terms of how the craftspersons working there articulated with their greater political system.

Middle Sicán Long Distance Trade

Given that one of the primary items deployed in Middle Sicán elite burials at Sicán are long distance trade items, this might be one focus of where Earle's concept of wealth finance is potentially edifying for understanding Sicán political economy. Gold, one item probably worked at Huaca Sialupe, was a material imported to the area, and given that it was a focus of the Sicán elite deployment of the 'aesthetic locus' in their tombs, it is important to consider how this controlled item made its way to Sialupe (Shimada 1985). Prior to the excavation of the elite tombs in the Sicán main precinct, Shimada (1985) argued that the great fluorescence of Middle Sicán productive activity, heralded by the monumental architecture and the long-range distribution of Sicán style ceramics, was built on the development of their extensive trade network. The idea was that copper alloy and blackware ceramics were leaving the region to be traded for imported goods, feathers, gold, gemstones, shells, etc. It was clear at this time, however, that gold/tumbaga sheet metal work was being produced for consumption by people of the La Leche River Valley, in particular elites buried in and around the precinct of Sicán. Understanding the relationship between the trade goods and the gold artifacts is one key to understanding the political economy of the Middle Sicán.

In 1982 and 1983 Shimada's excavations at Huaca del Pueblo de Batán Grande and Cerro Cementarios/Cerro Huaringa indicated that mining and smelting operations for the base materials of the copper alloy work found in the region were being locally mined and produced (Shimada 1985). At these sites large supplies of charcoal fuel would be required to meet the smelting needs. Shimada's 1985 work traces a series of artifacts known as axe monies or *naipes* throughout the North Coastal Andean region. By comparing the chemical composition of *naipes* found throughout the region with those from excavated contexts in the Sicán main precinct site of Huaca Las Ventanas, and from looking at copper-arsenic ore deposits local to Batán Grande, it became clear that *naipes* were largely produced in the La Leche River Valley. In trying to interpret the meaning of these artifacts Shimada (1985) argues that these materials served as equivalence trade or money items. Holm (1966) made similar claims for a comparable class of materials recovered from C.E. 600-Colonial period from South Coastal Ecuador. The *naipes* from Batán Grande in Peru are composed of 96% copper and 1-6% arsenic (mean= 3%; Shimada et al. 1994); those from Ecuador are made of 98-99% copper, but arsenic levels were not calculated in the analyses performed (Shimada 1985). If arsenic were present in 1-4% then it would likely indicate that Batán Grande, or at least Sicán polity metal sources were the origins for the Ecuadorian *naipes*. Based on style and other congruities of the artifact assemblages, Shimada argues that the wide distribution of these materials indicates that there was at least political, if not commercial, contact between the two regions (Shimada 1985).

In addition to *naipes,* the presence of *Spondylus princeps* and *Conus fergusoni* shells at Sicán also demonstrates a degree of contact with Ecuadorian peoples (Shimada 1995; Shimada 1985). *Spondylus* sp. and *Conus* sp. shells, native to coastal Ecuador, are some of the most preferred Andean sumptuous goods and were used as offerings to the supernatural from before the Sicán period continuing well into the Colonial period. It appears in a variety of archaeological contexts throughout Andean prehistory, both before and after the Middle Sicán. Shimada indicates that these shells, as well as precious stones, e.g, emeralds, both seen in Sicán period tombs, were notably, through their presence in Middle Sicán contexts looted and unlooted, worth trading for and Ecuadorian sources were potentially key to their procurement. Again, the indication is that an extensive external trade network, principally based on Sicán movement of copper alloy goods, was based in the La Leche Valley during the Middle Sicán.

Shimada (1985) also indicates that there was trade between Sicán and the eastern highlands in the Upper La Leche River Valley and into the eastern slopes of the Marañon drainage. Specifically, access to copper ore sources and, most importantly, gold sources would have been key for Sicán elites during the Middle Sicán. While local, the copper ore deposits were being exploited in the Lower La Leche Valley (Merkel et al. 1994), access to gold, from an elite perspective, was clearly a priority. Middle Sicán ceramics found in the highland Marañon drainage sites indicate contact with these upper areas. The presence of smelting and working sites like Huaca del Pueblo de Batán Grande and Sialupe demonstrate that processing these materials and, potentially creating the tumbaga, however, were under local control in the Lower La Leche Valley. Theoretically, the collection and allocation of fuels was dependent on local labor and trees to meet production demands. Shimada et al. (1982) indicate the presence of at least six different local smelting sites, likely overseeing mining ore to ingot production, in and around the Lower La Leche Valley, the zone surrounding the site of Sicán.

Our understanding of the importance of trade items increased with the opening of the East and West Tombs at Huaca Loro and the South Tomb of the Southern Sector of Huaca Las Ventanas in the main Sicán precinct in 1991-2 and 1995-6. These excavations increased our knowledge of the diversity of trade items used by elites in their mortuary practice, and dramatically altered our contextual understanding of their use. Not only were emeralds, *Conus* sp. and *Spondylus* sp. shell present, but amber, rainforest bird feathers, amethyst, and sodalite were also recovered, demonstrating the extensive range of Middle Sicán trade (Shimada 1995). The amber has been traced to loci as far as modern Colombia, and the feathers to the Peruvian Amazon basin (Shimada et al. 1997). Clearly, the recovery of these exotic items solely from elite burials demonstrates that these commodities were somewhat controlled; moreover, their diversity illustrates the extensive reach of Middle Sicán elites over an economy that commanded ties to a broad regional base encompassing the northwest corner of the South American continent (Shimada 1995).

Political elites at the ceremonial center of Sicán had access to and likely controlled a vast trade network that may have extended as far south as the Central Peruvian Coast, as far North as Colombia, as far East as the Amazon basin. This network was likely no small feat to maintain. The impetus for export was driven by the development of alloying technologies of copper ores, as seen in the *naipes*, and access to a broad network of trade resources. Trading partners abroad wanted access to the metalwork. In exchange, materials not available locally flowed into the center at Sicán. Paramount among the imports was gold for the important alloys deployed in burial contexts. These products were skillfully made through local expertise, to serve the religious and symbolic needs of Sicán elites (Shimada 1984). One point that remains unclear, however, is how, once trade items enter the La Leche River Valley, they are redistributed. Given the model from monumental construction and early colonial records, we might expect a merchant class to be emergent in Middle Sicán society, composed of mid-level elites. Additionally, if gold was an import, how does it get from being an imported object, part of what appears to be an elite controlled economy, to the workshop, a site like Sialupe, where goods are being produced? This point, however, remains unclear.

The Sicán deity portrayed on the blackware bottles and tumbaga metal masks and other elite ceremonial regalia recovered from elite tombs at Sicán was partly the currency of the elite. Its representation, as we will see below, was used in practice by artisans at Sialupe, evidenced by ceramic molds, potsherds, and technological features recovered from the site. Middle Sicán elites, manifesting power through construction labor, regional trade goods, and the harnessing of local crafting prowess, demonstrate that the artisans working and living around a location like Huaca Sialupe knew the elite's needs or desires. Yet, the potential for a middle management or even hands-off model of production at the local workshop is more realistic. Given the argument above regarding monumental construction and early colonial records, it seems unlikely that people working at the Huaca Sialupe workshop felt much pull from above to continuously be working to satisfy elite needs. On the other hand, artisans at Sialupe produced items that occur as external trade items, e.g., gold, copper alloy, blackware ceramics. So it is likely that some mechanism was in place that articulated with the needs of the political elites either remotely or closely. Given the 'aesthetic locus' principle accorded the Sicán deity on precious metalwork, and that these symbols were a component of production at Sialupe, to what extent were the desires of the political elite felt as a reality in terms of how production was maintained at Sialupe? In looking at how resources were used between the different kinds of production at the site, we might gain some further understanding of the domestic and political influences on production.

In terms of returning to the basic framework for political economy laid out by Earle (1997), taken from Marx (Tucker 1978), it would seem that given the information about Middle Sicán elite activity that some influence would have been felt at the site of Huaca Sialupe. From the input standpoint, resources reaching the site, precious metals were supplied to make metalwork at the site potentially to serve elite needs. These inputs were probably regulated by elites (Shimada 1985). The various extensions of the trade systems employed by Sicán elites controlled precious metals, and clearly involved Sialupe to some extent in that system. From an output standpoint, the denizens of Sialupe and its surrounds were engaged in production of funerary and utilitarian ceramics for local and regional consumption. Clearly one externality of production was completely borne on the backs of the producers at Huaca Sialupe themselves: the cost of maintaining the subsistence system and social reproduction through agriculture and the supply of fuel and people to meet production demands.

In terms of Earle's (1997) staple and wealth finance model, it is not clear from either the ethnohistoric evidence nor the archaeological evidence, illustrated above, the extent to which people, even craft producers, were directly controlled or overdetermined by political affairs by Sicán elites. There is indication the intermediaries functioned to control goods extraction from places like Sialupe in the ethnohistoric period (Ramírez 1996), hence the pull of a wealth finance system may have been felt only upon occasion. As far as staples are concerned, there is no indication that staple finance was the backbone of the Sicán elite. Rather trade and ideological power through construction of a sacred precinct was clearly central to elite control. It is similarly unclear to what level activities at Sicán impacted the affairs at Huaca Sialupe, as there is little evidence to connect the two (Shimada and Wagner 2007). In either case, wealth or staple finance, the case cannot be made for how low-level craft producers, far from the center at Sicán itself articulated with those structures of power at the center. In fact, it seems that production of ceramics and metalwork at Sialupe had some flexibility in terms of how the 'aesthetic locus' was used and distributed, apart from the myriad other items generated and consumed at the site (Shimada and Wagner 2007; Taylor 2002).

Not only do the ceramics at Huaca Sialupe indicate the deity's common usage in the industry, but metalworking features also point to the possibility that the finery of *tumbaga* sheet metal garments and masks were manufactured at this site. In effect, the production of the two primarily visible sets of ritual materials present in the elite tombs at the Sicán center is implicated in Sialupe's production potential. This situation begs the question, how then, given the model of the political system elaborated for the Middle Sicán, were the artisans integrated into this economy? Were they low-level administrators? Were they at the bottom tier of society, merely responsible for pooling together goods required for elite ceremony and ritual to be shipped to and used in elite-contexts? To assess these possibilities we must first understand something about Middle Sicán production technology, and the organization of the site of Sialupe itself.

Holism in Reconstructing Middle Sicán Technology

Shimada and Wagner (2007) recently put forward a programmatic statement regarding the SAP's holistic methodology that has been a part of the project from its inception (Shimada 1981; Shimada 1985; Shimada 1994b; Shimada 1998; Shimada 2000). The principle is simple: combining excavation of production loci, materials research, and experiment to reconstruct ancient technology to provide insight into understanding cultural and economic complexity. Shimada and Wagner's (2007) is to reconstruct craft production *en toto* starting with the conception of raw materials and products and encompassing its technology, organization, producer, cultural, social, and other factors that influence production directly and indirectly, vocally and non-vocally. Others have argued for similar attention to be paid to the holism implied by Shimada and Wagner's (2007) conception, e.g., Costin (2000). Yet, the reality of Shimada and Wagner's (2007) practice requires long-term laboratory and field research that continues to this day.

Based on this principle of holistic study the SAP has come to understand a great deal about different facets of material culture for the Middle Sicán and its use as the economic currency for this period. As argued above, in terms of construction labor was controlled through at least a three-tiered system, whereas control of trade goods is less clear. While imported trade goods may have been controlled through a top down relationship to non-elites, local resources, drawing on inferences from the 16th century (Netherly 1984; Ramírez 1996), may have been less important than the delivery of goods mediated by middlemen from local production to elites. Monumental architecture labor organization and the exterior trade economy, however, only provide a limited picture of the political economy during the Middle Sicán; the production of ceramic, metal, and other valued goods must be addressed to provide a more holistic account.

For this reason, I turn to a more bottom up approach to understanding resource use based on the Huaca Sialupe dataset, to help illustrate how the domestic economy at a site like Sialupe may have articulated with the greater political economy. Clearly, based on the remains of metalwork and blackware production at Sialupe there is a relationship between these two different levels of Middle Sicán society. What is implied by Earle's idea of the materialization of ideology on the landscape indicates that regardless of the type of power dynamic between these different levels of Sicán society, artisans at Sialupe must have left a mark on the landscape that tells us about how resources were controlled to either meet the demands of their daily lives and any other potential pressure from above. I have chosen to use charcoal assemblages from the technological features, where the important symbols of the Middle Sicán economy were made, to look for such an imprint from the landscape. As I indicate in this section, on one level an investigation of carbonized archaeobotanical remains implicates the economic activity of artisans and permits the reconstruction of the technology used along holistic lines (*sensu* Shimada and Wagner 2007). On another level it permits a look into the domestic economy of ancient craftspersons, their methods of social reproduction and subsistence, and lastly a frame for reconstruction their relationship with their ecological community. In the end this permits analyses of both their co-residential mode of production, and the landscape around them where they were actively integrated.

Analyses of archaeological charcoal assemblages offer a resource for reconstructing technological choices made by artisans, past environments, and forest structure (Fritts 1976; Newsom 1993; Schweingruber 1996; Smart and Hoffman 1988; Stahle 1996; Thompson 1994). From analyzing the charcoal from production contexts at Huaca Sialupe I can gather information regarding resource use, the basis of all economic activity, specifically fuel that we can discern from these data. Lemonnier's ideas concerning technological choice are some of the few to clearly articulate the cultural importance that choice, selection, and use of resources have for imbedding cultural and experiential information into the technological production of goods. As the majority of the contexts that form the dataset from Huaca Sialupe are composed of wood fuel remains, I begin with a review of some ways that archaeological charcoal has been interpreted. I then turn to Lemonnier's approach to craft production and in particular, how he interprets the selection of materials and use of resources as both culturally and ecologically salient. From there I suggest that a production model alone does not allow for a holistic enough understanding of production activities at the site. In particular I discuss how finding non-ceramic and non-metal production remains in these contexts implicates that a domestic economy or household approach more accurately represents the activities taking place at the site.

Archaeological Charcoal and Resource Use Theory

Theoretically, making the leap between charcoal identifications from archaeological contexts to reconstructing human activities is a challenge, but not impossible. In 1988, Johannessen indicated that although there are always biases inherent in archaeobotanical datasets, it does not preclude the utility of these data for use in activity reconstruction. In particular, she argues the potential for paleoethnobotanical analysis to be extremely effective in both reconstructing the ecology of human-plant interactions and identifying economic resources. Reconstructing resource selection and the variable proportions of plant resources in different archaeological contexts can highlight the roles plants played in broader economic systems. To be useful, however, archaeobotanical data must be systematically collected *en masse* across a site and from features and floors that are of comparable cultural and temporal contexts (Johannessen 1988). Her position focuses trends in archaeology that integrate social and economic analysis (Ingold 2000) and avoid artificial divisions of datasets relevant to both social and ecological reconstruction, as is commonly done in materialist enterprises, such as anthropology and ecology (Foster 2000).

Huaca Sialupe meets Johannessen's requirements for establishing a dataset that can address both social and ecological questions through archaeobotanical remains, and thus provides an excellent locus for investigation and analysis. The Peruvian North Coast, with its limited rainfall regime and tropical climate, maintains organic materials in both quantity and quality, and their associated archaeological contexts are often similarly well preserved. Additionally, the horizontal excavations of synchronic contexts performed at the site enhance the precision with which we can look at economic activity of plant use across contexts for the Middle Sicán. Due to the recovery strategy of organic remains at Huaca Sialupe, I generated a robust archaeobotanical dataset goes further than simply note the presence or absence of a certain taxon at the site, but permits me to use these data descriptively to illustrate how certain features were used and interrelate.

The archaeobotanical data recovered for the purposes of this study come directly from contexts where fire was employed to create specific artifacts, e.g., blackware, redware, and metalwork, as well as food stuffs, e.g., fermented beverages and meals. Our ability to distinguish between these contexts by activity enables us to assign meaning to specific fuel use patterns. Although we cannot assign *a priori* categories to all fire-use features found at the site, we *can* make clear archaeobotanical distinctions between them. Ethnoarchaeological studies of different production contexts provide categories of feature use, which I use to compare with unknown features from the Middle Sicán contexts; given the physical constraints of ceramic firing and metallurgy, similar fuel components can suggest similar activities. These contexts and their contents add further to the social and cultural content of reconstructions via the 'holistic' research design at Huaca Sialupe that incorporates comparative, ethnoarchaeological, and social/ecological approaches (Shimada and Wagner 2007) that are essential to understanding the dynamic and systemic relationship between humans and their environments.

Smart and Hoffman (1988) argue that archaeologists frequently turn away from analysis of charcoal, claiming it is an ephemeral class of remains most productively used for Carbon isotope dating. The common dismissal that charcoal deposits cannot represent in any way the fire as it was set or constructed disregards the immense potential of these remains to identify human behaviors, such as resource decision-making and pyrotechnology. Indeed carbon isotope dating is a critical innovation in modern archaeology and has made carbonized plant remains far more conspicuous in the field component of archaeological methodology. As a result, archaeologists are more and more often plucking charcoal from their archaeological contexts as chance finds that may aid in the relative dating of their excavation strata.

Yet, Smart and Hoffman (1988) and others (Bohrer 1986; Lopinot and Woods 1993; Minnis 1985; Newsom 1993) indicate that charcoal, when recovered in association with discrete archaeological features, provides a window into resource use; even so, charcoal's full potential is rarely exploited in archaeological reconstruction of the past. Newsom (1993) takes this one step further, arguing that, in fact, apart from resource use, we can examine both resource ecology and quality. These innovative researchers recognize the importance of the complementary qualities of charcoal as an archaeobotanical dataset:

1. Charcoal is the product of the conversion of plant cellulose, the product of plant growth, to a stable and recoverable physical matrix; the fact that charcoal is the result of plant growth as well as human action, highlights its importance for understanding plant growth and ecology in archaeological analysis.
2. Plant materials used for fuel are selected and brought to a site through the perception of individuals following culturally normative behaviors formed around their understanding of the surrounding environment and the technologies employed.

As a result, these two avenues form the main theoretical channels I pursue in my interpretation of charcoal in the archaeological contexts at Huaca Sialupe, those dictated by the biology and growth of the plant and the conscious selection and gathering of materials by humans for particular activities.

Charcoal Analysis: Examples Beyond Isotope Dating

Identification using modern materials allows for comparisons between ancient and modern forest compositions (Coles and Coles 1986; Lopinot and Woods 1993; Johannessen and Hastorf 1990; Newsom 1993). Charcoal macroremains are a data source for reconstructions of ancient environments and forest ecosystems, and push the potential of this data set beyond dating purposes and questions of the presence or absence of taxa (Newsom 1993; Smart and Hoffman 1988). Smart and Hoffman (1988) present that charcoal assemblages from fuel-use contexts can, in cases of well defined features, be interpreted as synchronic indicators of cultural preferences for materials. Similarly, shifts in the appearance or disappearance of certain taxa in the archaeological record aid in the recreation of forest succession and interpretation of diachronic vegetation change, at least as far as human exploited taxa are concerned (Smart and Hoffman 1988:192-4).

There are issues of sample bias from both human selection processes in antiquity and during excavation, not to mention physical degradation of the burned materials (Smart and Hoffman 1988; 180-4). But in instances where pollen counts and dendrochronological series do not exist, as is the case for north coastal Peru (Weir and Eling 1986; Craig and Shimada 1986), the ability to compare different contexts for burning wood, and to track species presence for a given time period, allows us to model vegetation change in the archaeological record (Smart and Hoffman 1982).

These analyses have been employed successfully at a number of archaeological sites around the world to analyze sustainability, e.g., the Mississippian site of Cahokia (Lopinot and Woods 1983), Bronze Age Thailand (Thompson 1994), Neolithic Turkey (Asouti 2003), and Bronze Age Iran (Miller 1985). Additionally, Newsom's (1993) work, which includes both a functional and ecological anatomical approach to wood exploitation, offers a fine grained analysis of resource quality that allows for the identification of resource management, e.g. coppicing, as interpreted from rapid growth anatomies visible in the examined charcoal (Newsom 1993). Newsom's approach looking for functional and ecological clues to interpreting the wood and non-wood charcoal assemblages that will be used here in this study to maximize the interpretive power archaeological wood remains. From this perspective, a more complete understanding of past ecological interactions between humans and forests can be established, and past resource availability becomes less unknown.

For the central Andes, only a handful of similar studies exist in the recent literature. Weir et al. (1988) reconstruct subsistence pattern changes at the three Formative period sites on the Peruvian Central Coast. Their work indicates that there was potential resource competition between wood used for firewood and wood used in construction (Weir et al. 1988). Pearsall (1983), like Thompson (1994), classed different types of wood species together in terms of their ecological niches and was able to reconstruct a general trend in changing wood resource exploitation. She demonstrates a shift in trees exploited from dry forest *Prosopis* sp. to moist ecological niche trees during the Valdivia period (3500-1000 B.C.E) at sites in south coastal Ecuador (Pearsall 1983). Although this shift was interpreted as an indication of overexploitation, the results were inconclusive (Pearsall 1983). Despite the innovative nature of these studies, however, both pay little attention to functional and ecological variation within the archaeobotanical samples and have difficulty relating concretely the archaeological contexts for resource use to specific instances of human resource exploitation. At Huaca Sialupe, however, 55% (n=26/47) of the fire use contexts are tied to features that have experimentally been demonstrated to be either metal or ceramic production related, permitting a more extensive analysis of fuel distributions (Shimada and Wagner 2001).

The most successful study in the central Andes to use carbonized fuel remains was carried out by the Mantaro Valley project looking at the Late Intermediate through Late Horizon to colonial period (C.E.400-1532) periods (Johannessen and Hastorf 1990). In the Mantaro Valley case, extensive ethnohistorical data detailing the forestry practices in the Late pre-Hispanic period exist (Ansion 1986; Sherbondy 1986). When the historical data were compared with charcoal remains from domestic hearths, patterns of fuel use were reconstructed, and researchers were able to trace the sustainability and continuity of certain management practices into the colonial period (Johannessen and Hastorf 1990). This study examines the variety of combustible materials used in specific contexts, as well as the ecological requirements of selected fuels over time. It also demonstrates the complexity of fuel use phenomena, such as the potential for fuel importation, raising woodlots, and the supplement of non-wood materials as coping mechanisms for depleted fuel resources (Johannessen and Hastorf 1990). The intercontextual approach of this research project enables a detailed analysis of fuel economics; moreover, it clearly benefited from the sustained research agenda in the region that facilitated a diachronic approach to resource use.

None of the studies described above, however, have been able to examine the impact of multiple industrial activities, along with domestic fuel use, on local forests. The Middle Sicán data set from Huaca Sialupe presents a unique opportunity for research of this type. An analysis of charcoal data recovered from production contexts at Huaca Sialupe in conjunction with the contemporary environmental setting provides the opportunity to tie cultural practices to the ecology of the Middle Sicán. As I argue below, fuel use implicates not only cultural choice, but ecological reality; the biological factors of tree growth are equally important to reconstructing fuel collection and use strategies, a topic that the above studies fail to address and that this dissertation attempts to reconcile.

The main variables affecting wood growth include factors that chemically alter how food is consumed and energy used in cell construction, e.g., nutrient deficiency, available groundwater, metabolic stress, and physically affect growth, e.g., breakage and physical strain. In both instances climate plays the significant role in how stress is mediated in the Huaca Sialupe research area. Dramatic climate shifts, especially manifested in the El Niño/Southern Oscillation event, dramatically alter groundwater availability that force shifts in metabolic rates, dramatically affecting either the acceleration or retardation of plant growth (Schweingruber 1996). These same events are responsible for the soil nutrient recharge in the lowland valley areas where the Huaca Sialupe site is located keeping them verdant and fertile (Delvaud 1984). Additionally, the liquefaction of clays, massive inundation of floodwaters, and the subsequent overloading of fruits and new growth in the post-ENSO period place extreme physical strain on plant growth (Delvaud 1984, Pastor Rozas 1993).

As a result (noted in Chapter 2), ENSO events play a dramatic role in forest regeneration, growth and vascular cambium activity, which leave a distinct mark, patterns of cell growth in xylem generation, on long-lived plant organisms in the survey area. Additionally, seed germination, sapling growth, shade and sun tolerance are all effected by the periodic ENSO events in concert with the annual stresses from the dry forest's ecological community, including humans. The results of these events are visible in the only record of plant growth that we have for the Late Intermediate Period, the xylem of plants currently growing in the survey area. Fortunately, a portion of this record is also available to us in the preserved charcoal remains at Huaca Sialupe.

Technological Choice of Fuel Use

The idea of fire construction as a cultural artifact has its place in studies of Andean ethnography and prehistory. Early on in the studies of South American Anthropology fire construction was seen as a salient avenue for research (Cooper 1946). However, most of the focus was on fire drills and tools for initiating fire, rather than how resources were gathered and utilized. Yet, it appears that along with other topics in technology and cultural studies in Anthropology, a desire to follow or elaborate this kind of research disappeared (Lemmonier 1986). In fact, it is in the study of modern peoples and in the 'development' literature where we see a greater focus on the elements of fire construction, especially as it relates to renewable resource use and exploitation (Adegbehin 1994, Agarwal 1986, Bogach 1985, FAO/UN 1983). This topic has appropriately come to the attention of the development community, as wood is still the most utilized fuel for energy needs in the world, and a growing human population has put an increasing amount of stress on the world's remaining forests. For the most part, however, these two avenues of study have attempted to interpret anthropogenic fuel use in specific historical and ecological instances; they rarely examine how fires are actually constructed from natural resources, e.g., the ordering of wood types, appropriateness of fuels, etc.

The charcoal dataset from Huaca Sialupe is the result of a series of human selective processes enacted upon plant life that formed a vital part of the artisan's cultural and ecological community. Lechtman and Steinberg (1979) present these decisions as being part of their framework of technological style. Where these decisions offer a chance to understand how fires were made and partially reconstruct the decision processes to use these fuels. Yet, how were these decisions made? Lemonnier (1986; 1992) presents a compelling theoretical model for the production of artifacts. He argues that artifacts are expressions of the accepted norms and values of a given society: 'technological choice' represents a compromise between constraints of production and the effects of the social and natural worlds (Lemonnier 1992). Lemonnier accepts that artifact design has the potential to be arbitrary in its construction, yet he asserts it is constrained by cultural norms and ecological practicality. Yet, the way that the ecology is approached is through a cultural frame. As a result, each artifact is the product of five main factors: the matter that composes it (materials), the energy that transforms it (work), the objects that contribute to its form (tools), the gestures that make an object (skills and actions), and the specific knowledge that allows the combination of the first four factors (Lemonnier 1992:5-6). Although many aspects of the Huaca Sialupe workshop could be assessed based on these five principles, my research focuses on the understanding the fuel loads used in fire features as artifacts. In this case, referring back to the idea that human economic activities create ideological landscapes (Earle 1997), the way that fire resources are extracted from nature leaves an imprint on the local ecology. This imprint and the fuel use features recovered are composites of the values and equivalences for how fuel use is mediated at the site. These values are based on cultural attitudes and social conceptions of what is acceptable as fuel and what can be conceived as fuel from the artisan's environment.

For Lemonnier (1992), every artifact takes a *particular* form that is the manifestation of the acceptable within a society. Under this circumstance, all objects, or parts of a technology are interrelated. For instance, while a knife can be used to cut cheese, bread, open a wine bottle, cut aluminum foil, or whittle; it is not, however, exclusive to any one of these functions. Then again the resources used to make the knife define not only what the qualities of the knife are, e.g., to cut butter or wood, but are held in contrast, conceptually by the maker, relative to the other materials that are *not* used. In the instance of fuel use, this contrast sets up a conceptual logic, regarding what can be used as fuel and in which contexts. On the other hand, the knife must be seen in relation to other tools in a particular context, for instance, at the dinner table, where a knife is not appropriate for eating soup or a stew; for this we have a spoon. Knives do not function well with foods contained in bowls. In terms of the fire-use features at Sialupe, each feature has generally the same purpose, to contain burning organic matter. Yet, the organic matter used as fuel is dictated by a set of cultural norms and technological needs that are actively used, and sometime passively ignored, by the artisans at Sialupe.

In the case of the metalworking and the ceramic kilns, however, there are certain ideal physical characteristics required to produce the artifacts in question. Leaving a whole set of valuations of resources that are either used, unused, or ignored, in the case of fire production for ceramics and metal work. These distinctions are both social as well as physical distinctions that are made by the artisan. From this point I consider that all of the fire-use features at Sialupe have an implicated social relationship. Additionally, as these features all used wood and dung fuels, I assume that there is an ecological, as well as cultural relationship between all of these features. Thus, comparing the contents of these different contexts permits us to reconstruct s elements of this social relationship as the different organic materials represent cultural distinctions made by the artisans as far as what is acceptable. Furthermore, the fuel contexts, as elements selected from the ecosystem, permit ecological interpretations of the communities used to serve the fuel use needs of artisans at the site.

Lemonnier does argues that there are always physical limitations in any technological system, e.g., eutectic temperatures of alloys, sintering temperatures for ceramics. Technologies, however, take distinct forms in response to *cultural* as well as physical factors of production. Lemmonier (1992) illustrates the power of this idea in light of the physical constraints of flight. For instance, modern airplane wings are designed in myriad ways, based on a variety of potential construction materials; all of these are viable under the physical needs for flight, lift, torsion, strain, stress, etc... Modern aircraft engineers choose, however, to produce airplanes with a restricted range of forms (Lemonnier 1992:66-77). This concept is widely known or considered as the notion of 'functional equivalence' in the social sciences. In the instance of fire making and fire feature production, the ways that fire-use features are prepared for firing, e.g., the fuels used and their design, are culturally mediated by their social and ecological contexts. Therefore the types of fuels procured are not always equivalent. In some cases a particular fuel might serve well in all instances of fire use, and in others not at all.

Huaca Sialupe is an archaeobotanical data set from a series of different fire-use features. In the case of this site, each feature type represents fuel collected for a fire and choices made regarding ecological resources. Yet, from the perspective of the arbitrary nature of the form of technology that Lemonnier describes, the physical constraints of the fire required as well as acceptable materials for each instance can differ. The way that they differ, for instance in the different woody taxa used as fuel, is the result of decisions made by artisans as to what kind of fire they wanted to create. If the surrounding area of Huaca Sialupe constitutes the catchment of fuel for the site, and different kinds of production were taking place at the site, then there should exist patterns in the dataset that indicate culturally and ecologically mediated decisions that the artisans made throughout the site. The full range of arbitrary decisions possible for fuel use exist within the context of the forests surrounding the site in the region in the 10th century C.E., and the fuels collected from them. These decisions made by artisans at the site, are made in context of their co-residential economic system (Wolfe 1997) and are the forces that create the landscape modifications, arguably Earle's (1997) ideological landscape.

Culturally, following Lemonnier's ideas about technological choice, the species or collections of materials burned in each instance represent not only what is 'available,' but also most importantly what is acceptable for production in each fire use feature. Following this model, the remains of each fire represent decisions made by artisans in the past based on their cultural conceptions of their environment, a 'chain of operations,' a linguistic idea taken and applied to material culture by Leroi-Gourhan (1945). Every artifact is constructed from a series of actions that are performed on the natural world to transform natural resources into material culture. This model is holistic in its interpretation of artifacts as it sees the gestures, motivations, labor, resources and form of the object created as inseparable from the cultural reality of the end result. While the outlook is similar to Lechtman and Steinberg's (1979) 'technological style.' I appreciate that Lemonnier highlights the interaction between humans and the environment not simply as the perception of limitations by the artisan. At Huaca Sialupe, the construction of fire and fire-use features is a single technological chain embedded within the overall process of metalwork production that extends outside the site itself, beginning with the ore in the mine, moving to the gathering of fuels, and ending in the final use of the product created; for instance, a tumbaga ceremonial headdress buried with a Sicán lord.

Fire Use at Co-Residential Workshop Units

One important aspect to Wolf's (1997) co-residential unit is that people, in my case artisans, make decisions at the local level about what resources to use to meet their subsistence and production needs, often uncolored by the direct pull of a dominant political economy. As we have seen above, one of the key local resources was fuel. Since early conceptions of understanding ancient ceramic and metal production, this idea of humans mediating resources through cultural frames is well known (Smith 1965). In the archaeological reconstructions of pyrotechnology and fire use in production contexts Matson (1989) and Wertime (1964), present relevant observations about how co-residential workshops use and share fuel. The thrust of most of Wertime's work is aimed primarily at interpreting the ascendancy of western civilization in the Levant and its technologies of iron, glass, copper alloy, and ceramic productions. Although his work acknowledges the importance of fire, the concept that fire in itself constitutes a technology and artifact is not recognized (Wertime 1982). Matson (1989), on the other hand, offers important observations regarding the compatibility and overall interdependence of metal and ceramic fuel use and the potential implications of fuel catchment to the development of both industries (Smith 1965). As with Wertime (1982), Matson (1989; 1966) focuses on clay mixtures, refractory materials, and the thermal processing of clays and metals. Matson (1989) departs from Wertime (1982), however, when he observes from Levantine ethnoarchaeological data that ceramic production is the most elemental process in non-food material transformation through fire.

For Matson, ceramic technology is the focus or main contributor of all pyrotechnology at a site where multiple economic processes are taking place. Matson (1966) includes household productive activities, e.g., farming/pastoralism and food production, in his assessment of ancient pyrotechnology. Matson also reasons that all pyrotechnological industries, apart from subsistence fuel needs, require expertise in the handling of high temperatures and material transformation, and require steady and reliable sources of fuel catchment. He continues to argue that ceramic industries are the base of all pyrotechnologies found at a production site. Metalwork, glasswork, etc., never appear in isolation without the presence, on some level of ceramic production (Matson 1989). Essentially, Matson (1989; 1966) describes the situation at Huaca Sialupe.

Matson's (1966) argument, at first glance is an initial presentation of the importance of fire in its own right. Second, however, it illustrates the reality that fuel materials used in all pyrotechnic settings at a workshop and co-residential unit implicate different ecological and economic processes related to not only production, but also social reproduction and subsistence. Matson's (1966) interest is in non-wood fuels that contribute to the workshop's energy quotient. These come from the subsistence portion of the domestic economy at the site farming economies, through fodder refuse and animal dung, and subsistence garbage, and are substitutes, functional equivalents, for wood fuels. He argues that in many cases of ancient technology we see that humans are very resourceful, tending to burn anything as fuel that can be burned (Matson 1966:151-2). These non-wood or non-forest products are key to interpreting the way that pyrotechnology can articulate with other economic, political, or domestic processes in any archaeological or contemporary context. This possibility echoes Lemonnier's (1992) idea that cultural factors are clearly at work in production sites in terms of resource apportionment. Following from these studies, the fuel remains from Huaca Sialupe represent much more than the residues of two sophisticated technologies. They provide a general description of the overall resource base, as well as the ways in which fuel was understood in different use contexts at the site. The charcoal deposits from workshop contexts provide a window into the localized economic activity and cultural practices at the workshops of Huaca Sialupe.

Households and Palimpsests

Huaca Sialupe's workshop complex is one small part of a five-mound complex that makes up the larger site. As discussed above, however, excavations are not feasible due to modern settlements atop the site. Additionally, we know that the production debris stretch across the five-mound complex, it just happens that we located what is likely the core of production at Mounds I and II (Shimada 2000). Yet, Matson's (1966) implication about fuel use catchment and the integration between all other economic processes and with artisanal production, suggests that I consider the implications of the Mound I and II workshop relative to the adjacent settlement remains. Additionally, Lemonnier observes that cultural factors, including attitudes towards the local ecology, tend to weigh heavily in how technology is performed. It seems that whatever the activities at the nearby structures, their combined effect would have been felt, and perhaps left a mark, in the Huaca Sialupe workshop contexts. Again, this leads back to Wolf's (1997) concept of the importance of co-residential units as primary economic movers in pre-capitalist systems. This approach to the data at Huaca Sialupe is appropriate for my analysis, given the scale and quality of the features implicating a household, co-residential, or domestic economy at the site.

Archaeologically, households have been dealt with slightly differently given the impossible situation of being unable to interview or measure household activities in the past. The have mostly focused on architecture and extrapolating kinship dynamics onto past populations (Ashmore and Wilk 1988) or differentiating the use of space in synchronic contexts (Gero and Conkey 1994). Karl Polanyi (1944), in his development of ideas related to substantivist economic theory in the social sciences proposed the existence of the household as an economic unit. Key to this idea of the household is that where people live exist conjunctions between several different cultural institutions. Additionally, in instances where non-industrial and non-urbanized economic settings are being studied, the household is often the focus of production, consumption and reproduction activities associated with kin and other social affines (Polanyi 1944: 53 as quoted in Halperin 1994). Households are where several institutions cross, conflict, and resolve their different needs (Halperin 1994:143). As a result they are the crossroads for many different economic activities where the people associated with the household are influenced by their overlap. Halperin's (1994: 145) point that is key to my study is that households are integrative processes where farming, food production, saving and hoarding, distribution, cooking, consumption, and any other form of productive activity are all taking place at once. In the case of Huaca Sialupe, then, we can argue that apart from ceramic and metal production, other productive processes left their mark in resource use at the site, and fuel use would be a good place to look for the conjunction of the production of artifacts and other household institutions like cooking, construction, food processing, etc.

Multiple anthropologists have dealt with the concept of domestic economy from the point of view of the reality of how households, groups of cohabitating people, make decisions (Folbre 1986; Halperin 1994; Hart 1992; Moore 1992). In each of these experiences, however, the common sense approach derived from these investigators is that theorizing the household, in terms of prioritizing individual institutions functioning at the household level, is near to impossible. Folbre (1986) indicates that both substantivist (Polanyi 1944) and formalists (Herskovitz 1952) tend to miss the point on households in that the institutions that form the household, specifically, at the domestic level, cannot be teased apart or individually examined with much success without an ample dataset and good formal interviews of living people. This limitation is due to exactly the reality of domestic activity that is best described as a palimpsest of superimposed activities, residues, and negotiations made between the people occupying the household's space. Halperin's (1994) discussions of the palimpsest work well in their application for modern people's, where one can identify and ask for descriptions of institutions directly from living informants. They permit the investigator to describe with detail the institutions that exist. In archaeology this is slightly more difficult.

At Huaca Sialupe, in the instance of the 100 years that make up the Middle Sicán levels, they are the results of using the workshop over time. While I am not able to discern the order or number of instances that certain fire-use features were in production, I can say that they collectively represent the intersection of various institutions related to the household, writ large, of Huaca Sialupe during the Middle Sicán. At least two of these institutions, metal and ceramic production, and their technologies are well reconstructed by the SAP. The other activities that undoubtedly influenced the lives and development of the site, such as farming, food consumption, and herding, also left their mark at the site. Interpreting them adds a whole new dimension to how we understand the co-residential production unit, their resource use decisions, and their ecological community in the Middle Sicán. The fuel use contexts at the site leave little choice but to be interpreted in this way, as domestic institutions that overlapped as a palimpsest of social activities, domestic economy, and resource apportioning at the site.

I believe that the previous SAP work (Cavallaro and Shimada 1988; Shimada 1984; Shimada 1995) and work by ethnohistorians (Netherly 1984; Ramírez 1996) looking at the construction of political economy in the Middle Sicán indicates that elite presence as a driving force at the site might not be visible archaeologically, or even present at a site like Sialupe. As a result, here I consider that models of householding allow me to interpret the conjunctions between the different features and fuel apportioning at the Huaca Sialupe workshop as a palimpsest of activities and intersection of institutions. In returning to the opening concepts of Earle (1997), it may be that the datasets so far developed for interpreting Sicán period political systems are not yet sufficient to understand or address the full spectrum of power relationships expressed by Middle Sicán political economy. Instead, the Middle Sicán activity residues of overlapping social, cultural, and ecological decision-making processes, are better suited for interpreting workshop level and other institutional activities participating directly in the site's development. In this instance, I employ the fuel use contexts to interpret the presence of these conjunctions and reconstruct how their use of available natural resources was interdependently structured by their production and consumption practices at the site of Huaca Sialupe, Mounds I-V, as a whole.

Theoretical Interpretations of Charcoal at Sialupe

The fire-use features at Huaca Sialupe represent the remains of choices made by artisans based on their ecological and cultural perceptions of what fuel resources were available and adequate for use. Although this study cannot provide a complete understanding of all cultural and ecological constraints that shaped fuel procurement and apportionment, it can reconstruct general patterns reflecting the past environment. Each context of fire use is a complex overlay of the selection and use of specific combustible materials. These materials were drawn by people, with their own social needs and ecological outlooks, from local resources in light of production fire requirements. From these data, I interpret the remains of fuels that were not fully consumed in the combustion process as the results of cultural decisions made about combustion. I see the remains as the results of overlapping domestic and production working at the site, and they permit me some qualitative understanding of what the ecological community looked like where fuels came from. Lastly, the fuel remains permit me to interpret to some extent the relationship between artisans and the living features of their environment, the trees, herbs, and other plants in their world.

At the broadest scale of analysis, the fuel use contexts at Huaca Sialupe can enhance our understanding of how the Middle Sicán political economy may have influenced activities at the site. In particular, we may see a preference for fuels in certain contexts, as implied from our ethnoarchaeological experiments inn 2000, where certain technological demands were necessary or cultural attitudes necessitated use of some fuels over others in particular contexts. At a more refined level of analysis, fuel apportionment across the site allows us to interpret which elements from the environment served what distinct needs in the Middle Sicán. At the most refined level of analysis, the individual charcoal remains, we can reconstruct fuel selection and identify the elements that composed the artisan's conceptual ecological community.

From the charcoal remains, I envision the location, gathering, and relocation of materials to a specific kind of hearth and we can determine that it was burned with a certain end in mind. Each gathering episode terminates in a constructed artifact of the laid fire: constellations of resources gathered in a specific combination, with a specific intent for a desired end, similar to the production of 'traditional' artifact classes. If production of these goods was a priority, on the basis of some perceived elite demand, we might see the intensive use of local resources on an extractive level indicative of a command type relationship. More likely, based on the models reviewed above, the pattern is likely to be generalized with a pattern of shared fuel use across the site, perhaps centered around ceramic production, and mediated by local co-residential economic activity. In particular the essential resource for production in the case of Huaca Sialupe was wood fuel. The patterns of use of certain wood fuels, at the expense of others, might indicate some kind of primacy over different types of production. Likewise, this patterning may suggest that craftspeople were forced to prioritize elite requirements over more immediate and local economic and cultural demands, such as domestic economy, relation to local ecology, etc.

CHAPTER 5
FIELD AND LABORATORY METHODS

The paleoethnobotanical analysis executed at Huaca Sialupe combines traditional paleoethnobotanical approaches of species identification via seed morphology with the determination of woody plant species through wood anatomy; archaeobotanical remains are compared anatomically to modern materials and available identification keys to determine their plants of origin. The goal is to reconstruct ancient resource use and availability in an archaeological site relative to its surrounding ecology to understand how people apportioned organic resources at this ancient craft production site, specifically relating to fuel use. By analyzing fuel types in different contexts in the workshops at Huaca Sialupe, we can identify how local production and daily life intersect with the elite political economy; moreover, determining their distribution across the site can provide insight into labor and activity organization and artisan's role in the domestic economy.

Given this focus, my research concentrates on the patterning of arboreal plant species across the sites of ceramic and metal craft production at Huaca Sialupe. I completed this investigation in three stages: the development of a comparative modern plant collection, archaeobotanical recovery from craft production contexts, and laboratory analysis. To perform this work efficiently and in line with demonstrably effective methodologies, I constructed the appropriate reference collection of both seeds and wood materials appropriate to the environmental, historical, and cultural tradition relevant to the research question (Pearsall 2001; Wheeler and Baas 1998). To ensure that I was collecting fuel resources salient in the area I compiled a typology of fuels and list of potential fuel resources with the help of local informants. In conjunction with the SAP's excavation crew, I implemented and developed a strategy for the field recovery of archaeobotanical remains and then executed this methodology during the 2001 field season. Finally, I analyzed the archaeological and the modern collections. My aim was to identify the modern materials from the local materials based on modern field botany keys for the area. I tried to determine the genera or taxa represented by the archaeological plant materials either through visual comparison with modern materials in the case of the non-wood remains, and through comparison with modern wood sections from my modern collections and wood anatomy keys for the wood remains.

Reference Collection

I spent four months of the 2002 field season surveying the plant communities in the lower La Leche River Valley. Permits were acquired from the Peruvian Department of Natural Resources (No003-2002-INRENA-DFFS) under the Department of Agriculture for the collection of all plant specimens. A file is on report with this organization in Peru detailing the outcome of the 2002 season. The goal of my botanical survey was to locate and collect samples of any and all potential fuel resources present in the research area today (see below for how fuel resources were determined). While this survey focused on the collection of arboreal plants, I also collected woody herbaceous vegetation, those with xylem or simply reinforced primary cell walls (collyenchyma). Developing a reference collection included:

1. Identifying the modern plant specimens and collecting botanical voucher specimens of all woody vegetation in the area
2. Collecting wood specimens from known trees in each ecological zone (more than one where possible) that represent a variety of age grades within the sample
3. Collecting wood specimens of each of these trees from a variety of locations in the tree, where possible, to interpret functional variation of the cellular anatomy by part of the tree.

Ostensibly, the woody vegetation in the area is used for two main purposes, fuel and construction. Trees are also used as sources of renewable resources apart from these materials, such as fruit for human consumption, bark for lashing, thorny scrub for living fences, and animal fodder. My observations of experimental ceramic production in the 2000 SAP field season, 1999 fieldwalking and visiting of Huaca Sialupe denizens, and my 2002 fieldwork all indicate that non-woody herbaceous vegetation often serves to initiate fire, e.g., preheat ceramics, initiate kilns (Wagner et al. 2001). So, I was certain to collect herbaceous vegetation during my field collection. Based on concerns with the variation inherent in woody plants due to age-grade and ecologically conditioned anatomical changes in the wood collected (Panshin and DeZeew 1980, Wheeler and Baas 1998, Dawson and Lucas 2005), an effort was made to collect modern specimens from the three different ecological zones with visibly different vegetation within the La Leche Valley (Standing Dry Tropical Forests, River Bottom Areas (Riparian Forests), and the Dry Lower Foothill Slopes; Chapter 2).

Site Selection and Botanical Sampling

Initially, I selected two sites for ecological prospecting and sampling of the regional flora (Table 5.1, Figures 5.1 and 5.2), based on the following criteria. Zone 1 borders the National Historical Sanctuary, Bosque de Pomác, the largest dry forest preserve in South America. The Historic Preserve is protected by the National Institute of Natural Resources for Peru (INRENA) the National Institute of Culture (INC) and is the only place where active forest cutting and cultivation clearance has been effectively prohibited on the North Coast (See Discussion in Chapter 2); it also encompasses the large archaeological ceremonial precinct of Sicán. INRENA, however, did not permit access to this area for the sampling purposes of this study due to its short tenure as a protected area (since 2000), the existence of a limited previous survey conducted by a Peruvian non-governmental organization (NGO), and my status as a

Zone	I.G.M. Quadrant Carta Naciónal del Peru 1:100,000	UTM
1	Jayanca	E ^{6}44 to E ^{6}56 N ^{92}82 to N ^{92}92
2	Chiclayo	E ^{6}12 to E ^{6}24 N ^{92}64 to N ^{92}74

Table 5.1 UTM Coordinates for Botanical Sampling Zones 2002 Field Season

graduate student. A local NGO, Proyecto Algarrobo, *lit trans.* the algarrobo project, conducted a limited floristic survey and forest inventory in this area (Proyecto Algarrobo 1997, Zevallos P. 1986). Sampling of the plant materials, however, was restricted to plants larger than 50cm high; in addition, only two 100m x 2m transects were completed within National Historical Sanctuary, Bosque de Pomác, which only covers one of the four main ecological zones in the area. The survey also did not collect wood samples or complete vouchers. Some of their materials were helpful and used for basic identification purposes. Their collected materials, however, from their research did not generate a comparative collection of wood deposited in a xylarium necessary for this research. So I had to select locations and collect my own wood and voucher specimens.

Fortunately, the process of establishing the National Historical Sanctuary, Bosque de Pomác raised awareness among the local population of the importance of co-existing with forest resources. During two weeks of preliminary fieldwork in 2000 and 2001, I learned that some landowners were opting not to cut down older forest growth as a conscious decision towards protecting the last of the forested areas in this portion of the La Leche River Valley. This shift in mentality is largely responsible for the survival of forested areas contiguous to the park and upstream along the La Leche River Valley that provided an alternative region for our botanical survey. The local sugarcane company, once a cooperative and former hacienda, stationed at Batán Grande, however, was in dire economic straits and the easiest source for supplementary income by people who are occasionally employed by the cooperative is to cut *Prosopis* spp. and *Acacia macracantha* trees in the area

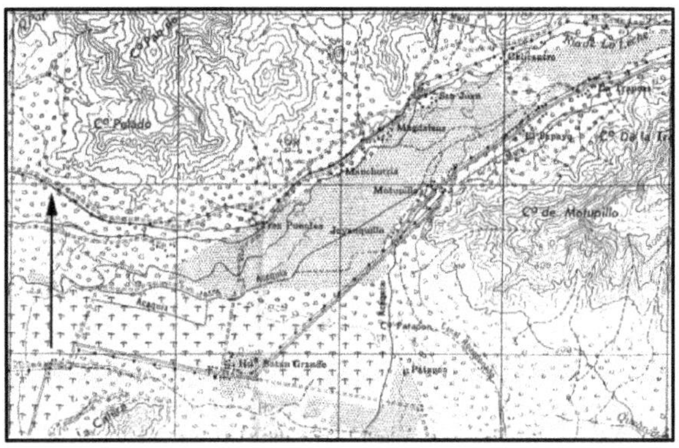

Figure 5.1 Zone 1-Batán Grande Region

to sell as charcoal when there is no cane work. This economic situation in 2002 led to a condition where some landowners were actively protecting their land from intrusion by wood cutters, others were cutting their trees as fast as they could, and some had internal family struggles where some household members were interested in cutting while others wanted to protect their trees.

These species constitute the primary forest cover and are the backbone of the forest regeneration dynamic in the area; under these economic circumstances their wood is converted to charcoal for sale in the regional and local markets, threatening the local ecological balance and heightening the urgency of our 2002 survey. The areas to the west of the National Historical Sanctuary, Bosque de Pomác, open onto a combination of heavily cultivated and/or desertified territory, and the northern and southern boundaries are mountainous with very little vegetation above 30 cm high. The eastern most boundary of the National Historical Sanctuary, Bosque de Pomác, however, provides a comparable sample of woody vegetation; I chose this region to replace the original Zone 1 of the National Historical Sanctuary, Bosque de Pomác.

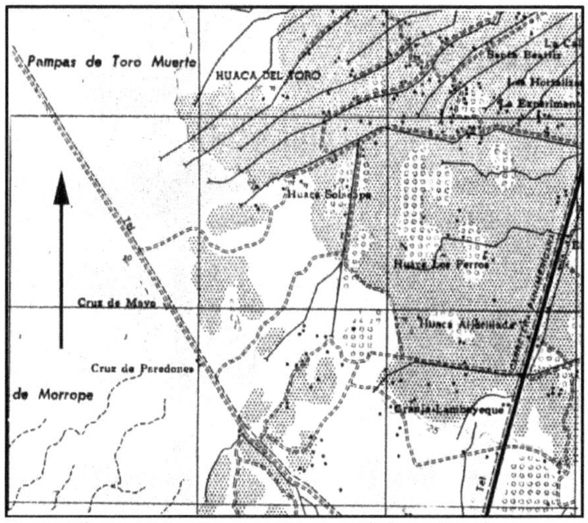

Figure 5.2 Zone 2-Huaca Sialupe Region

Zone 2 is located at the perimeters of the archaeological site of Huaca Sialupe, the object of this investigation; its degraded ecological state, however, presented additional methodological difficulties. During the 1999 and 2000 field seasons at Huaca Sialupe I began a preliminary collection of samples from local woody vegetation and recorded the dominant flora in the area. I noted that trees, regardless of genera, had an estimated density of only seven to ten (Diameter Breast Height (DBH) 50 cm < 10 cm) per 100 m^2. In 2002, fieldwalking confirmed that farming disturbance over the past 100 years or more had heavily deforested the area. Locating a good reference sample, representative of a variety of age grades and ecological zones for each species, for either the anatomical or floristic data was impossible. Not only was the diversity of the area's woody flora scarce in number, I did not locate a single tree with a probable age over 40 years of age within a three km radius of Huaca Sialupe. Given the damaged character of this zone,

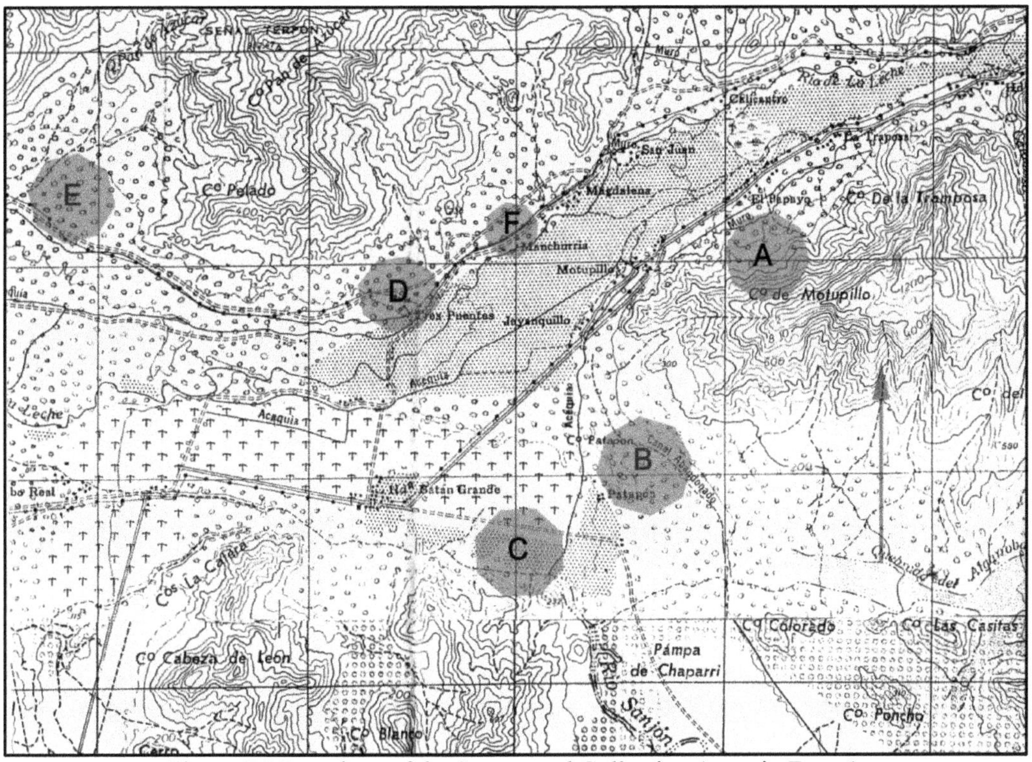

Figure 5.3 Locations of the Survey and Collection Areas in Zone 1

potentially in a state of deflected succession, I shifted my focus back to Zone 1: I decided that the more or less uniform ecological niche in Zone 2 (savannah/agricultural fields under irrigation), as recognized by INRENA and Proyecto Algarrobo (1993) in their survey of the Department of Lambayeque, could be sampled from similar ecological niches as they manifest themselves in Zone 1. Spending more time in the area was not likely to reveal new information or useful comparative materials, so I abandoned Zone 2 for survey and the collection of reference material for this study.

Zone 1, then, became the primary area of concentration for reference collection sampling and floristic survey. Certainly, deciding to move away from Zone 2 created a less than ideal situation for performing survey work directly associated with the site of Huaca Sialupe. Unfortunately, due to the level of deforestation and my desire to understand dry forest ecology, there was little choice. As a compromise I focused on Zone 1. My team and I made two six-week forays into this area in May-June 2002 and again in August-September 2002 to collect during both the Spring and Fall flowering seasons to ensure that I compiled as many vouchers with flowers or fruit as possible.

Zone 1 was divided into 6 areas for intensive sampling (Figure 5.3). These areas can geographically be distinguished and located as:

A. The Northwestern slopes of Cerro Motupillo;
B. Cerro Patapón and its surrounds;
C. The forest held by the Guevarra Family ESE of Batán Grande;
D. Both sides of the La Leche River near the settlement of Tres Puentes;
E. The plain to the West of Cerro Pelado;
F. Above the flood plain of the La Leche near to Manchuria.

Each zone exhibits different kinds of ecological constraints and poses different challenges to plant growth and regeneration (Table 5.2). My collection strategy was geared towards covering as much ground as possible in the six different survey areas to best understand the makeup of each ecological community. Sampling was largely opportunistic based on the vegetation that was present, correspondence to what I collected in other areas, and to take care not to destroy resources that were unique in each place. We first located species of interest based on my ethnotaxonomy of fuel materials (see below) and then searched for specimens to be sampled across the different collection zones. UTM coordinates were taken with a Garmin GPS72 unit for each sample.

I attempted to collect multiple samples from each species from each of the three different ecological zones present in the lower La Leche River Valley as defined by Proyecto Algarrobo (Standing Dry Tropical Forests, River Bottom (Riparian) Areas, and the Dry Lower Foothill Slopes; 1997). The aim was to record the differential effects of environment on the growth of each species. To do this I collected the same parts of trees, from the same species, growing in different environments. For example, I collected tertiary, secondary and primary branchwood from specimens of *Capparis angulata* in four of the six different ecological

ZONE	SOIL WATER INSOLATION CONSTRAINTS	DESCRIPTION OF FACTORS FOR GROWTH INHIBITION
A. The Northwestern slopes of Cerro Motupillo;	Dry, Well Drained Rocky Soils, Shaded 4 hrs/day, Sloped 5% grade	Water is Most Limited Resource, Human Predation for Fuel Wood
B. Cerro Patipon and its surrounds	Dry, Clay/Sand, Sun 8 hrs./day, Flat, Irrigation Canals	Water is Most Limited Resource, Sun is Excessive. Unless Canals/ Irrgation is Functioning
C. The forest held by the Guevarra Family to the east-southeast of the town of Batán Grande	Shaded, Cool, Full Cover Forest, Flat, Irrigation and Subterranean River	Human Predation for Fuel and Farmland. ENSO Activity Floods River and Damages Trees
D. Both sides of the La Leche River near the settlement of Tres Puentes	Dry, Eroded Sandy Soils, Floodplain, Sun 12hrs/ day, Flat	ENSO Activity Floods River and Damages Trees. Roots Unstable
E. The plain to the West of Cerro Pelado	Dry, Well Drained Rocky Soils, Limited Shade, Sun 8hrs/day, 4% grade	Water is Most Limited Resource, Goats Eat Young Trees, Selectively Denegrated Landscape.
F. Above the flood plain of the La Leche near to the settlement of Manchuria.	Dry, Irrigated Fields and Settlement Area, Human Disturbed, Clay/Sand, Limited Shade, Sun 8hrs/day, 3% grade to Flat	Human Predation for Fuel and Farmland. ENSO Activity where Survey Area Meets Floodplain

Table 5.2 Survey Areas in Zone 1 and Factors Limiting Tree Growth

zones where the plant grows. For each specimen, I documented important ecological statistics and GPS coordinates; these data were recorded on forms and entered into a Microsoft Access database (Appendix A).

I also collected samples of each species for as many of the different age-grades of the plant in ecological zone. For instance, I sampled *Prosopis* spp. several times in the same ecological zone to represent different ages of growth; I did this for each zone where the plant was present. By these means I developed a comparative sample that can account for changes in wood anatomy based on ecological and functional differences for each species throughout their life span. Branchwood samples were cut from standing trees. In a few instances wedges were cut from the boles of standing mature stature trees as a standard increment borer could not breech the wood of any of the species in the area. This situation is ostensibly due to the hardness and extremely high density of the wood species included in the survey. Leaf, fruit, and flower vouchers were cut with a pair of pruning shears and pressed in the field between newspaper and standard unbleached blotter in a field press. Vouchers were dried in the courtyard of one of the storage facilities at the National Sicán Museum.

For each species in this investigation I collected and compiled five copies of a comparative leaf and flower voucher specimens and wood collections. Four of these collections were donated to the following institutions, while the fifth resides with the researcher:

1. The project archive at the National Sicán Museum, Ferreñafe, Peru.
2. The xylarium and herbarium at the Department of Forest Science, UNALM, La Molina, Peru.
3. The herbarium of the Field Museum of Natural History, Chicago, Illinois (vouchers), and the xylarium of the Forest Products Laboratory, Madison, Wisconsin (wood).
4. The National University of Pedro Ruíz Gallo, Lambayeque, Peru, (still at the National Sicán Museum awaiting acquisition)

All materials are stored on archival card (vouchers) and in polyethylene bags with acid free paper (wood) identification tags. A digital record of the collected ecological and reconnaissance data for each specimen in a Microsoft Access database accompanies each set of samples.

In summary, my crew and I identified by common name 192 different plants potentially used in fire construction. Appendix B presents the relevant botanical data collected during the four months of botanical sampling. These plants represent 131 recognized genera within 52 botanical families. Of those plants, I cut 115 wood sections from the 63 woody (those with secondary xylem) genera for anatomical identification and comparison; at least four different tree parts were sampled, for a total over 1,000 individual wood specimens. I determined all Latin name designations and systematic classifications for all identified plants through comparisons with herbarium specimens from the National Agricultural University at La Molina, Peru, in conjunction with available keys/surveys for woody vegetation and regional flora (Gentry 1994, Sagastegui Alva 1973, Brako and Zarucchi 1993, Weberbauer 1945). A series of floristic inventories and voucher specimens of the major woody vegetation of the area, generated by INRENA and Proyecto Algorrobo's recent interests in the sustainability of dry coastal forest in the Batán Grande region, provided an alternative resource (Proyecto Algarrobo 1997; Cuba Salerno et al. 1998, INRENA and Proyecto Algarrobo 1993). I solicited and received additional determination information from Dr. Carlos Reynel, Head of the Department of Forest Sciences, UNALM, Peru, Lisc. Biologist Tina Lerner, Lima, Peru, Lisc. Biologist Anibal Calderon, Chief of the Interpretive Center, National Historical Sanctuary, Bosque de Pomác,

Common Name	*Scientific Name*
Hualtaco	*Loxopterygium huarango* Spruce ex. Engler
Palo Santo	*Bursera graveloens* (H.B.K.)Triana and Planch
Vichayo	*Capparis avicennifolia* H.B.K.
Palo Verde	*Cerciduim praecox* subsp. *praecox* (R. and P.) Harms.
Sapote	*Capparis angulata* R.and P. ex DC.
Overo	*Cordia lutea* Lam.
Faique	*Acacia macarantha* H. and B. ex Willd.
Algarrobo (macho)	*Prosopis pallida* (H. and B. ex Willd.) H.B.K.
Cuncun	*Valesia glabra* (Cav.) Link.
Pajaro Bobo	*Tessaria integrifolia* R. and P.
Alcapparillo	*Cassia bicapsularis* L.
Angolo	*Pithecellobium multiflorum* (H.B.K.) Benth.
Cabelludo, Palo Verde	*Parkinsonia aeuelata* L.
Pacay or guaba	*Inga feuillei* DC.

Table 5.3 Names of 14 Major Wood Species Sampled for Anatomical Description from Research Area (Brako and Zarucchi 1993; Zevallos P. 1998)

Peru, and Dr. Michael Dillon, Department Chairman, Division of Botany, Field Museum of Natural History.

All of these samples can serve as proxies for understanding growth, community dynamics, and th effects of ecological stress on the woody vegetation of the region; for this research; however, I focus on the growth qualities of fuel resources to gain insight into Middle Sicán collection strategies. Additionally, they provide a baseline from which to evaluate and/or estimate ecological change in the forest over time (Fritts 1976; Schweingruber 1996). Overall, I expect that future students working in the area will have access to these materials and develop new research strategies that look at the functional responses, that is the cellular changes in anatomy, due to ecological realities of plant growth in terms of archaeological assemblages and the forces that created them.

Ethnobotanical Terms for Fuel Identification

The primary portion of the study was conducted using local names for the various plant types. I used this terminology to facilitate communication with my crew, all local farmers/residents of the area. My crew consisted of Ismael Llauce and Rafael Benitez, both local farmers/laborers, Juan Carlos Santoyo, a local beekeeper, tour guide, and longtime member of the SAP, and Don Julio Navarro, a local herbalist/*curandero*. All crewmembers have spent their whole lives in the region of Zone 1 and, collectively, retain generations of ecological and botanical knowledge of the area. Using these terms allowed me to incorporate farmers' knowledge into the goal of identifying plants with desirable or potential fuel characteristics. I refined the terminology from my field notes and informal discussions with local farmers and my field crew during fieldwork. While collecting specimens, we used three general terms to classify the species local people use as fuel: *leña*, *combustible*, and *poña*. *Leña* and *combustible* are both Spanish words; *poña* is a local term derived from the now extinct *Muchik* language, once indigenous to the area.

Leña, lit. trans. wood, refers to any durable plant part that can be harvested and measures larger than approximately two cm in diameter. The uses of *leña* range from fire building to house construction. *Leña*, however, is further distinguished by the part of the tree that it comes from, e.g., *brote* (branch), *raíz* (root), *trunco* (trunk), *corteza* (bark): this series of sub-classifications is used to describe this type of fuel. Once the *leña* is selected for a particular use, prior to and throughout cutting or processing, farmers often replace the term *leña* with the name of the item the wood is to become, e.g., *viga* (house post), *puerta* (door), *lumbrales* (window frame supports), *cuartones* (corner wall supports), *alambre* (binding), *batán* (anvil for some leather and wood working). As the material, *leña*, is transformed through its use, it no longer is referred to as *leña*. In the case of firewood collection, *leña*, is conceptually transformed into *combustible*, fuel, through the prospecting and harvesting process.

Combustible, lit. trans. fuel, spans a variety of flammable materials, some of which are renewable plant resources (also encompasses fuels made from refined petroleum). In addition to *leña*, small herbaceous plants and lianas, products of furious growth spurts associated with ENSO rain events and plentiful on the soil surface in the region as dried plant matter, are the most common non-woody sources of *combustible*. Persistent grasses and herbaceous plants associated with fallow fields in irrigated agriculture systems, which abound in the area, are also frequent choices for *combustibles*. One type of *combustible*, critical to this study, is *carbón* (charcoal) that is produced from *leña* (Chapters 2-3). Charcoal is manufactured by burning wood, *leña*, in the absence of oxygen; the anaerobic environment reduces the carbon-based cellulose molecular structure that is the backbone woody plants. Generally, these reactions take place at high temperatures, over long periods of time, and result in the volatilizing of trace elements and non-carbon based molecules (FAO/UN 1983; Rosotti 2002). The result is a physically stable, light, and portable *combustible*.

The last term used to refer to potential fuel resources in this study is *poña*. This local term for kindling does not have a direct Spanish translation. It may have its root, however, in the extinct local language of *Muchik*. Salas' (2002:31) compilation of *Muchik* vocabularies identifies the terms *pong* or *pongiio* [poŋ, poŋjo] and also *poñ, -eió* [poŋ, -ejo] as phonetic matches to the modern *poña*. Carreras' (1644 as reported in Salas 2002) *pong* closely approximates the definition my crew utilized. Salas (2002) traces shifts in the definition from 'a series of items grouped by dozens, horses, rocks, cane, that are not money or fruits' by Carerras (1644 in Salas 2002) to more recent definitions of *pong* as 'a rock' (Brüning 1928 in Salas 2002). Nevertheless, it seems that the term *pong* maintained its collective meaning as 'an aggregation of items in dozens, multitude of things' (Salas 2002:31). The significance of the term for this study is that *poña* is never applied to a specific plant or single type of fuel. Instead, it refers to a *combustible* gathered as a mass of burnable, and even some non-burnable, materials used to start a fire.

In collecting kindling materials for the experimental kiln research carried out in 2000 and in the experimental charcoal furnace constructed in 2002 (see Chapter 2), *poña* collection was a general grab of proximate dry plant materials near to the fire construction site. In 2002, *poña* was collected with a shovel from underneath trees in the area, taking advantage of animal defecation in the shade of trees and leaf fall. Shimada (personal communication 2007) indicates that 4 grown men can collect about 50 Kg of *poña* in about 30 minutes. During my 2002 botanical survey and collection of potential fuel plants, I often asked my crew if they recognized *poña*-specific plants. In fact, *poña* is such general a term that they rarely identified a plant's use as being specifically for *poña*. This general sense of the term can be applied to anything dry, plant or otherwise, to be used as kindling. The term has also been recorded as being used for describing organic material collected as fertilized (Nordt et al. 2004).

For the most part my crew, and others that I spoke with, did not distinguish between woods based on a particular fire use. In other words, fuel was seen as fuel. What they were aware of is that some fuels burn hotter than others, *Prosopis* spp. being at the top of the list, as well as a series of other relative differences between woods, some were harder, wetter, softer, longer burning, smokier, etc, than others. Utilizing these local terms enabled me to develop a general system of use classification for fuel resources and to locate, sample, and categorize plants based on these distinctions. Seeing as the modern landscape cannot stand in for the ancient one in terms of diversity, species, or land use (Hather 1994), I do not assume my collection to be exhaustive or complete relative to the historic landscape. It is, however, the best approximation available for the purposes of reviewing the archaeological material from Sialupe. By outlining the terms and the plants collected by my crew and myself, I am attempting to convey the sense of holism involved in generating this comparative plant database. The collection of botanical and ecological information for reference materials was carried out with technical precision; equally important, this strategy also maintained a sense of how people live and consume resources in the research area.

Excavation Protocol and Archaeobotanical Recovery

The arid environmental conditions of the North Coast of Peru present exceptional conditions for the recovery of archaeobotanical specimens (Chapter 3). With this in mind, the protocol for sample recovery during the excavation of Huaca Sialupe in 1999 and 2001 relied largely upon dry sieving. I developed a sampling protocol based on visual examination of specimens and on-site activity during the 1999 and 2001 seasons. The following features were classified and recognized by the excavation team (see Chapter 3): *hearths* (cooking fires), *ash lenses* (temporary occupation fires), *ceramic preheating fires*, *blackware ceramic kilns*, *oxidation kilns*, and *metalworking furnaces*. The feature typology that I used was defined largely based on the 2001 excavation season coupled with my understanding of the morphology of feature that we tested in our 2000 experimental work (Chapter 3).

During the 1999 season, I spent two weeks on-site evaluating the quality of the organic materials recovered from the features and floor contexts being excavated. This assessment demonstrated that both carbonized and dry remains constituted the bulk of the recovered organic materials. Large areas of burning and distinct fire-use features provided the best contexts for the recovery of organic remains. Floor contexts, on the other hand, while yielding good overall organic material recovery, were fraught with other taphonomic problems. Within the excavation, floors were only partially extant, and often the workshop-related features did not have recognizable intact floor contexts associated with them. This situation is mostly due to the floors' composition of nothing more than tamped or compressed earth.

After the 2001 excavation season this phenomenon was attributed to the relatively short occupation of the site during the Middle Sicán, the post-abandonment phase related to the ENSO inundation and erosion processes, and later construction on the site (Shimada and Montenegro 2002). The preservation of dry and desiccated materials in floor contexts was also probably limited due to incidents of inundation that caused periodical fungal growth within the contexts over the past 1000 years. Fungal growth is visible in the wood, dried organic matter, and charcoal materials from the Middle Sicán levels at the site in the form of fungal spores recovered during the microscope analysis. Additionally in 1999, I observed that animal dung made a substantial contribution to the fuel used at the Middle Sicán workshop. By the end of my stay during the 1999 season, I could see that the carbonized material from the features, as far as wood species determination was concerned, was in excellent condition. I saw the potential to reconstruct fuel resource consumption at the site from the charred remains, as well as provide a baseline of Middle Sicán ecology at Huaca Sialupe.

A sampling strategy was necessary, however, that would permit a significant amount of material to be recovered for analysis. On the other hand, the desiccated wood was less useful in terms of identification and utility in the study that I conceived. The overall preservation of all organic materials, whether carbonized or not, however, indicated that all contexts and floors needed to be systematically sampled to assemble the dataset for the analysis proposed by my research. Without systematic sampling, where known volumes of each sample are compared, one cannot compare feature contents at the site. Up to that point the SAP strategy for collecting charcoal samples was not systematic in this way, instead opportunistic sampling of some of the features was made when the excavation crew thought it necessary.

In 2001, excavators made systematic efforts to sample the entire contents of all pyrotechnology features recovered at the site. Additionally, they collected a minimum of four-liter samples per one m^2 quadrant of floors. Feature sampling began from the top down, removing the entire contents in layers: from the top surface, layer by layer (where present) was removed with a trowel and placed in polyethylene bags left open to ventilate any accumulated humidity. Once the relative humidity of the bag contents appeared stable, i.e. no condensation visible in the bag, we sealed the bags. Excavators sampled the floors using a trowel to scrape up the entire floor surface, lifting it in clumped wedges, where possible. This protocol was followed in an effort to sample materials ground into or otherwise incorporated into the floor matrix during its use.

In some of the excavation areas these protocols could not be followed either due to excessive sample volume or difficulty in defining the boundaries of features or floors. In these instances, we recorded the estimated geometrical forms and dimensions of the feature, e.g., spheroid, ellipsoid, rectangular prism, so that the relative volume of the sample could be compared to the feature dimensions. Similarly, we documented floor dimensions and measurements relative to the size of the sample removed. All features and floors from the site were individually recorded and sketched on forms (Appendix C). For the Middle Sicán levels, at least 75% of the features' contents were sampled (Table 6.1); we managed to get at least one sample from every feature and floor Middle Sicán level, even in cases that the entire floor of feature was not sampled due to preservation problems. This sampling provides a data set that can distinguish between feature and contiguous/contemporary floor components, as well as between features. All data was compiled in a Microsoft Access database for easy sorting and data manipulation.

Laboratory Analysis and Dataset Determination

In March and April of 2002, I screened and rough-sorted all of the 1999 and 2001 Huaca Sialupe samples at the newly opened National Sicán Museum, Ferreñafe, Peru. Screening protocol required sieving of Middle Sicán context-specific samples representing 41 contexts (n=56; 239.51 Kg ;240.06 L) through a standard series of geological sieves (4mm, 2mm, 1mm, and 0.5mm mesh sizes). I individually weighed and measured the volume of each sample and sub-sample by screen size; I also weighed and measured each individual fraction. For each sample, a 250ml sub-sample of the sediment smaller than 0.5mm was taken for future microanalysis, e.g., pollen, phytolith, or soil trace-elements. These microsamples were separately stored in an archival quality polyethylene bag at the National Sicán Museum, Ferreñafe, Peru.

The identification of charcoal relies on a clear view of three-dimensional structure of the wood. Species determination is speculative at best in samples less than 2mm square. Even much of the 4mm is difficult to determine. This is especially the case in ancient charcoals, as they are materials that have undergone a series of post-depositional environmental changes. Hence, separating the 4mm and larger charcoal materials from the samples for species determination was one of the goals of sorting the archaeological matrix. As my primary interest in this study is the determination of the wood fuels used at the site, I removed all of the charcoal from each sample from the 4mm and 2mm fractions of each sample. I scanned all four fractions of each sample for non-wood botanical remains and artifacts using a Meiji Technologies EMZ-8TR Zoom incident light zoom microscope (7-180x magnification), in the Museum's laboratory. Two halogen light desk lamps were used together for illumination. One hundred percent of all 4mm, 2mm and 1mm fractions were scanned under microscopy. Fifty percent of each 0.5mm fraction was scanned for botanical remains. Non-wood remains were packaged together with the wood remains in archival polyethylene bags for export to the United States. All materials were exported under permission from the Peruvian National Cultural Institute to Dr. Izumi Shimada, and I was responsible for packaging and shipping the materials to the United States.

Sample Preparation and Determination: Non-Wood Materials

Once in the United States, I prepared the seed and wood reference collections in the laboratory for comparison with the archaeological materials. Initially, a collection of seeds and fruits was made from the voucher specimens collected during the Zone 1 botanical reconnaisance. These materials were removed, numbered, and catalogued based on their survey information. Each specimen removed was placed in a labeled, numbered archival-grade polyethylene bag or scintillation vial. I took digital photos of all seeds and constructed a digital seed atlas of the fruits and seeds recovered in my botanical survey (Appendix D: unknowns only). I added digital photos of other known reference materials to this catalogue, including seeds housed by the University of California at Berkeley's Paleoethnobotanical Laboratory Andean seed reference collection. These references, in combination with textbooks for seed identification (Martin and Barkley 1961; Corner 1976; Gunn 1972; Sagastegui Alva 1973; Vozzo 2000), provided the comparative keys and visuals for the determinations of the seed materials found at Huaca Sialupe. Due to the large quantity of dung observed within the remains, I generated a small reference collection of Camelid dung from *Llama glama* and guinea pig dung from *Cavia porcellus*. These references were used to distinguish between the two species on visual identification (see Chapter 6).

All non-wood remains were examined using a Meiji Technologies EMZ-8TR Zoom incident light zoom microscope (7-180x magnification). Two halogen light desk lamps, and a Leica incandescent spot were used together as light sources. Each non-wood remain was classified to botanical genus or class of material (e.g., fish bone, ceramic, dung, metal, slag) and placed in an archival grade polyethylene scintillation vial with an acid free label inside each vial with the determination and context written on its exterior in permanent black ink. In instances where the recovered materials did not fit inside of a scintillation vial, a small polyethylene bag was used. All information was recorded on paper forms (Appendix E). All determined and non-identified materials, in their respective packages, were gathered and stored in their own individual bag by archaeological context for later reference.

Sample Preparation and Determination: Modern Wood Samples

I sampled 14 species for anatomical description. One from each of the major woody species collected from Zone 1 that produce secondary xylem, wood (Table 5.3), to create a set of comparative reference slides of local woods in the region of investigation. Additional samples were sectioned of each species of reference material on an *ad hoc* basis to locate specific morphologies related to age-grades and organ-based differences in each plant. All samples included at least four different tree parts: primary branchwood, secondary branchwood, tertiary branchwood, and trunkwood. These 14 descriptions and slides constitute the core reference samples that correspond to the most common woody plant species in the area today. These were used in conjunction with computerized and printed wood keys for tropical woods to facilitate the species determination of the archaeological wood recovered from Huaca Sialupe (Wheeler et al. 1991; Record 1942-8; Record and Hess 1972). These keys have the additional advantage of allowing me to cast a wide net of most Neotropical tree species, this permits the possibility to identify tree species that do not grow in the area at present. I also used wood anatomical descriptions and keys that I located through library research and data resources available through the UNALM Department of Forest Sciences, La Molina, Peru (Ocampo 1991, Acevedo M. and Kikata 1994).

For each of the 14 different samples, each specimen, representing the various parts of the plant sampled, was sectioned and mounted. First, each specimen was rough cut on a freestanding bandsaw with a 2.5 cm wide blade in the three directions of analysis, i.e., tangential, radial, and cross sections, to expose the diagnostic anatomies. Next, I manually cut the specimen with a disposable microtome blade; the thin section was mounted on a glass microscope slide in glycerol with a cover slip. Each cover slip was sealed with a viscous solution of cellulose nitrate in amyl acetate (commercially available fingernail varnish) after a 48-hour period allowing for the sample to settle and air bubbles to evacuate. I identified each slide with a self-adhesive backed white label filled out with permanent black ink, identifying the species, provenience information, sectional view orientation, and part of the tree represented. All slides were viewed using a Fisher Scientific Micromaster CK Transmitted Light Microscope (CK 100, 200,400, and 1000x objectives). Under magnification, I used a paper schedule to record the anatomical characteristics of the modern comparative specimens within the initial 14 woody species (Appendix F). The schedule is the one used by my advisor, Dr. Lee Newsom, The Pennsylvania State University, and conforms to the standard anatomical features for hardwood trees established by the International Association of Wood Anatomists (IAWA Committee 1989), and the features used by the InsideWood (2004) on-line wood anatomy database housed at North Carolina State University. This particular anatomical survey is slightly more specific than either Inside Wood or the IAWA schedule; highlighting a wider variety of patterns expressed in the axial parenchyma and takes more specific notice the cell composition of radial parenchyma (Elizabeth Wheeler personal communication 2005). Both the radial and axial parenchyma are often critical distinguishing factors in tropical woods.

To familiarize myself with the qualities of carbonized wood, I examined the remains from the experimental firing of full-scale replica kilns and metalworking furnaces (from SAP 2000 season) and the experimental charcoal furnace (from my 2002 fieldwork) under magnification. This allowed me to practice, under the experimental conditions of knowing what specific woods were present in the case of both experiments, the sorting and identification procedures followed with the archaeological materials.

Sample Preparation and Determination: Archaeological Wood

The archaeological features recovered from Huaca Sialupe and used in this study consist of 61 samples. Carbonized remains compose the majority, by weight (8.21 Kg), of the dataset utilized in this dissertation. All charcoal remains larger than 4.0 mm were examined for determination from each context. Wood species determination for each individual charcoal specimen is based on anatomical features recorded in cross (transverse), radial, and tangential sections, under magnification, up to 180x, using the incident light Meiji microscope described above (see Wheeler and Baas 1998). I also consulted a series of reference materials for tropical woods (InsideWood 2004; Ilic 1991; Moutarde 2005; Acevedo M. and Kikata 1994) and seed identification (Martin and Barkley 1961; Corner 1976; Gunn 1972; Vozzo 2000). Specimens were fractured by hand in each plane of view and examined under magnification to locate salient anatomical features for each piece of charcoal. For each archaeological sample I recorded, when possible, I also described a qualitative assessment of any anomalous cellular anatomy present in the wood charcoal and part of the plant, e.g. branch-, root-, or trunk wood. Initially, the anatomies of each different recognizable type of wood was described and outlined using the morphology schedule described above for modern wood reference samples (Appendix F). Appendix G is a report of the wood anatomy of the 14 major species in the La Leche Valley region the focus of this dissertation.

Basic Wood Anatomy

Plant materials in archaeological contexts provide a critical link between the living ecology of the Huaca Sialupe area and human activity at the site in the Middle Sicán. Each sample represents an aggregation of organic materials selected by individuals from their surrounding environment for a specific purpose. The organic materials recovered from Huaca Sialupe's contexts, however, are emblematic not only of human selection and technological choices (Chapter 4), but also represent the results of ecological growth and environmental dynamics.

Because some plants generate secondary xylem in species-specific patterns, we can identify wood remains by looking at the three-dimensional structure of the secondary xylem. To this end wood anatomists have compiled extensive databases of modern or ancient comparative specimens and developed keys for the identification of wood from observations made of its structure in three dimensions (Acevedo Mallque and Kikata1994; Ilic 1991; InsideWood 2003; Wheeler and Baas 1998). Plants are distinct from mammals, a common point of view for humans when considering growth, in that they grow structurally throughout the course of their lives. This growth is particularly significant in tropical areas where seasonality is often less pronounced, or influenced by two peak periods of moisture, leading to year-round growth apart from dormancy periods of a up to few months. Wood charcoal, the majority of the archaeobotanical finds from Sialupe, serves as a record of the growth state of the plants used for fuel, which can provide insights into the health the forest. For example, in instances where tree growth has been inhibited or affected by climatic or human interventions (Baas and Vetter 1989; Fritts 1976; Schweingruber 1996), distinctly marked anatomical structures occur at the cellular level, e.g., scarring or false rings. In some instances these distinct anatomical patterns can be correlated with ecological processes observed in modern tree growth (Baas and Vetter 1989; Newsom 1993; Schweingruber 1996). Below is a brief explanation of how woody plants generate secondary xylem to explain the environmental import of wood charcoal.

Botanically, the complex plant cell structure, identified commonly as wood, is known as secondary xylem. The lignified woody cells collectively comprise the secondary xylem tissues of the plant's construction, i.e. that interior to the cambial zone, and are resistant to most forms of organic and physical decay as long as the living plant's immune response and other living physical features remain resilient to incursion. These cells are responsible for the physical and metabolic, e.g., water conduction and exudate storage, support of the plant during its lifetime, and form the bulk of the resource known to humans as wood.

At the microstructural level, certain dicotyledonous plants generate both primary and secondary xylem through the basic metabolic process of consuming sugars and combining their byproducts to form lignin, cellulose, hemi-cellulose, and a variety of secondary metabolites, e.g., oil, gum, and sap. Cellulose, lignin and hemi-cellulose, however, are the basic construction material of plant cells (Panshin and DeZeew 1980; Raven et al. 1999). In the case of secondary xylem the primary cell walls, referred to as S1, are reinforced with two additional cell walls, the S2 and S3 that both contain cellulosic reinforcement and polymer infusion of lignin, an organic polymer highly resistant to hydrolysis. Cellulose itself, while providing the carbon polymeric structure of incremental plant growth, is not resistant to organic breakdown. In woody plants an additional set of polymers that result from sugar to energy conversion, compounds hemicellulose and lignin, provide the strength and resistance for wood to be resistant to physical and biological deterioration.

Secondary xylem is generally a non-living vascular tissue component of the woody plant composed of cells with a thrice-reinforced cell wall, known as the S1, S2, and S3 cell walls. These cells consist of cellulose structures reinforced with layers of the polymeric compounds cellulose, hemicellulose, and lignin (Panshin and DeZeew 1980). While the S1 and S2 layers contain unidirectional microfibrils, more or less forming a textile like structure, the S3 layer is an additional layer that has a chaotic network, almost felt like, of interlocking and similarly reinforced microfibrils. Together they create a strong and impermeable seal to the cell's interior. The secondary xylem contrasts starkly with the phloem region of the living portion of the tree stem at the outer periphery of the plant's stem, exterior to the vascular cambium. The secondary xylem also contrasts starkly to the networked rows of living parenchyma cells that are imbedded within the secondary xylem tissues. These parenchyma play a significant role in regulating water management, nutrient transport, and emergency response within the xylem. All of the cells in the secondary xylem are all formed in the living part of the tree at its outer periphery, at the phloem-xylem junction known as the vascular cambium. Once formed, with their S1, S2, and S3 layers intact and their forms matured, the secondary xylem effectively dies and the living components are evacuated from the cell leaving behind a reinforced superstructure of cellulosic material. Essentially the secondary xylem provides piping for water and nutrients to complement the living parenchyma structures within the secondary xylem. Crucial to understanding wood anatomy, the vascular cambium is constantly generating xylem material (Wheeler and Baas 1998). For some species there are periods of slowing down or even stoppage during periods of stress, for example when nutrients are scarce (Baas and Vetter 1989; Baas and Miller 1985). These shifts in metabolism leave distinguishing patterns in the xylem in terms of overall cell size, distribution of specialized cell types, and their frequency. These cellular characteristics provide essential testimony to the ecological and environmental dynamics of a tree's life (Schweingruber 1996).

As a result, I describe in Chapter 6 not only the genera present within the archaeological samples, but also what the recovered materials indicate about the growth of those trees. Once I have an idea about how these different plants were growing, I can then relate this information back to the ecological description of the local forests as set out in Chapter 2. Ideally this level of interpretation permits a discussion about the strategies of fuel gathering and labor organization used at the site, as well as some notion of the transmission of knowledge of long-term ecology employed by Huaca Sialupe's artisans.

Conclusion: Methods and the Resultant Datasets

Due to the incidental weight for most of the non-wood remains, e.g., seeds and fruits, these materials were scored as present in the dataset as counts. Wood material, however, was scored by weight. For wood samples, counts are less likely to represent actual pieces of wood due to fragmentation effects in the features examined, including the depositional (burning and conversion to ash) and post-deposition taphonomic processes and breakage due to overburden and extinguishing fires. The methods that I used at the site resulted in the development of two datasets from one set of archaeological data. Quantitatively, these are dealt with separately in the analyses of Chapter 6. Both wood and non-wood archaeological remains were counted and/or weighed for comparisons of the data sets across the site and among the contexts under investigation. All counts and weights for each archaeological sample were scored on the same forms, by hand, using paper data recording forms (Appendix E; Appendix H), again copied from Dr. Lee Newsom, Pennsylvania State University. All paper forms are filed in my personal archives. Data corresponding to the wood and non-wood determinations appear in a Microsoft Excel spreadsheet in the Appendix I.

Although extrapolating the existence of modern plant species and communities into the past presents theoretical difficulties (Harris 1989; Hather 1994), the use of modern comparative materials to determine wood species from the cellular structure remains an effective method (Wheeler and Baas 1998). The case is similar for seed morphology (Pearsall 2001). Due to one major evolutionary constraint, however, it is critical that the methods used here are understood as determinations and not identifications. Hather (1994) articulates what is perhaps the most important principle in paleoethnobotany: archaeobotanical datasets must be regarded with caution in terms of making identifications of species based on comparisons with modern materials.

Chief among Hather's (1994) reasoning is that plants are part of a dynamic ecological and evolutionary system, and although some of them are heavily dependent on humans, others are not. His concerns develop from the application of homologies in paleontology to use present environments for patterning our interpretations of ancient ecology. In the case of archaeology, the human use of similar plants in the past does not indicate that they represent the same species. Even those species that are classified as domesticates create problems when we extrapolate the presence of the same species of the present into the past. There are exceptions, however, to this rule: for example, in the case of maize, *Zea mays*, there is only one organism, with a demonstrably huge amount of variation that corresponds to its seed, fruit, and overall morphology. Additionally, the same is likely true for other potential domesticates emblematic of an extremely specialized relationship with humans, e.g., *Erythoxylon coca* used as a ritual plant and stimulant throughout the Andes (Henman 2005). Archaeologists are confident, when dealing with organic macroremains, talking about *Z. mays* and *E. coca* in the past. In the case of a plant like the potato, *Solanum tuberosum*, however, that is demonstrably in constant hybridization with ambient weeds and cultivars where it is grown (Brush 1992), the identification of plant macroremians from this taxon in the past to the species level is suspect, based on its known ecology. The best that paleoethnobotanists can do, in most cases, is argue for identification to genus. As a result, my family, species, and genus classifications must be regarded as determinations and not identifications.

In the case of the wood anatomy, however, where species designations are used, these are indeed determinations, not full on identifications, signaling distinct similarities or correspondences with identified species from the modern era. For instance in the case of *Prosopis* spp., as described in Chapter 2, there are at least three different species in this genus present in the area. Current scholarship indicates that there is quite a good amount of hybridization between these species, as is indicated by the ethnotaxonomy in Chapter 2 for *Prosopis* spp. In the case of *Acacia macracantha*, there is only one species present in the modern era for comparison, additionally, this species is well represented along the entire Pacific Coastline from Ecuador to Northern Chile, and while it is, in the same sense as *Zea mays*, likely the tree I here only make a distinction to genus. I cannot rule out the possibility that another *Acacia* species was present during the Middle Sicán or that overall *Acacia* genus diversity was not higher along the South American coastline in antiquity. Additionally, as is discussed in Chapter 6, the genus designation is used when the wood taxonomy fits what I have identified as a salient distinguishing trait between *Acacia* and *Prosopis*. Where these two determinations cannot be distinguished I have used a family based Fabaceae designation, indicating that it is in the family and could be either of these. If no distinguishing characteristics could be assigned, then the wood went into the unknown designation.

With the methods used in this study outlined, the theoretical orientation described, the historical, technological, and ecological background explained, we turn now in the following two chapters to the analysis of the data. It follows in two parts, a general overall site view of fuel, a mound-by-mound view, and finally a specific fire-use feature description of the different fuel contexts. The methods, from fieldwalking to talking with dry-tropical forest residents, and lastly a biological look at determination of the remains, pull this dataset together for reconstructing human activity and a general view of the environment. The results of this overview are discussed in the concluding Chapter 8.

CHAPTER 6
SITE-WIDE COMPARISONS OF ARCHAEOBOTANICAL DATA

Paleoethnobotanical Analysis and Interpretation

Throughout the Sialupe workshop we encountered exceptional preservation of features and some living/working surfaces from the Middle Sicán. The floors, consisting of tamped dirt, are often not as well preserved as the features; for this reason, I focus on the features in this analysis and utilize the remains from floors as a general comparative dataset. The features, kilns, hearths and furnaces, consist of carbonized or burned materials, and sometimes a mix of burned and unburned materials. As explained in the previous chapter, dry sieving was used to recover all materials for analysis after which I sorted and determined materials under incident light microscopy. To recover the greatest detail of fuel resource consumption, I separated seeds and wood for different methods of determination: seeds were determined via a comparative method and wood through pattern recognition, anatomical comparison with modern reference material, and anatomical keys.

The diversity of materials that compose this archaeobotanical data set require that different types of data are, to some extent, quantified separately. Wood charcoal composes the bulk of the weight of the overall data set (8222.73g); seeds and other botanical remains, however, have negligible weights, even when grouped by taxon and context. As a result, these remains cannot be expressed analytically in the same manner. For the purposes of this analysis, I use wood *weights* per sample volume for context description and comparison between areas of the site and between features. This measurement is effectively a density, weight per volume. All wood

Figure 6.1 *Llama* sp. dung

Figure 6.2 *Cavia* sp. dung

remains are classified in the database as 'stem,' in keeping with the overall plant organ categories recognized in plant biology (Raven et al. 1999). Seeds and other botanical remains are considered separately as *counts* per sample volume. Both datasets are analyzed per sample volume to account (to an extent) for the differences in sample sizes across the site. These units, weight/volume of sample and count/volume of sample, measure the density ratio of a given taxon or class of material, linking these two datasets together into a single analytical framework. Further refinement, a category called 'Item Description' uses more qualitative terms for both count and weight data, e.g., branchwood, cob, 12-row cob, etc.

The most ubiquitous material found at Huaca Sialupe, specifically in fire use contexts, is not wood or another type of botanical remain per se; dung served as an important fuel source during the Middle Sicán and must be considered when examining fire-use features. I distinguished two classes of dung in this analysis. One is dung from *Llama* sp. (found in 100% of 56 samples) features), characterized as plump bullet-like, compact feces, between 8 and 20mm long (Figure 6.1): the second comes from *Cavia* sp. (42% of 56 samples) and has an elongated, cigar-like form ranging from 4-12mm long (Figure 6.2). Based on morphology alone, there is the potential for confusing these two types of dung, regardless of the relative sizes of the two mammals. For these cases, I created a third category of unidentified Dung (11% of 56 samples; Figure 6.3) to include those that were neither clearly *Llama* sp. nor *Cavia* sp., were partial feces, or fell somewhere in between the typology I used to distinguish between them.

Like wood charcoal, I often recovered camelid dung in large enough quantities to be weighed; *Cavia* sp. dung, however, was less prevalent and is reported as count data in all contexts. In burning features, *Llama* sp. dung was weighed; where it is not carbonized or appears in small quantities, it has been counted. To convert between counts and weights for the benefit of analysis, I used an estimation expression for *Llama* sp. dung (1 dung count =

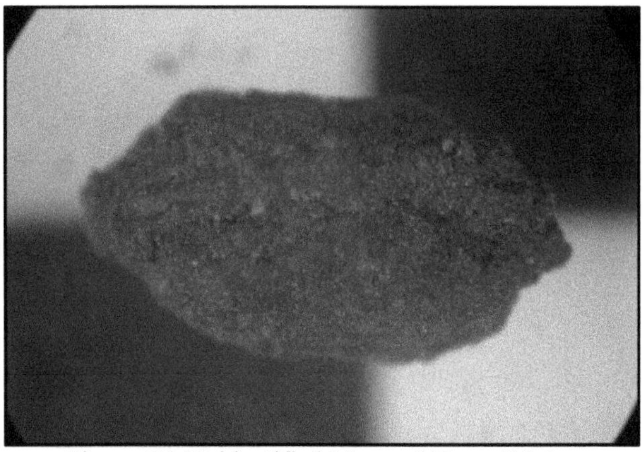

Figure 6.3 Unidentified Mammal Dung (10x)

0.05g, based on a mean of all counted and summed weights for camelid dung). For *Llama* sp., the weights that are estimates appear in the wood weight data tables in italics (Appendix I). I did not make weight estimates for *Cavia* sp. dung; it only occurs in the count tables. Unidentified dung appears as counts.

The use of animal dung as fuel in human society is quite common. It is an incredibly efficient way to harness energy ancillary to animal production that is otherwise lost to human consumptive needs (Miller and Smart 1984; Sillar 2000; Winterhalder et al. 1974). Ruminants (Ruminantia) and camelids (Tylopoda), for example, have a four-stomach digestive tract forcing the fermentation and regurgitation of fibrous cellulosic material until it is sufficiently disintegrated and passable. By the time that much of the herbaceous and tree species plant matter has made it through the digestive tract and into the intestinal system the meals of these animals have been compacted and ground into a fine particleboard-like mass. In the case of both camelids and *Cavia* sp. these animals produce resilient, portable, relatively odorless, compact pellets of cellulose particles that are an excellent fuel source, maintaining their compact form and insulating the combustion process (Rosotti 2002). This quality allows the combustion to be near complete, making these two dung sources excellent fuels.

The majority of the features at the site represent loci where resources were gathered and destroyed (i.e. fire-use features), hindering our ability to reconstruct entirely the original assemblage of fuels from examining the remains. Popper (1988) asserts that as human and natural taphonomic processes are both at work to differing degrees with archaeobotanical data, it is essential to use forms of descriptive statistics, other than straight counts or weights, in our analyses. Adjusting count and weight data for sample size is one quantitative method to reconcile differences in sample, feature, and taphonomic variation. Much of the analysis that follows relies on the comparisons of taxon densities within the samples and across the site.

Scales of Analysis

The archaeobotanical materials recovered from excavation can address fuel resource use at a variety of scales. At the broadest scope, I look at the site-wide prevalence of different materials to generate a general description of the catchment and subsistence base for the site. For this scale of analysis, I use taxon ubiquity, total number of incidents of a single taxon/total number of features and floors at the site (Popper 1988), i.e. proveniences were analyzed separately. The raw data-- wood weights and taxon counts-- and their densities, counts, or weights adjusted for sample volume, provide overall relative frequencies for the different taxa encountered at Huaca Sialupe.

At the next level of analysis, I compare the distributions of fuel resources between the two craft production areas of Mounds I and II at Sialupe, Mound I being used for a variety of purposes, most ostensibly ceramic production, and Mound II being used exclusively for metal production (Shimada 2000; Shimada and Montenegro 2002). Here, I employ the same descriptive statistics used for the site-wide comparison, ubiquity and taxa densities, to further refine and qualify resource remains in each part of the Sialupe crafting complex. To identify the fuels most commonly used for combustion at the site, I examine the overall contribution of any one taxon to the constellation of plant materials used: a simple diversity ranking compares the total density of each taxon to the total incidences of all taxa at each mound. Expressed as a percentage of all taxa found in that mound, this ranking, expressed as a percentage, indicates the relative contribution of any single taxon to the overall assemblage at each mound. Differences in overall contribution to diversity, where the same number of taxa are present, will distinguish differences in taxa use at each mound. I will use the same measurement to analyze the features and floors, in the next level of analysis covered in Chapter 7.

Lastly, and perhaps most critical to the analysis of this research, is to establish a background for reconstructing catchment and subsistence for room-by-room activities. Because the Middle Sicán period contexts at Sialupe were excavated using cultural levels, we are able to examine bounded activity areas as defined by features within rooms (Shimada 2000; Shimada and Montenegro 2002). Identifying fuel resources present in each context permits me to interpret where Middle Sicán artisans employed specific taxa for some of the identifiable production activities at the site. As mentioned in Chapter 3, previous SAP experimental archaeology and ethnographic research allows me to assign use to some features at Huaca Sialupe, particularly kilns and metal furnaces, based on morphology. The fuel resources associated with these known contexts will clarify and refine our current understanding of metal and ceramic production. Moreover, I may be able to assign use to features (i.e. metal or ceramic production) that exhibit distinct morphologies, yet share similar archaeobotanical assemblages. With more generalized fire-use features, such as hearths, identifying fuel components can highlight the presence of other activities taking place within the workshop contexts, *a la* household or domestic production behaviors. More importantly, it allows us to identify how craft production intersected with other types of ongoing production.

Comparing fuel resource use across the site, between mounds, and between features enables me to identify 1) the ecological and environmental conditions affecting resource collection, 2) the organization of metal and ceramic production, and 3) the integration of the political and domestic economies (i.e. craft production with other daily activities).

Site-Wide Comparisons of Archaeobotanical Remains: Ubiquity of Taxa

To begin, I present the results of the ubiquity scores of both the wood/weight data and non-wood/count data. This portion of the analysis uses site-wide ubiquity, the total number of incidents of a taxon, i.e. features that include the taxon, over the total number of features (n=41). Here the analysis, as we are only looking at incidence of taxa in the data, the two datasets are reported together. As noted above, the individual incidences of seeds and non-wood plant remains are based on counts; wood remains are examined as weights. In some instances, in this section of site-wide comparisons, I use their densities (weight or count per feature volume) to qualify and interpret the finds that occur most frequently and in abundance at the site.

There are 41 features and floors present in the entire dataset recovered from Middle Sicán levels from both Mounds I and II at Sialupe. Fifty-six samples were taken from these different features and floors and constitute the dataset that represent 240 L of collected and hand sieved soil, with a total weight of 239.8 Kg (Table 6.1). The non-wood remains represent 143 taxonomic distinctions or categories from Mounds I and II; 33 categories were identified for the wood taxa (Appendix J). Combined, only 33 of the taxa determinations were more than 18% ubiquitous across the site (non-wood, n=26/143 or 18%; wood n=8/56 or 19%; Table 6.2). Only about 20% of the different taxa recovered are present across the site. Site-wide comparisons of the archaeobotanical materials, based on the datasets described above, provide the most general picture of fuel resource use at Huaca Sialupe. Most of the wood charcoal and desiccated wood remains from Sialupe are juvenile stems or tree branches, 2-5 cm diameter, and exhibiting anywhere between 3-10 years of growth (Figure 6.4). This high frequency of branchwood complicates the determination portionof the analysis, as it is difficult to tell the difference between young taxa based on anatomical characteristics or criteria; this is significant when considering three of the most common taxa, *Prosopis* sp., *Acacia* sp., and Fabaceae. For the purposes of this analysis, where *Acacia* sp. and *Prosopis* sp. could be distinguished, based on the anatomical criteria described in Appendix G, they were categorized into their separate taxa; where they could not, they were placed in the category Fabaceae juvenile. In addition, some occurrences of taxa (e.g., *Prosopis* sp.) appear in the data set in multiple ways: as fruits, stems (wood), seeds, leaves. In Table 6.2, I report the specific plant part/organ present in the contexts; for the purposes of this analysis, I classified 'stems' as wood charcoal. Unidentified 2mm and 4mm charcoal categories include those materials thatcould not be identified due to their small size, incomplete anatomy, or their deteriorated state.

Mound	Feature	Floor	Sample Analyzed Weight (Kg)	Volume (ml)	Percent of Actual (Sample V/ Feature V)
1	49	1	4.462	3900	23%
	63	1	5.062	4050	100%
	67	1	1.244	975	100%
	68	2	6.900	9840	100%
		4	1.991	2925	100%
	81	2	1.900	3975	14%
	83	2	3.062	2450	1%
	84	1	1.400	1975	59%
		2	2.362	3700	87%
	96	2	2.062	2225	2%
		4	4.900	6000	7%
	97	2	2.762	3670	41%
	98	2	2.462	2350	100%
	99	1	5.362	3625	100%
	100	2	4.562	4375	78%
	101	2	5.062	4215	100%
	103	1	2.962	3130	89%
	106	1	21.000	24975	100%
	109	4	3.400	4950	6%
	110	4	5.900	7875	18%
	111	4	2.300	2850	9%
	117	4	11.800	10850	24%
	122	4	1.200	3200	26%
	123	4	1.600	3450	18%
	131	4	1.500	1300	2%
	Floor	1	34.948	21925	100%
		2	4.612	3000	50%
		3	14.622	10370	100%
		4	7.000	7125	44%
Mound 1 Total			168.406	165250	
2	8	5	5.711	8175	11%
	9	5	10.808	11950	42%
	11	2	7.425	5680	16%
	12	5	5.862	4950	46%
	14	2	6.009	7300	35%
	15	5	1.062	3475	100%
	16	2	4.562	3495	28%
	19	5	6.418	7350	71%
	20	2	4.240	7025	54%
	21	2	5.000	3570	38%
	Floor	2	3.600	2725	6%
		5	10.400	9111	77%
Mound 2 Total			71.100	74806	
Total Analyzed			239.507	240056	

Table 6.1 Overall Context Statistics 2002 Excavations

As only 20% of the taxa determinations were more than 18% ubiquitous across the site this leaves over 70% of the different taxonomic distinctions restricted to certain areas. Although the total assemblage for the entire site is quite diverse, certain taxa dominate the examined contexts. Artisans clearly selected specific fuel resources from a broader universe of potential materials: only 27 wood and non-wood taxa are present in a majority of the contexts, representing 11 plant families, at least 3 animal families, and a few unknown or unidentified materials (Table 6.2).

My primary goal in examining site-wide ubiquity, Table 6.2, is to provide a general description of the most prevalent materials that are common to fuel resource use

Figure 6.4 Fuel Remains Demonstrating the Predominance of Branchwood

at the site. Given Matson's (1966) observation regarding workshops as intersections of domestic activities and production, I additionally aim to generate a list of materials present across the site that represent connections between the workshop and other economic activities occurring at the site. Once I have reviewed the wood and non-wood remains based on site-wide ubiquity, I use these data to illustrate how humans were drawing on their environment for primary subsistence and catchment and its impact on local ecology.

Ubiquity of Taxa: Dung and Wood Fuel Remains

The most ubiquitous material at the site and a major component in nearly every charcoal assemblage is dung from camelids, probably *Llama* sp. Keep in mind that a high ubiquity score does not necessarily implicate use across the entire site with the same intensity. The more ubiquitous a certain taxon is, however, the more likely it is a certain taxon represents a generalized process of use and activity pattern. In the case of dung, its ubiquity and the quantity that we recovered in fire-use features indicates that dung was a fuel used in almost all contexts of fire use at the site, regardless of any consideration of the different wood fuels being used. As the importance for coastal herding of camelids is well known in archaeological record, Huaca Sialupe is clearly no exception (Shimada and Shimada 1985).

Figure 6.5 Beetle Larva Eaten Wood

In addition to dung, immature branchwood comprises the other major element of fire-use features. focus their fuel gathering and use of fuel on young, immature resources. This topic is explored throughout the analyses described here. Despite the greater energy The predominance of small branches is critical as it represents a series of decisions and 'technological choices' (Lemonnier 1992) made by artisans at the site to potential of mature wood, the large proportion of juvenile branchwood highlights a cultural preference for younger wood, or that constant harvesting limited/stimulated growth making this the most available fuel source; this becomes even more apparent when you consider that these young branches are smaller and if only a small portion of the overall fuel assemblage, would likely be consumed in a fire attaining 1200°C. By weight, mature wood accounts for only 409.42g of the actual charcoal weight, about 5% of the total wood assemblage at Huaca Sialupe. Large mature tree boles or timber do not appear to have been used intensively as fuel at the site; there is little or no indication that large pieces of trunk wood from high stature trees were used for fuel. In addition, wood boring insects (Figure 6.5) and wood fungal spores (Figure 6.6) are present in all of the major wood taxa recovered, regardless of age grade of the materials recovered at the site, indicating that collected dead wood was the primary source of branch wood used at the site (see below). In chapter 7 a more detailed analysis of wood fuel taxa provides additional insight into artisans' preference at Huaca Sialupe.

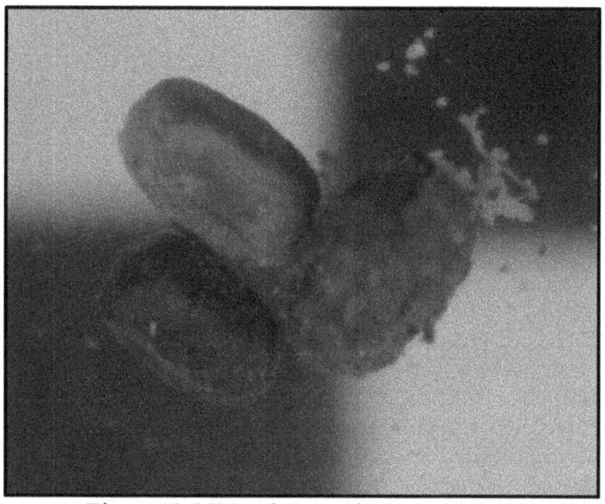

Figure 6.6 Wood Fungal Spores (15x)

The most ubiquitous fuel woods consumed at the site include *Prosopis* sp. (98% of 56 samples: Table 6.2), *Acacia* sp. (64% of 56 samples), Fabaceae juvenile wood (62% of 56 samples), *Capparis angulata* (39% of 56 samples), *Tessaria integrifolia* (37% of 56 samples), and finally *Capparis ovalifolia* (25% of 56 samples). All of these species are present in the local flora today; given their significant percentage of the composition of woody materials at the site, their presence as part of the local ecology during the Middle Sicán is also likely. It is important to note, I discuss below, that not only is the wood from *Acacia*, *Capparis*, and *Prosopis* present, but so are the seeds, fruits, juvenile branches, spines, leaves and flowers. The presence of other plant parts from these

Family	Determination	Organ/Type	Total Taxon Incidence	Site-Wide Ubiquity (n=56)
Other	*Llama* sp.	Dung	56	100.00%
Unid	2mm Charcoal	Stem	55	98.21%
Fabaceae	*Prosopis* sp.	Stem	55	98.21%
Poaceae	*Zea mays*	Seed	52	92.86%
Verbenaceae	*Lippia* sp.	Seed	47	83.93%
Fabaceae	*Prosopis* sp.	Seed	47	83.93%
Unid	4mm Charcoal	Stem	41	73.21%
Fabaceae	*Acacia* sp.	Stem	36	64.29%
Fabaceae	cf. Fabaceae	Seed	36	64.29%
Fabaceae	Fabaceae	Stem (Juvenile)	35	62.50%
Poaceae	*Zea mays*	Fruit	34	60.71%
Unid	Unid.	Seed	32	57.14%
Fabaceae	*Acacia* sp.	Seed	31	55.36%
Other	Shell 4mm	Shell	29	51.79%
Chenopodiaceae	*Chenopodium* sp.	Seed	28	50.00%
Boraginaceae	*Heliotropum* sp.	Seed	26	46.43%
Caviinae	*Cavia* sp.	Dung	24	42.86%
Fabaceae	*Prosopis* sp.	Leaf	24	42.86%
Capparidaceae	*Capparis angulata*	Stem	22	39.29%
Asteraceae	*Tessaria integrifolia*	Stem	21	37.50%
Fabaceae	cf. Fabaceae	Fruit	19	33.93%
Euphorbiaceae	*Euphorbia* sp.	Seed	18	32.14%
Other	*Donax* sp.	Shell	18	32.14%
Verbenaceae	*Lippia* sp.	Fruit	18	32.14%
Malvaceae	*Gossypium* sp.	Seed	17	30.36%
Other	Fish	Bone	15	26.79%
Capparidaceae	*Capparis ovalifolia*	Stem	14	25.00%
Solanaceae	*Nicotiana* sp.	Seed	14	25.00%
Other	Slag	Metal	13	23.21%
Other	Wood Fungi	Spore	13	23.21%
Solanaceae	*Capsicum* sp.	Seed	12	21.43%
Solanaceae	cf. Solanaceae	Seed	12	21.43%
Fabaceae	*Acacia* sp.	Flower	12	21.43%
Unid	Unid.	Fruit	12	21.43%
Amaranthaceae	*Amaranthus* sp.	Seed	11	19.64%
Other	Mammal	Bone	11	19.64%
Fabaceae	*Acacia* sp.	Fruit	10	17.86%
Fabaceae	*Prosopis* sp.	Fruit	10	17.86%
Other	Ceramic Redware	Ceramic	10	17.86%
Erythoxylaceae	*Erythoxylon coca*	Seed	10	17.86%
Poaceae	*Zea mays*	Stem	9	16.07%

Table 6.2 Most Ubiquitous Taxa Recovered Site-Wide

genera further reinforces their presence in and around where fuel used at Sialupe was coming from. Additionally, the category Fabaceae juvenile wood is ubiquitous at the site; as mentioned above, this category is a catchall for carbonized stem/branch remains that could not be classified as either *Acacia* or *Prosopis* and could represent either of these or another species present in the Fabaceae family. This point clearly indicates that small branches were the most utilized fuel source at the site.

The presence of both *Prosopis* sp. and *Acacia* sp. in large numbers across the site provide some clues to the local ecology during the Middle Sicán. The most ubiquitous tree taxa present at Sialupe, (*Prosopis* sp.= 2630.05g; *Acacia* sp.=711.26g), they occur in most of the features. As noted in Chapter 2, the relationship between *Acacia macracantha* and the *Prosopis* spp. trees in the region of study is symbiotic and potentially coevolutionary. *Prosopis* spp., well known today as a source of wood charcoal (Chapter 2), figures prominently in the local ecology and history as an important cultural and biological resource; *Acacia macracantha*, a sun-tolerant and cohabitant successional, or pioneer, species often appears as new growth in previously disturbed or abandoned areas, providing shade to other saplings. The presence of both taxa in high numbers, usually as juvenile branches, demonstrates that part of the fuelwood catchment area at the site was amply disturbed enough that *Acacia* sp., a successional species, was allowed to grow back in areas of fallow or recent disturbance, such as clearings for agriculture, maintained as a living fence system, or by ENSO related effects. This evidence also suggests that mature stands of *Prosopis* sp. were populous enough to be considered a locally available fuel source from scavenged dead branches or structurally compromised boles or branches that were adventitiously pruned. Although there is evidence in some contexts for the presence of wood chips from metal tools associated with *Prosopis* sp. (Chapter 7), it is clear that dead branch wood, either culled from trees or collected from the ground surface, accounts for the majority of this species use at the site.

There are two fuel woods recovered and identified from the Family Capparidaceae, *Capparis angulata* (39% of 56 samples) and *Capparis ovalifolia* (25% of 56 samples). Again, both plants grow in the area adjacent to the site today, and both are commonly used as fuel with their fruits occasionally used as fodder; humans occasionally consume the fruits from *C. ovalifolia.* In the case of both woods, they have a similar calorific output (Table 2.1), as either of the two Fabaceae genera, but their grain is less compact, making it light, easy to cut and work using metal tools, and easy to burn. Experimentally, both species were used as expedient local fuels for initiating combustion in our 2000 field firing experiments. The areas below the branch coverage of these trees are also choice locations for collecting *poña,* kindling materials (Chapter 5). Trees in these genera have a heavy waxy coating on their branches and leaves, a common adaptation of trees to arid environments, and this wax is adventitious in catching fire while initiating combustion. These trees, like some dry tropical species are drought deciduous, dormant and losing their leaves at the end of the wet season, and their dried broad waxy leaves and occasionally broken limbs are sought for kindling and are common sites for gathering *poña.*

Tessaria integrifolia (37% of 56 samples) is in the Family Asteraceae and is a wood consistent with contemporary traditional ecology and cultural use in the area. The plant currently grows in and along canal systems, irrigated areas, and river beds especially in the flat low lying coastal areas of Lambayeque. It is a favorite building material for roofing poles and stockade-like *quincha* (pole and daub; Figure 6.7) wall supports, similar to those that would have likely surrounded the main firing areas in room 39 of Mound I and the large open metal furnace area of Mound II. Although *Tessaria integrifolia* is ubiquitous in the assemblage it comprises less than 1% of the overall wood weight recovered from the site. When compared to other taxa like camelid dung, *Prosopis* sp., and *Acacia* sp., this taxon is only present in ephemeral quantities. Its ubiquity and presence by weight at the site is more akin to other potential fuel additives like *Zea mays* (16% of 56 samples; 3.55g), discussed below, or *Gynerium* sp. (3% of 56 samples; 53.77g) a local tall wetland grass used in construction that will be discussed more below.

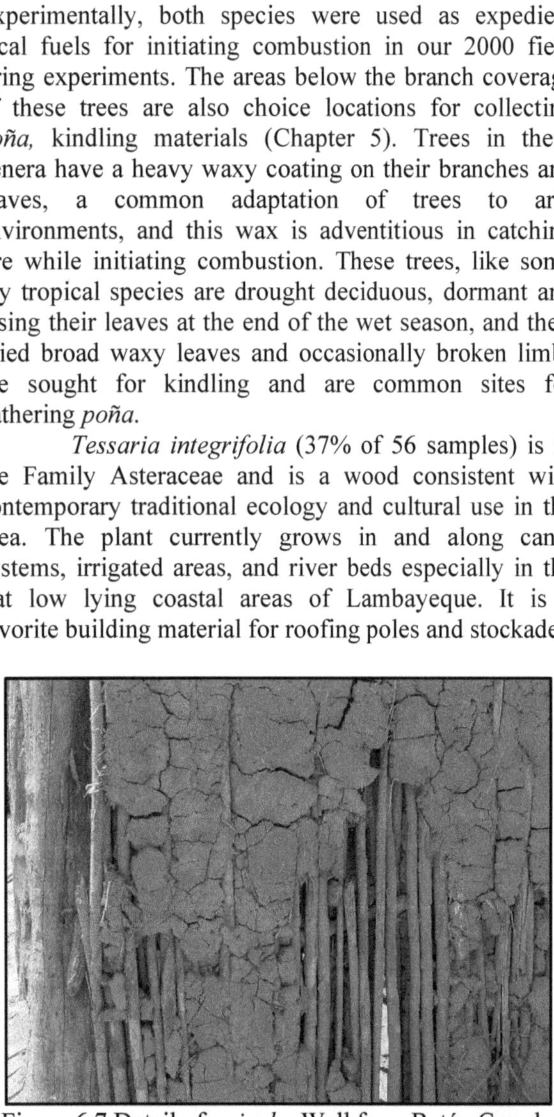

Figure 6.7 Detail of *quincha* Wall from Batán Grande

From the wood and dung fuel data, we are able to distinguish the major components of fire features, as well as a general picture of the local ecology during the Middle Sicán. A site-wide analysis of non-wood fuel complements this dataset, as well as begins to illuminate the relationship between local ecology, craft production, and cultural dynamics at the site.

Ubiquity of Taxa: Non-Wood Remains

The most ubiquitous non-wood material is seed from *Zea mays* (92% of 56 samples). At least two different cob types are present in the assemblage indicating that at least two different varieties of *Z. mays* were grown near to or adjacent to the site, a 12-row (12.5% of 56 samples) and a 16-row variety (7% of 56 samples). The significance of the corn remains will be discussed further in Chapter 7. For the most part, the seeds are best described as flour type (13% of 56 samples), indicating a preference for large starchy corn seeds consumed at or near the site. Varieties with flour type seeds are commonly used today for making *tamales* or *humitas*, reconstituted dried or fresh maize flour fashioned into stuffed and steamed or boiled dumplings. In addition to the presence of seeds throughout the assemblage, there are also *Z. mays* stalks (15% of 56 samples), cobs (16% of 56 samples), cupules (19% of 56 samples), and glumes (5% of 56 samples) within the material at Sialupe. It is important to note that even as a cultivar, it is more ubiquitous than weedy species present at the site. The recovery of all parts of a domesticated plant at a site indicates that there is likely a close relationship between those planting, harvesting, and consumption of a resource at a site. The presence of all *Z. mays* parts, those consumed (seeds and fruits) as well as those that are normally discarded (stems), demonstrates the important role that corn played in daily life at the site as a means of subsistence (i.e. food) and perhaps also as a concurrent economic concern (i.e. agriculture, foddering animals). Although ubiquity, essentially presence across the site, is not comparable with overall abundance or cultural significance, it is clear that activities related to *Z. mays* production and consumption were generally tied to the craft activities at Sialupe, a topic that I will revisit in subsequent sections.

Across the site, *Lippia* sp. seed (83% of 56 samples) and fruit (32% of 56 samples) remains are the second most ubiquitous materials recovered. Based on an evaluation of the modern environment, the presence of *Lippia* sp. is not surprising: within the immediate surrounds of the modern habitation sites visited at Sialupe, *Lippia* sp. is a dominant herbaceous species. In particular, most people use it as an herbal tea infusion in hot water consumed at many meals or as a cure for sore throats and colds. One informant described the herb as the key to making people of the upper La Leche River Valley strong because they drink the tea everyday. More importantly, *Lippia* sp. is a disturbance taxon, directly associated with human intrusion and settlement in the area.

Within my own botanical reconnaissance, I noted its presence along avenues and particularly in animal pasture directly adjacent to houses. In areas where grazing is limited, the plant can grow to a height of 0.5m and has a slightly woody stem. Although camelids are not present in the modern setting of Huaca Sialupe, goats, pigs, and other grazing animals often feed on these taxa. Some of the camelid dung remains recovered from features at Huaca Sialupe included partially digested seeds within it from what appeared to be *Lippia* sp.. Additionally, we recovered fruits, flowers, leaves, and seeds of *Lippia* sp. from many different features across the site. It is likely that this plant, in areas adjacent to human settlement at Sialupe, was a good potential kindling material during drier periods. Its presence in features at Sialupe, where ceramic and metal production occurred, is likely due to either its direct use as kindling or indirectly, as fodder for grazing camelids, whose dung was later used for fuel. The presence of *Chenopodium* sp. (50% of 56 samples), *Heliotropum* sp., (46% of 56 samples), *Euphorbia* sp. (32% of 56 samples) and other ephemeral weedy species and disturbance taxa at the site also probably owe their ubiquity to movement and use of camelid dung or dried tinder across and throughout the site. The overall presence of a plant like *Lippia* sp. and its companion disturbance taxa indicates that although wood and dung were sought as main sources of fuel, weeds related to human habitation and settlement were easily accessible and played a supporting role as fuel, likely for starting combustion.

Seeds from *Prosopis* sp. (83% of 56 samples) and other Fabaceae species (64% of 56 samples) form the third most ubiquitous non-wood remain. As noted in Chapter 2, and demonstrated below, the presence of *Prosopis* sp. seeds (83% of 56 samples; 867 seeds), leaves (42% of 56 samples; 177 leaf pinnae) and fruits (17% of 56 samples; 76 fruit remains) indicates that fuel was collected for the workshop from a dry tropical forest. The fact that all of these many plant parts are present at the site indicates to me that the site was at least partly surrounded by trees of this species, and it likely served as the dominant fuel wood. Apart from being a choice animal fodder, seeds from this plant are often used to make fermented beverages and it is recognized as a famine food in both modern and historic populations. Given the probable abundance of *Prosopis* sp. in the past, the occurrence of this taxon highlights its importance as a local resource for grazing or collecting practices of the Sialupe locals. It also demonstrates that these primary forest taxa, *Prosopis* sp. and other Fabaceae species, were utilized in a variety of ways apart from fuel for craft production: they appear to have been integrated into the broader base of local resources that kept Huaca Sialupe up and running. Their utility was not only for fuel, but some indications demonstrate that Prosopis sp. in particular was also used for ritual purposes at Huaca Soledad at the site of Sicán during this period (Shimada 1986).

As mentioned above, *Acacia* sp. seeds (55% of 56 samples), flowers (Figures 6.8 and 6.9) and other plant parts associated with this genus were recovered throughout the site. As noted above, wood from *Acacia* sp., the dominant and perhaps only *Acacia* species in the area, is the second most ubiquitous by weight as a fuel source in the fire-use features. Ecologically, this plant offers two fruiting seasons annually and is a preferred fodder source among modern grazing animals, as it was in the region historically. As with *Prosopis* sp., the ubiquity of this largely successional species within the Sialupe assemblage for the Middle Sicán indicates that the successional processes similar to those of that I observed today in and around Batán Grande categorize the ecological processes dominant in the site catchment area.

Interpretations of the presence of *Acacia* in the fuel assemblages are two fold. First, it indicates that fuel catchment was related to a forest that was, to some extent, successional, as *Acacia* is the predominant sucessional species. When paired with the presence of *Prosopis* sp., however, it indicates that there was a mix of disturbance and primary forest tree taxa. This situation may represent an ecology, one in dynamic flux, similar to that recorded in parts of the 2002 survey area. Second, based on *Acacia*'s clear advantage as a fodder (i.e. always fruiting abundantly), the quantity of the plant, in conjunction with the ubiquity of *Llama* sp. dung at the

Figure 6.8 Carbonized *Acacia* sp. Flowers

Figure 6.9 Modern *Acacia macracantha* Flowers

site, may indicate a relationship between *Acacia* and animal herders: herders working in and around Sialupe may have tended this fodder source to provide food for their animals. The people of Sialupe may have even encouraged *Acacia* growth for a number of reasons, perhaps, most importantly, to maintain a constant source of camelid fodder; or as a living fence. In turn, both extraneous *Acacia* branchwood and camelid dung provided an ample source of fuel for workshop production. Additionally, the plant in the modern day is encouraged throughout coastal Peru for its rigid and numerous hard woody stems with thorns that serve as living fences and expedient fencing material as dead wood, around habitation sites, animal corrals, and trucks.

Apart from identifying the role of food production and herding at the site through the fuel remains, additional non-wood botanical materials point to the occurrence of other cultural activities across the site. For instance, *Gossypium* sp. (cotton) seeds (30% of 56 samples) and *Erythoxylon coca* seeds (17% of 56 samples) are present at both Mounds I and II, which may reflect the importance of integrated site-wide social activities, ancillary or concomitant with craft production. Although *E. coca* is not traditionally chewed socially on the Andean coastline today, it certainly was an important part of domestic and daily life in antiquity (Henman 2005). While it grows only nominally on the North Coast today, in previous centuries it was cultivated, serving as a locally and regionally traded good (Henman 2005; Rostworowski de Diez Canseco 1989). Additionally, the cleaning of *Gossypium* sp. seed hairs from cotton seeds is depicted in photographs of daily life in Lambayeque during the recent historic period, where it appears to have been a social and domestic activity (Schaedel 1988; Vreeland 1986). The ubiquity of these materials, coupled with the variety of *Z. mays* products present in the archaeological record across the site, suggests the integration of daily household activities, not normally associated with specialized craft production, filtering through the background of ancient Huaca Sialupe workshops; an analysis of context-specific fuel use assemblages, however, provides a more detailed study of the intersection of these production activities.

Overall Ubiquity and Subsistence

The above analysis (Table 6.2) of the taxa ubiquitous in 20% or more of the Sialupe contexts provides a broad picture of the ecological and economic setting of the site. Table 6.3 presents the material based on the botanical families and genera present among this top 20%. Additionally, I have added a column that demonstrates how my work crew, consisting of local residents and subsistence farmers, indicated their classification when I asked them to distinguish between three categories of plants; weed (*maleza*), domesticate (*planta domesticada*), and wild (*planta del bosque/planta salvaje*). I present the growth form that the plant exhibits as a part of the local ecology as surveyed in 2002. The idea behind Table 6.3 is to understand the kinds of space, farmland, forest or human disturbance area, the artisans at Sialupe collected their materials. By recognizing the kinds of plant present, understanding their growth form and ecological niche, I can describe the general qualities of the local ecological community that artisans were exploiting for fuel, and qualify, to some extent, the management practices that dominate that landscape. The local dry-tropical forests are represented in a variety of forms (fruits, small stems or branches, flowers) acknowledging the life cycle of these plants. The representation of the entire life cycle of these plants is a likely indicator that easily scavenged resources, e.g., *poña*, leaves and seeds, in the assemblage) that encompass the entire life cycle clearly played an important role in fire use and craft production at the site. *Prosopis* sp. and *Acacia* sp. were coming from an area where these plants had coverage. The pruning of young branches, and moreover, the collection of dead, infested and decomposing branches, demonstrates access to a living forest. The decompositional as well as new growth cycles were in place, as well as successional species, indicating that some land disturbance and maintenance, perhaps agricultural fields surrounded by a living fence, best describes the local landscape at Huaca Sialupe in the Middle Sicán. As demonstrated from Chapter 2, it is likely that these materials composed part of the fire materials at Huaca Sialupe due to the expedient access to these resources in local forests. Notably, only five of the 14 core taxa recognized in my 2002 botanical survey, however, are present in the fuel assemblages. This presents that fuel selection, and the kinds of fuels preyed upon for human energy consumption was limited. Unfortunately, it is unclear if artisans were selectively using on these materials due to their cultural/technological preference and/or their relative abundance, e.g., because other trees species were not present on the landscape. The large presence of *Prosopis* sp. and *Acacia* sp., as well as a number of non-successional species, in fuel contexts across the site points to a local environment affected to some extent by natural or human disturbance with some forest cover.

From this evidence, it may be that farming practices in the area required a degree of land clearance: the ubiquitous *Z. mays*, appearing in fuel contexts as cobs, cupules, grains, and other plant parts, was likely an important cultivar at Huaca Sialupe that required tended fields. Fallow fields would have provided an opportune location for a variety of herbaceous and arboreal sucessional taxa to flourish, providing an ample source of fuels in the catchment area. Land disturbance allowed for a successional boundary of *Acacia* sp. around the perimeter of the domestic and farming area, accounting for the young branches and flowers present in the fuel remains.

From the ubiquity statistics alone, we can see that these craft producers were clearly integrated with a local farming economy that likely included maize, camelid, and cotton production. The fact that *Llama* sp. dung is found in every context at the site in one way or another, both in floors associated with artifact manufacture and also food production indicates that free access to this quality fuel source was a given at the site. Its presence in all fire-use features probably implicates camelid herding at a profound level in the lives of the artisans at the site. *Z. mays*, as mentioned above, is also present in a variety of forms, not only as fruit and seed fragments, the

Family	Plant Organ	Determination	Growth Form	Weed/Domesticate/Wild
Amaranthaceae	Seed	*Amaranthus* sp.	Herb	Wild
Apocynaceae	Stem	*Vallesia glabra*	Shrub	Weed
Asteraceae	Seed	*Ambrosia* sp	Herb	Weed
		Bidens sp.	Herb	Weed
		Taraxacum sp.	Herb	Weed
		UKN#50-142		
	Stem	*Tessaria integrifolia*	Shrub	Weed
Azioiaceae	Seed	*Trianthema* sp.	Herb	Weed
Bixaceae	Seed	*Bixa* sp.	Tree	Domesticate
Boraginaceae	Fruit	*Cordia* sp.	Shrub	Wild
		Heliotropum sp.	Shrub	Weed
	Seed	*Cordia* sp.	Shrub	Wild
		Heliotropum sp.	Shrub	Weed
Brassicaceae	Seed	*Lepidium* sp.	Herb	Weed
Cactaceae	Seed	cf. Cactaceae	Cactus	Wild
		Opuntia sp.	Cactus	Wild
Capparidaceae	Flower	*Capparis angulata*	Tree	Wild
		Capparis sp.	Tree	Wild
	Fruit	*Capparis ovalifolia*	Tree	Wild
		Capparis sp.	Tree	Wild
	Seed	*Capparis crotonides*	Tree	Wild
		Capparis ovalifolia	Tree	Wild
		Capparis sp.	Tree	Wild
	Stem	*Capparis angulata*	Tree	Wild
		Capparis ovalifolia	Tree	Wild
Caryophyllaceae	Seed	*Silene* sp.	Herb	Weed
Chenopodiaceae	Seed	*Chenopodium* sp.	Herb	Weed
		Suaeda sp.	Herb	Weed
	Stem	*Chenopodium* sp.	Herb	Weed
Convulvulaceae	Seed	cf. Convulvulaceae		
		Ipomoea sp.	Herb	Domesticate
Cucurbitaceae	Fruit	*Lagenaria* sp.	Herb	Domesticate
	Seed	Cucurbitaceae	Herb	
		Cucurbita maxima	Herb	Domesticate
		Cucurbita sp.	Herb	Domesticate
		Lagenaria sp.	Herb	Domesticate
		Cyclanthera sp.	Herb	Domesticate
Cyperaceae	Fruit	UKN#49-142	Herb	Weed
		UKN#64-120.1	Herb	Weed
	Seed	*Cyperus* sp.	Herb	Weed
		Scirpus sp.	Herb	Weed
		UKN#45-126	Herb	Weed
		UKN#49-142	Herb	Weed
Erythoxylaceae	Seed	*Erythoxylon coca*	Tree	Domesticate
Euphorbiaceae	Seed	*Euphorbia* sp.	Herb	Weed
Lauraceae	Seed	Persea sp.	Tree	Domesticate
Nyctaginaceae	Fruit	UKN#41-121.1	Herb	Weed
Fabaceae	Flower	*Acacia* sp.	Tree	Wild
	Fruit	*Acacia* sp.	Tree	Wild
		cf. Fabaceae		
		Prosopis sp.	Tree	Wild
		UKN#43-121.1	Herb	Weed
	Leaf	*Acacia* sp.	Tree	Wild
		Prosopis sp.	Tree	Wild
	Seed	*Acacia* sp.	Tree	Wild
		Arachis sp.	Herb	Domesticate
		Caesalpinia sp.	Tree	Wild
		cf. Fabaceae		
		Neptunia sp.	Herb	Weed
		Parkinsonia sp.	Tree	Wild
		Phaseolus sp.	Herb	Domesticate
		Prosopis sp.	Tree	Wild
		UKN#43-121.1	Herb	Weed
	Stem	*Acacia* sp.	Tree	Wild
		Fabaceae		
		Prosopis sp.	Tree	Wild

Table 6.3a Taxa Recovered from Huaca Sialupe Excavations 2001 by Family, Plant Part, Determination, Life Form, and Ethnographic Category (Based on 2002 Reconnaisance)

Family	Plant Organ	Determination	Growth Form	Weed/Domesticate/Wild
Malvaceae	Seed	cf. Malvaceae	Herb	Weed
		Gossypium sp.	Shrub	Domesticate
		Malva sp.	Herb	Weed
		Sida sp.	Herb	Weed
Myrtaceae	Fruit	cf. Myrtaceae	Herb	Weed
	Seed	*Psidium* sp.	Tree	Domesticate
Oxalidaceae	Root	*Oxalis* sp.	Herb	Domesticate
	Seed	*Oxalis* sp.	Herb	Weed
Piperaceae	Flower	cf. Piperaceae	Tree	Domesticate
Plantaginaceae	Seed	*Plantago* sp.	Herb	Weed
Poaceae	Fruit	*Zea mays* (12-row	Grass	Domesticate
		Cenchrus sp.	Grass	Weed
	Leaf	cf. Poaceae	Grass	Weed
	Root	cf. Poaceae	Grass	Weed
	Seed	*Andropogon* sp.	Grass	Weed
		Bromus sp.	Grass	Weed
		cf. Poaceae	Grass	Weed
		Cenchrus sp.	Grass	Weed
		Panicum sp.	Grass	Weed
		Zea mays (Flour)	Grass	Domesticate
	Stem	Gynerium sp.	Grass	Weed
		Zea mays	Grass	Domesticate
	Corm	cf. Poaceae	Grass	Weed
Portulacaceae	Seed	cf. Portulacaceae	Herb	Weed
		Portulaca sp.	Herb	Weed
Rubiaceae	Seed	*Gallium* sp.	Herb	Weed
Solanaceae	Seed	*Capsicum* sp.	Shrub	Domesticate
		cf. Solanaceae	Herb	Weed
		Nicotiana sp.	Herb	Domesticate
		Physalis sp.	Herb	Weed
		UKN#53-147	Herb	Weed
Ulmaceae	Seed	*Celtis* sp.	Tree	Wild
Verbenaceae	Fruit	*Lippia* sp.	Herb	Weed
	Leaf	*Lippia* sp.	Herb	Weed
	Seed	*Lippia* sp.	Herb	Weed

Table 6.3b Taxa Recovered from Huaca Sialupe Excavations 2001 by Family, Plant Part, Determination, Life Form, and Ethnographic Category (Based on 2002 Reconnaisance)

obvious focus of human consumption, but also as stems used for fuel. Again, this evidence indicates that not only were the harvested and processed products being consumed in the workshops, but also the stems, the products left in the fields or stockpiled for fuel and/or fodder, were burned.

The artisans of Huaca Sialupe were clearly tied to the status economy through the metal and ceramics goods they produced; they were also closely linked, however, with local and regional economics that permitted their acquisition of *E. coca* and their fabrication of cotton goods. The distribution of cotton seeds throughout the site indicate that either refuse associated with cotton cleaning or the cleaning itself was tied to crafting activities at both ends of the site. A later, intrusive burial placed inside of Room 39, consisting of a 15-20-year-old woman (Burial '01-8) and her complete backstrap loom and spinning and sewing utensils, including wooden and copper needles, copper and ceramic spindles, and worked bone and stone implements adds support to this supposition (Klaus 2003: 188-197; Shimada and Montenegro 2003).

In the end, it appears that the workshop areas functioned in close association with a community that locally, was engaged in land clearance, subsistence farming, and the extraction of a variety of forest materials; on a large scale, evidence from the fire-use features points to an interconnectedness between the local and regional economies. Moreover, the selection and utilization of resources at the workshops demonstrates Lemonnier's (1992) 'technological choice' that highlights an intimate relationship between the artisans, their needs, and their local ecology. Despite the clear advantages of mature wood for fuel, the craftspeople focused largely on fallen or dead branchwood, or even small saplings, even when the dominant tree species in the area, *Prosopis* sp., could have been at a mature stature. From here, I continue my analysis by comparing the fuel remains assemblages of Mounds I and II. This next way of partitioning the data for analysis gives insight into the particular needs for fuel at each side of the site relative to the evidence for the goods that were produced there.

Comparing Mound I and Mound II: Distinguishing Workshop Spaces by Archaeobotanical Data

The ubiquity data listed in Appendix K present a general impression of fuel resources consumed in the two different work areas at Sialupe, Mounds I and II. As noted in Chapter 3, these distinct sections of Huaca Sialupe represent the two major divisions of the site during the Middle Sicán, metalworking and ceramic production. Based on the features and technological remains recovered there, e.g., ceramic kilns, mold remains, and worked clay, Mound I is considered a locus of mold-made ceramic production. Mound II, is regarded as the locus of Middle Sicán metallurgy at the site, based on the recovery of furnaces used for working metal and some small slag and corroded copper. In both cases, it is clear from Appendix K that although the majority of the taxa designated in this study are present at both loci, there are several taxa that appear exclusively only at one or the other.

Of the 143 distinctions of the materials present at the site, 31 of them are exclusive to Mound I; 35 are only found at Mound II. Seventy-nine of the taxa occur at both Mounds. First I will discuss the materials that are common to the fuel remains at both sides of the site. Second, I will provide an analysis of the materials found exclusively at each Mound.

Mounds I and II Shared Fuels: Wood Fuel and a Close Look at* Prosopis *sp.

Although *Prosopis* sp. wood is present throughout the site, it appears in every feature and floor of Mound I, but not of Mound II. In fact, the density of *Prosopis* sp. wood at Mound I (5147.8 g/ml) is more than twice the density for *Prosopis* sp. wood at Mound II (1946.46 g/ml). The densities for the *Prosopis* sp. seed and spines (leaves) at both mounds represent similar trends of these non-wood materials in the dataset. The densities of *Prosopis* sp. fruits, however, present an opposing trend: more fruit remains are found at Mound II (n=71, 201.51#/ml) than at Mound I (n=5, 27.92 #/ml). The presence of more fruits at one side of the site may indicate that the fuelwood at Mound II was collected during a season when the plants are fruiting, e.g., Andean summer, or that they are using fodder litter as *poña*. Additionally, the location of the large camelid dung deposit adjacent to the metal workshop area may indicate that camelids were corralled and fed here, making it unnecessary to transport dung to the Mound II furnaces. In fact an area to the NW of the Mound II excavation area is where Shimada and Montenegro (2002) report the recovery of a camelid dung deposit just beyond the quincha wall boundary of the metalworking area. A similar pattern of *Acacia* sp. stems, flowers, fruits, and seeds demonstrates that its plant parts are present in different quantities in the metal and ceramic workshops (Appendix K): flowers, leaves, and seeds are present in higher numbers at Mound I, and fruits in higher numbers at Mound II. Again, this may relate to camelids being fed and corralled near to the Mound II workshop. In the case of the *Acacia* and *Prosopis* materials, they clearly play a major role in the fuel base at both mounds. Their relative quantities and the parts of the tree present in the features, however, may indicate that they are used in different proportions based on the technological need at each separate part of the workshop; this possibility offers the potential that individual features contain distinct fuel assemblages, which will be further explored in Chapter 7.

As there are many different taxa and parts of taxa to be examined in this analysis, in an attempt to make the data more manageable, I turn to Table 6.4. This table presents a comparison of only carbonized materials from the features between the two sides of the site and are expressed in terms of their density measurements, weight of taxon recovered per sample volume. They are represented by the percentage they contribute to the total adjusted weight of charcoal recovered (density) from each mound. These numbers are also expressed as percentages of the total charcoal density, including both identified and unidentified charcoal, and as a percentage of only 4mm identified charcoal. To highlight the importance of taxa use for fuel in this part of the analysis, the table excludes materials recovered from floors. I also excluded unidentified materials to highlight the significance of individual determined taxon at each Mound: 64% of taxa from the Mound 1 assemblage have been determined (72% of the 4mm Charcoal) and 53% of the assemblage from Mound II (70% of the 4mm Charcoal).

Looking at the individual taxon contributions to the proportion of the identified charcoal from both mounds, we see that *Prosopis* sp. wood is by far the most common item in features at both mounds. As noted above (Appendix K), however, it is less intensively utilized at Mound II than it is at Mound I: *Prosopis* sp. composes 65% of the overall charcoal matrix from features identified from Mound I, while it only makes up 40% at Mound II. The next most frequently encountered and most represented taxon at both mounds is *Acacia* sp., comprising 15% of the overall charcoal density at each mound. The remaining 20% of the charcoal remains at Mound I consist of nine additional taxa (11% collectively), plus a 9% contribution from Fabaceae juvenile charcoal, potentially a mix of *Prosopis* and *Acacia* that could not be determined due to their juvenile morphology. In contrast, the remaining 45% of the Mound II charcoal assemblage at Mound II consists of Fabaceae juvenile charcoal (15%), *Llama* sp. dung (9%), *Capparis angulata* (8%), *Capparis ovalifolia* (5%), and *Gynerium* sp. (3%).

The indications from this level of comparison are that both Mounds drew the bulk of their fuel from the same primary constituent *Prosopis* sp. and from a similar resource pool, a predominantly Algarrobo forest with some contribution from successional species. It appears, however, that the need for additional wood fuel and dung was greater at Mound II, the metal workshop. In this case, the species that compose the remaining 45% indicate a higher reliance on two species in the Capparidaceae family, which may illustrate that the combination of *Prosopis* sp. and *Capparis* spp. as primary taxa was particular to the kinds of production employed at both parts of the site.

In the case of fuel wood charcoal, the woody plant taxa present at Mound I are slightly more diverse overall than those used at Mound II. Several more unidentified taxa (Wood Types AA-LL; Appendix G) are present in the Mound I fire use contexts, which may demonstrate an ongoing need for expedient fuel resources when a load of *Prosopis* or *Acacia* was not available. In fact all of the

Family	Determination	Mound I Incidence of Taxon	Mound I Ubiquity (n= 38 samples)	Mound I Sum of Determinations g/ml (Density)	Mound I Density as Pct. of Total Charcoal Density	Mound I Density as Pct. of Identified 4mm Charcoal Density	Mound II Incidence of Taxon	Mound II Ubiquity (n=18 Samples)	Mound II Sum of Determinations g/ml (Density)	Mound II Density as Pct. of Total Charcoal Density	Mound II Density as Pct. of Identified 4mm Charcoal Density
Fabaceae	*Prosopis* sp.	31	82%	5087.17	41.93%	65.87%	15	83%	1928.49	21.41%	40.75%
Fabaceae	*Acacia* sp.	19	50%	1198.55	9.88%	15.52%	11	61%	739.64	8.21%	15.63%
Fabaceae	Fabaceae	16	42%	703.13	5.80%	9.10%	17	94%	746.78	8.29%	15.78%
Capparidaceae	*Capparis angulata*	14	37%	293.04	2.42%	3.79%	7	39%	384.20	4.26%	8.12%
Camelidae	*Llama* sp.	33	87%	167.52	1.38%	2.17%	18	100%	453.58	5.04%	9.59%
Asteraceae	*Tessaria Integrifolia*	11	29%	147.03	1.21%	1.90%	5	28%	8.30	0.09%	0.18%
Apocynaceae	*Vallesia glabra*	3	8%	53.28	0.44%	0.69%	2	11%	11.13	0.12%	0.24%
Capparidaceae	*Capparis ovalifolia*	4	11%	25.36	0.21%	0.33%	9	50%	245.34	2.72%	5.18%
Poaceae	*Gynerium* sp.	2	5%	0.59	0.00%	0.01%	3	17%	155.04	1.72%	3.28%
Poaceae	*Zea mays*	3	8%	3.74	0.03%	0.05%	5	28%	13.78	0.15%	0.29%
Poaceae	Wood Type FF	1	3%	2.31	0.02%	0.03%		0%		0.00%	0.00%
Poaceae	Wood Type KK	1	3%	0.10	0.00%	0.00%		0%		0.00%	0.00%
Chenopodiaceae	Wood Type LL	1	3%	0.51	0.00%	0.01%		0%		0.00%	0.00%
Unid	Wood Type AA	2	5%	13.29	0.11%	0.17%		0%		0.00%	0.00%
Unid	Wood Type BB	1	3%	5.89	0.05%	0.08%		0%		0.00%	0.00%
Unid	Wood Type CC	2	5%	21.38	0.18%	0.28%	1	6%	0.13	0.00%	0.00%
Unid	Wood Type GG		0%		0.00%	0.00%	1	6%	45.70	0.51%	0.97%
Unid	Wood Type JJ	1	3%	0.14	0.00%	0.00%		0%		0.00%	0.00%
Unid	4mm Charcoal	19	50%	3016.65	24.86%		12	67%	2057.49	22.84%	
Unid	2mm Charcoal	28	74%	1393.60	11.49%		16	89%	2218.73	24.63%	

Table 6.4 Fuel Charcoal Ubiquities and Densities at Mounds I and II (Just Features, Wood Only)

unidentified Wood Type taxa have included phloem and/or fiber bundles and are represent small diameter sizes indicating that they are largely from herbaceous or grass taxa. Two of these unidentified Wood Type materials, accordingly, fall into the families Chenopodiaceae and Poaceae, highlighting that non-wood fuels from herbaceous plants in the local environment represent a small proportion of the fuel remains found at the site. Granted these materials, if dried, would probably be mostly consumed as fuels and may not be fully represented in this analysis. Fortunately other plant parts, e.g., seeds indicate their presence as expedient combustion fuels. Additionally, there appears to be no crossover between the two mounds of the different unknown woody plant types. All of these variables suggest that they are weeds or on-site dried plant materials that were being thrown into the fire or used as materials to bulk out *poña* or kindling. Alternatively, their low overall ubiquity and small quantities raises the possibility that their non-determined status represents investigator error.

In the end, the comparison of the charcoal materials indicates that the two sides of the site, although drawing on a similar set of resources for work, were using them in different combinations and most likely for different ends. To some degree, this difference can be attributed to the two known productive activities in either part of the site. Mound I ceramic production required long-term burning with good fire stability. Our 2000 experimental kiln firings lasted between 12 and 17 hours each time we fired the kilns (n=5). These fires, once laid, are not opened up to add more fuel. The fuel burns from front to back through the reduction furnace; reopening or disturbing the fuel load would permit oxygen to enter the kiln, disrupting the reduced environment necessary for blackware. The large quantity of *Prosopis* sp. at Mound I corresponds with ethnographic and ethnoarchaeological evidence as a common fuel resource for ceramic production (Cleland and Shimada 1998): an excellent candidate for this kind of technology, this taxon takes up to seven days to reduce to charcoal in a small-scale kiln, all the while maintaining its overall branch structure.

Shimada and Montenegro's (2002) and the SAP's overall model for metal furnace use (Goldstein et al. i.p.) indicates that these were used most effectively when the on-shore breeze picked up in the afternoons, and that these furnaces were structured to take advantage of natural drafts. The draft itself would be responsible for the intensity of heat emitted from the furnace. The furnaces, based on our 2000 experience could attain, using *Prosopis* sp. alone, temperatures of 1200°C and higher without much trouble. Schmidt (1994; 1997) records a similar reliance on a single key tree species for production fuels in East African ironwork. This temperature, however, may have proved too hot for the needs of the artisans using the braziers, as we believe these to be used principally for annealing, which requires lower temperatures (Shimada et al. i.p.; Shimada and Wagner 2001; Shimada and Wagner 2007). Also, Mound II was mostly using the furnaces for shorter periods of time, with occasional starting and stopping of the fires, perhaps by sealing the draft hole and then removing the coals and saving them for later once the fire was extinguished (Chapter 7; Shimada and Wagner 2007). The variety of fuels present at Mound II may represent that the metalwork activities at Huaca Sialupe did not require the hot and consistent burning properties of *Prosopis* sp., and allowed for a more flexible fuel mix and less dependence on any one particular fuel.

Mounds I and II Shared Fuels: Non-Wood Remains

The non-wood materials recovered from both Mounds indicates that *Llama* sp. dung, *Z. mays* seeds and fruits (cobs), and *Prosopis* sp. seeds are, apart from wood, the most ubiquitous materials recovered (Table 6.2). By examining the densities of these materials, we can distinguish how resources were apportioned and utilized at both mounds. Although *Llama* sp. dung was in use at both parts of the site, we recovered a greater quantity from Mound II (total density = 588.0g/ml: 5.91% of the assemblage: contributing 14.07% to overall diversity) than at Mound I (total density=176.14g/ml: 1.49% of the assemblage: contributing 15.54% to overall diversity). Sampling error, however, may skew the results of this data and underestimate the importance of *Llama* sp. dung at Mound II: half as many features were sampled from Mound II than from Mound I. As a result, the amount of *Llama* sp. dung used for metalworking was probably greater than represented in this analysis. Comparing the two different diversity rankings, however, demonstrates that *Llama* sp. dung is responsible for 15% of the diversity within the Mound II assemblage; this may indicate that camelid dung was more frequently used in metal furnaces, an important differentiating factor between resource use at both mounds. As an easily stockpiled, lightweight, and predictable fuel additive to start or maintain a charcoal brazier, *Llama* sp. was essential to the metal work production at Mound II of Sialupe.

Like *Llama* sp. dung, *Zea mays* remains were also recovered from contexts of both Mounds; in this case, however, *Z. mays* appears in similar quantities throughout (Appendix K; Mound I total density=313.19 count/ml: 0.89% of the assemblage: contributing 6.48%; Mound II total density=330.20 count/ml: 4.38% of the assemblage: contributing 5.38%). Again, as there were more features recovered from Mound II, it may be that *Z. mays* played a larger role in the activities at Mound II than at Mound I; but because the density rankings are similar, this does not appear to be the case. The *Z. mays* fruit (cob) remains appear in similar concentrations with similar percentages, a pattern I will address with more detail in Chapter 7.

Mound I: Exclusive Wood and Non-Wood Remains

As mentioned in the discussion of common fuel resources between Mound I and Mound II, there is a list of materials that are found exclusively at each of the mounds. Table 6.5 lists the presence of materials that are only found at Mound I and Table 6.7 lists those for Mound II. In most of these instances, the overall ubiquity of these materials is based on an incidence of one, making it difficult to assess their presence in one locus over another. That said, classifying the materials by plant organ strengthens the presence of an item in one locus even when present in only a single feature or floor: for plant parts that are ephemeral or seasonal, for instance fruits or flowers, their presence may indicate the seasonality of the activities performed.

Of the materials present only at the ceramic production site of Mound I, most notable are flowers from *Capparis angulata* and *Capparis* sp. and several fruits from *Capparis ovalifolia*. *C. angulata* flowers most commonly in July and August, the heart of the Andean winter on the Peruvian North Coast. These are large, showy, and fragrant flowers that are a staple of the wintertime honey production season, along with *Prosopis* sp. sap which runs down the trunks during the winter. *C. ovalifolia*, on the other hand, flowers during the late Fall, with fruits available quite quickly into the Andean winter, July and August. These flowers are almost sickly sweet, attracting ants, bees and wasps. While the fruits of *C. ovalifolia* are edible, they are an acquired taste, and in my interviews I found that most people choose not to eat them. The fruits of *C. angulata*, on the other hand, are extremely poisonous and apparently enticing to grazing livestock. Donkeys, goats, and sheep, are attracted to them, but exhibit strange behavior, including attacking humans and sudden death, upon eating them (Egg 1999). Humans occasionally ingest them to commit suicide or as an abortifacient.

Table 6.5 shows that both the fruits and flowers of both *Capparis* species are present in the assemblage, as well as seeds. In both instances I interpret their presence not as a result of food production, but rather as relating to gathering activities. We see flowers as a product of fuel gathering with *Acacia* sp. as well. In the case of the flowers from *C. angulata*, their flowers are quite robust, and often remain complete long after they have become senescent and fallen to the ground. Additionally, they have rather woody sepals that desiccate rather than wilt, often finding their way into gathered *poña* materials. The

Family	Determination	Organ/ Type	Mound I Incidence of Taxon	Mound I Ubiquity (n= 38 samples)	Mound I Sum of Determinations count/ml (Adjusted by Sample Size (Density))	Mound I Pct. of Total Assemblage by Density (Density /Total Density all Taxa Mound I)
Capparidaceae	*Capparis angulata*	Flower	3	7.89%	9.72	0.03%
Capparidaceae	*Capparis* sp.	Seed	3	7.89%	27.04	0.08%
Cyperaceae	*Scirpus* sp.	Seed	3	7.89%	18.95	0.05%
Other	Burnt Clay	Ceramic	3	7.89%	41.33	0.19%
Cyperaceae	*Cyperus* sp.	Seed	2	5.26%	8.60	0.02%
Fabaceae	UKN#12-177	Seed	2	5.26%	5.92	0.02%
Malvaceae	cf. Malvaceae	Seed	2	5.26%	6.02	0.02%
Oxalidaceae	*Oxalis* sp.	Seed	2	5.26%	4.13	0.01%
Portulacaceae	cf. Portulacaceae	Seed	2	5.26%	4.13	0.01%
Cactaceae	*Opuntia* sp.	Seed	1	2.63%	3.13	0.01%
Capparidaceae	*Capparis crotonides*	Seed	1	2.63%	46.67	0.13%
Capparidaceae	*Capparis ovalifolia*	Fruit	1	2.63%	26.67	0.08%
Capparidaceae	*Capparis ovalifolia*	Seed	1	2.63%	15.00	0.04%
Capparidaceae	*Capparis* sp.	Flower	1	2.63%	2.02	0.01%
Fabaceae	*Caesalpinia* sp.	Seed	1	2.63%	7.69	0.02%
Malvaceae	*Sida* sp.	Seed	1	2.63%	2.72	0.01%
Malvaceae	UKN#44-126	Seed	1	2.63%	6.39	0.02%
Other	Lithic	(blank)	1	2.63%	3.01	0.01%
Other	UKN#13-177	Seed	1	2.63%	2.50	0.01%
Other	UKN#28-45	Seed	1	2.63%	2.34	0.01%
Other	UKN#53-147	Seed	1	2.63%	20.51	0.06%
Oxalidaceae	*Oxalis* sp.	Root	1	2.63%	10.26	0.03%
Plantaginaceae	*Plantago* sp.	Seed	1	2.63%	2.72	0.01%
Verbenaceae	*Lippia* sp.	Leaf	1	2.63%	8.33	0.02%

Table 6.5 Non-Wood Materials Present at Mound I/ Absent from Mound II

case of the *C. ovalifolia* fruits is similarly inconclusive. I recovered these fruits mostly as peduncles (stems) of the fruits, again, indicating that they are from either consumed or dried fruits, probably from *poña* as well. Unfortunately, flowers from *C. ovalifolia* are not as robust as the *C. angulata,* hence their absence from the dataset does not aid in interpreting the presence of the *Capparis angulata* flowers.

The concurrence of these plant remains tentatively suggests the use of the site during the Andean Summer, as both *C. angulata* flowers and *C. ovalifolia* fruits would be on the ground at that time (Table 6.6). This possibility conforms to Cleland and Shimada's (1998) ethnoarchaeological observations about appropriate times to be making and firing ceramics at the nearby modern town of Morropé. Summer is a key time for extra-productive activities, such as ceramic production, as the crops have been harvested, the air is dry and hot, and weather is predictable. Although these potential seasonal indicators do not prove conclusive in defining site seasonality, their presence does highlight that seasonality has the potential to be assessed in datasets similar to that derived from Huaca Sialupe. In fact, as will be addressed in the conclusion and on a feature-to-feature basis, there are some instances that permit us to interpret the data for seasonal activities.

The fruits and seeds from the two different genera in Cyperaceae are also present only at Mound I. These plants grow along the modern agricultural canals and irrigation networks in the area today. It is not clear when they go to seed, but it is certainly tied to the seasonality of either irrigation water or rainfall. Based on the local ecology of the site and what I assume to be its proximity to either irrigation canals or standing water, I would have expected within this dataset more of these kinds of materials. As is evident, however, we see very little influence of this family in the archaeobotanical data for fuel. Their presence in small numbers may indicate that fuel use at the site was focused on forest resources and disturbance taxa close to home, rather than materials associated with agricultural fields, deltaic swamps, or irrigation systems.

In terms of materials associated exclusively to Mound I, I turn to inorganic remains. I recovered several pieces of burnt clay, most of which include finger impressions, probably thumbs. I assume that these were used as spacers in the kilns and may explain their presence here at Mound I and not seen at Mound II (Arnold 1985). The presence of fingerprints was also noted on a number of bricks and ceramic mold fragments, and is indeed significant as it opens up entirely new avenues of research in terms of understanding the identity of craft producers in the archaeological record (DeBoer 1990; Kamp 2001). Note here, however, that redware or blackware ceramic sherds, and metalworking debris, e.g., slag or scoria, are not exclusive to either Mound I or Mound II. These spacers, if that is what they are, however, are only found here at Mound I. This occurrence may indicate that metal production had more interplay with ceramic production, than vice versa. This supposition is supported by Matson's (1989) position that ceramic production, in all cases of ceramic and metalwork multi-crafting situations, is the anchor process on which all other production relies and feeds off of for innovation. From a material culture standpoint, the necessity for ceramic vessel construction for ingot molds and vessels for making the furnaces themselves, both items recovered from the site, indicate that ceramic production was a primary production concern. Further discussion of the inorganic materials will be included in the feature-to-feature comparisons below.

Species	Season of Flowering/Fruit (Southern Hemisphere)
Cordia lutea	Midsummer/Early Fall
Bidens sp.	Midsummer/Late Summer
Acacia macracantha	Early Winter/Late Winter (lesser) Early Spring/Midsummer (greater)
Lippia sp.	Early Summer/Midsummer
Cucurbita sp. *Lagenaria* sp.	Mid-Spring/Late Summer
Capparis angulata	Midwinter/Early Summer
Capparis ovalifolia	Late Winter/ Midsummer

Table 6.6 Fruit and Flower Seasonal Indicators Recovered from Huaca Sialupe

Mound II: Exclusive Wood and Non-wood Remains

Table 6.7 illustrates the taxa that are present only at Mound II. The most ubiquitous and robust of these is the exclusive presence of *Lagenaria* sp. seeds (n=80). The seeds are the results of the cleaning of gourds, locally known as *mate*, that are used most commonly as serving vessels for fermented beverages, soups, and food in general. Remains of the gourds (fruits) themselves are not exclusive to either side of the site. In many cases, however, seeds from the family Cucurbitaceae (Actual Counts:n=101 overall; Mound I=22; Mound II=79) do not survive as entire seeds in the archaeological record. For that reason, when their specific morphology did not allow me to determine the genus, I placed them in a general Cucurbitaceae taxon (actual counts:n=14 overall; Mound I=2; Mound II=12). Taking this into consideration, it is clear that *Lagenaria* sp. seeds are not well represented at Mound I. Part of the reason for this may have to do with the role that domestic garbage played at the site and how charcoal materials were shared between the two sides of the site, ideas that are better explored in the feature-to-feature comparisons in Chapter 7.

Lagenaria sp. seeds comprise more than half of the overall Cucurbitaceae seed assemblage, a striking finding. As gourds take a relatively long time to fruit and dry out, they are harvested during the late summer; often cleaned and prepared for use when harvested, the presence of seeds may relate to the seasonal pooling of garbage at Mound II (Schaedel 1988). Perhaps the cleaning coincided with a period when the metalworking area was dormant, or the refuse was collected as domestic garbage to be used as fuel at during the active period Mound II. When taken together with the *Lagenaria* sp. data, the presence of other Cucurbitaceae seeds (*Cucurbita maxima*) and other seeds (e.g., *Bixa* sp., *Celtis* sp., *Neptunia* sp.) and fruits (e.g., *Cordia* sp., cf. Myrtaceae, *Cenchrus* sp.) may collectively point to another seasonal aspect of workshop production (Table 6.6).

Apart from the Cucurbitaceae data, there are several other seasonal indicators that occur in the taxa that are exclusive to Mound II. There are three types of seeds from the family Asteraceae present in the assemblage. Additionally, the fruit (Density: n=2.5) and seed (Density: n=2.96) of *Cordia* sp. (Boraginaceae) are present in the Mound II contexts. The presence of *Bidens* sp. seeds (Density:n=201.44) and *Cordia* sp. seeds and fruits are also potential indicators of seasonality (Figure 6.6). In 2002, I observed the presence of ripe *Cordia lutea* fruits, edible and tasty, in conjunction with the seeding of *Bidens* sp. at the end of the Andean summer, March to April. Most members of the family Asteraceae respond well to water for germination. In this case, the extremely rare summer rains or irrigation water from spring crops might have triggered growth leaving them to seed at the end of the summer. The co-incidence of these taxa and their exclusivity at the Mound II metalworking context may indicate that at least the collection of fire materials, if not the use of Mound II, occurred during the late summer-early fall.

I consider the gathered seed materials exclusive to this side of the site as constituting an assemblage of seeds collected during the off-season when these plants were growing, setting fruit, and going to seed. *Cucurbita maxima* (Density:n=78.79) has a long growing season. The presence of its seeds, together with *Cordia* sp. fruit and seed, with a short and more punctuated fruiting schedule, may be indicators that these materials were deposited in these contexts when the furnaces were not in use. If this were the case, then it may be that the furnaces were charged during the spring and summer to prepare for use during the late summer or fall. This part of the site, the metalworks, could have been the prime production locus during the winter and spring months, when ceramic production may have been temporarily halted. This scenario would mean that ceramic production

Family	Determination	Organ/ Type	Mound II Incidence of Taxon	Mound II Ubiquity (n=18 Samples)	Mound II Sum of Determinations #/ml (Adjusted by Sample Size (Density))	Mound II Pct. of Total Assemblage by Density
Cucurbitaceae	*Lagenaria* sp.	Seed	5	27.78%	80.40	1.07%
Asteraceae	*Ambrosia* sp	Seed	3	16.67%	27.83	0.37%
Boraginaceae	*Heliotropum* sp.	Fruit	3	16.67%	105.45	1.40%
Convulvulaceae	*Ipomoea* sp.	Seed	3	16.67%	18.11	0.24%
Cucurbitaceae	*Cucurbita maxima*	Seed	3	16.67%	78.79	1.04%
Cactaceae	cf. Cactaceae	Seed	2	11.11%	16.15	0.21%
Fabaceae	*Neptunia* sp.	Seed	2	11.11%	66.63	0.88%
Fabaceae	UKN#43-121.1	Fruit	2	11.11%	7.89	0.10%
Fabaceae	UKN#43-121.1	Seed	2	11.11%	32.84	0.44%
Asteraceae	*Bidens* sp.	Seed	1	5.56%	201.44	2.67%
Asteraceae	UKN#25-37	Seed	1	5.56%	1.42	0.02%
Asteraceae	UKN#50-142	Seed	1	5.56%	2.78	0.04%
Bixaceae	*Bixa* sp.	Seed	1	5.56%	1.65	0.02%
Boraginaceae	*Cordia* sp.	Fruit	1	5.56%	2.50	0.03%
Boraginaceae	*Cordia* sp.	Seed	1	5.56%	2.96	0.04%
Capparidaceae	*Capparis* sp.	Fruit	1	5.56%	1.36	0.02%
Fabaceae	*Parkinsonia* sp.	Seed	1	5.56%	2.50	0.03%
Lauraceae	*Persea* sp.	Seed	1	5.56%	2.63	0.03%
Myrtaceae	cf. Myrtaceae	Fruit	1	5.56%	3.03	0.04%
Piperaceae	*Piper* sp.	Flower	1	5.56%	11.43	0.15%
Poaceae	*Cenchrus* sp.	Fruit	1	5.56%	4.35	0.06%
Poaceae	*Cenchrus* sp.	Seed	1	5.56%	74.82	0.99%
Rubiaceae	*Gallium* sp.	Seed	1	5.56%	2.63	0.03%
Ulmaceae	*Celtis* sp.	Seed	1	5.56%	2.88	0.04%

Table 6.7 Non-Wood Materials Present at Mound II/ Absent from Mound I

held sway at the site during the spring and summer months. Ethnoarchaeology information collected at the nearby town of Morrope indicates that summer months are preferable for ceramic production due to clear dry and calm weather, without the potential for encroaching, damp and coastal fog (Cleland and Shimada 1998). Arnold's work (1985; 1993) that compares observations of several different ethnographic ceramic production examples notes these same seasonality concerns cross culturally among traditional ceramic production in modern contexts. A final possibilty is that these production processes took place simultaneously as some form of co-production (Shimada 2007).

Mound I and II: Patterns of Fuel Use

By refining the dataset and looking independently at Mounds I and II for comparative purposes I have tried to demonstrate a pattern of fuel use among the main taxa used exclusive to each side of the site. Both sides of the site, however, appear to have drawn on the same set of resources for up to 80% of their fuel needs. Additionally, while there are some instances of exclusivity among the finds at both mounds, there are few indicators that allow us to truly distinguish between the two sites without looking more critically at the actual features where fuels are used. An important exception, however, regards the remains that may demonstrate some rhythm of seasonality among the use or activity at the two sides of the site. These broad trends provide a potential avenue for exploring seasonality, potential work scheduling and planning, as well as suggest the possibility of some instances of resource hoarding and sharing.

I turn now to the specific fire-use features. Examining fuels on a feature-by-feature level illustrates how fuel use at each side of the site conformed or deviated from our reconstructions of the ancients technologies used at the site. Additionally I can better illustrate the different domestic or household processes, attendant to metal and ceramic production that occurred at Huaca Sialupe.

CHAPTER 7 PALEOETHNOBOTANICAL INTERPRETATION OF ARCHAEOBOTANICAL DATA: FEATURE-TO-FEATURE COMPARISONS

This chapter narrows the analyses begun in Chapter 6 to examine the fuel remains recovered from features associated directly with either metalwork or ceramic production. A qualitative description of the individual features establishes a pattern of fuel use at the different loci of production; these patterns are then compared to unknown fuel use contexts at the site to aid in their interpretation. This comparative analysis reveals an intersection of domestic and agricultural activities with the recognized craft production at the Huaca Sialupe mounds. The incorporation of domestic garbage and agricultural residuals into workshop features points to an integration of food production and consumption with ceramic and metal crafting. For the most part, the following discussion specifically examines the materials recovered from within the features. When useful, floors are compared to their associated features to clearly define activities taking place in certain parts of the site.

A Qualitative Look at Fuel used in Mound II Metal Production

Metal working furnaces at Mound II of Huaca Sialupe provide a detailed record of collected fuel materials used for metal fabrication. Many of the furnaces recovered during the 1999 and 2001 field seasons contained several liters of carbonized fuel remains, which consisted largely of wood charcoal from the Capparidaceae and Fabaceae families and camelid dung (Table 7.1). As mentioned in the previous chapter, the wood charcoal is noticeably and predominantly in the form of juvenile branch wood. The small and consistent diameter sizes, 2-5 cm, indicate that this sample represents wood from small branches. In all of these small twigs and branches we see the same anatomical characteristics. Evident in the cross sections of these specimens are two or three cycles of rapid growth: secondary xylem vessels occur in radial multiples, fiber cells are vastly outnumbered by parenchyma, and the final or last growth series begins to conform to species-based growth patterns.

Also I recovered 326 wood fungal spores from contexts throughout Mounds I and II, indicating that some of the fuel wood branches had undergone fungal decay precipitated by damp conditions. The presence of spores was noted throughout even when they were not collected and counted as part of the dataset, again highlighting their ubiquity across the site. Living woody species in dry tropical forests experience fungal decay as part of periodic inundation following destruction due to ENSO events and prolonged epochs of coastal fog; thus, the presence of these wood fungi on carbonized remains of metal furnace fuel indicates that wood was probably collected from the forest floor or from deador damaged branches. Also, fungal decay could set in during periods of wood storage at the site or elsewhere. Additionally, we see signs of infestation of wood boring insects, even in the instance of the smallest branches. With the exception of termites, few animals can consume woody fibers; instead, infestation often occurs when fungal growth is advanced and the insects consume the weakened infested wood (Allsopp and Seal 1986). Taken together, these observations indicate that the bulk of the wood materials being used as fuels in metal working contexts were collected materials, not actively procured from living trees. The branches were likely collected as dead or dried senescent juvenile growth from locally growing trees. The presence of carbonized insect larvae, beetles, and wood fungal spores may indicate that the dead wood was first collected and stored, converted to charcoal, and finally placed in the furnaces for combustion. It is additionally conceivable that branchwood was collected and stored well in advance of use. This dynamic of fuelwood collection will be discussed in Chapter 8.

Table 7.1 shows evidence of little or no mature stature trees of the main genera were cut or used in any significant way to fuel the metalworks at Mound II. As discussed in the previous chapter *Prosopis* sp. and *Acacia* sp. make up the majority of the fuels used, with the addition of branches from the two *Capparis* species. A closer look at the fuel used in this area demonstrates that two species in particular, *Tessaria integrifolia* and *Vallesia glabra*, occur in the metal workshop features listed as 'mature' wood. Unlike the primary tree species discussed in this dissertation, *Tessaria* and *Vallesia* are much less substantial: the growth form of these two species are shrub to low trees, with mature wood trunks or boles no more than 15 cm in diameter. In my dataset, the diameters of the wood recovered are considerably smaller. Due to their relatively smaller size, and accelerated growth, these plants, however, were likely harvested as mature plants in contrast to the other primary species recovered, e.g., *Prosopis* sp. or *A. macracantha*. *T. integrifolia* and *V. glabra* are both frequently used in *quincha* and roof beam construction and other organic parts of architecture that are replaced over time. In the present day their use as renewable construction resources is due to both their abundance and quick growth (Correa Álamo 2000). Their inclusion as fuel resources provides an additional dimension of 'technological choice:' renewable and relatively easy to procure, artisans included these species within an assemblage that consisted largely of dead juvenile branchwood, pointing to the possibility that craftspeople were aware of the important idiosyncrasies of local ecology. Alternately, they may have been recycling parts of old *quincha* walls or using stores of wooden staves as fuel in the instances of these species.

The following analysis of fuel resources by feature provides additional insight into human relationships with the environment, as well as the relationship between local artisans and the domestic economy of Huaca Sialupe.

Mound II: A Fully Charged Furnace, Feature 8-'01

Of the features recovered in the Mound II metal workshop only one conformed exactly to the morphology ascribed to the furnaces as recovered in 1999 and

Family	Determination	Mound II Location	Open Area						Room 3			Total Density g/ml	Taxon as Pct. of All Mound II Feat.
		Feature #	8 (Charged)	9A	9B	12	15	19	14	21	20		
		Feature Type (Est. Vol.L)	Porron (73.59)	Charcoal Hearth (37.66)	Charcoal Hearth (49.02)	Old Porron (16.03)	Old Porron (3.47)	Old Porron (10.23)	Old Porron (20.70)	Old Porron (9.41)	Old Porron (12.90)		
Fabaceae	*Prosopis* spp.	Branchwood	141.75	7.38	130.11	309.39		172.71	1136.92	3.28		1901.55	21.70%
Fabaceae	cf. Fabaceae	Juvenile	52.36		62.18	585.65			6.73	2.35		709.27	5.46%
Fabaceae	*Acacia macracantha*	Branchwood	119.67		249.50	198.37	0.55	37.46	60.76	34.43		700.73	8.09%
Dung	*Llama* sp.		3.86	0.80	1.95	2.35	434.45	0.26	4.52	1.06		449.25	4.67%
Capparidaceae	*Capparis angulata*	Branchwood	13.03		22.61	306.48		22.68	15.82	3.59		384.20	1.75%
Capparidaceae	*Capparis ovalifolia*	Juvenile	86.28		17.80	74.59			47.45	18.04		244.16	2.40%
Poaceae	*Gynerium* sp.						154.62		0.42			155.04	1.61%
Metal	Scoria	Amorphous	21.33					19.78	5.05			46.16	0.78%
Unid	Wood Type GG	Branchwood			45.70							45.70	0.65%
Fabaceae	cf. Fabaceae	Desicc.Juv.	9.66				10.27		1.76	4.31		26.00	0.28%
Poaceae	*Zea mays*		0.21		0.28	12.83						13.32	0.03%
Apocynaceae	*Vallesia glabra*	Juvenile	2.11		9.03							11.13	0.15%
Asteraceae	*Tessaria integrifolia*	Dessicated								4.99		4.99	0.05%
Asteraceae	*Tessaria integrifolia*	Mature		0.98			0.17			0.42		1.57	0.01%
Fabaceae	*Acacia macracantha*	Mature								0.42		0.42	0.00%
Unid	Wood Type CC	Branchwood							0.13			0.13	0.00%
Unid	2mm Charcoal	Branchwood	202.29		331.08	235.35		105.92	1331.55	3.98	3.16	2213.33	28.72%
Unid	4mm Charcoal	Juvenile			78.60							78.60	1.12%
Unid	4mm Charcoal	Branchwood	509.65		339.40	601.90		389.77	61.06	13.05		1914.84	22.51%
Grand Total			1162.18	9.16	1288.24	2326.92	600.06	748.57	2672.17	89.92	3.16	8900.38	100.00%

Table 7.1 Fuel Contents of Features from Metalworking Area Mound II (Density of Weighed Charocal g/ml)

experimentally reconstructed in 2000. This feature was systematically sampled during the 2001 season, Feature 8-'01 (Figure 7.1). Curiously enough, this was one of the best preserved features (Figure 7.2) in terms of its overall form, as well as in its charcoal fuel charge. This sample provides an unparalleled look into a complete metal working furnace from the site, for both overall feature morphology and its fuel components. Table 7.1 compares the charcoal fuel contents of Feature 8-'01 with the other *porrón* related features found at Mound II. As indicated in the previous chapter (Table 6.1), the array of fuels used in the metalworking portion of the site, Mound II, does not vary significantly from Mound I ceramic kilns in term

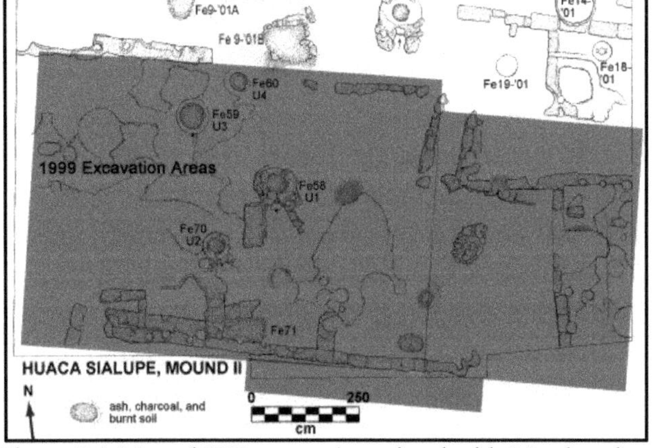

Figure 7.1 Mound II Features Associated with Metalwork Production

of the species represented. Instead the materials differ by the representative quantities between features at the two mounds. Feature 8-'01 is unique in that it is filled with a combination of both carbonized and uncarbonized materials (Table 7.1). The overall volume of the feature is approximately 74 L; we recovered about 8.1 L of charcoal from two different samples taken from the interior of the feature, about 10% of the feature's total volume.

The SAP's 2000 experimental metalworking furnace, based on a design that mirrors Feature 8-'01, was most efficient when filled with around 20 L of charcoal, which covers about three quarters of the air intake at the base of the furnace. One problem of correlation between our experiments and the ancient furnace is that we did not measure the overall volume of the experimental furnace making comparison difficult. I estimate, however, that the original volume of the experimental *porrón* used in 2000 was around 50 L, indicating that we filled it with 2/5 of its overall volume and achieved acceptable temperatures for annealing. Recall that a key feature of the *porrones* is their air intake hole, positioned to utilize the afternoon offshore breeze. In our experiments we found that covering the hole with the charcoal level maintained a consistent temperature of 750° C at the top of the furnace, a reasonably good temperature for annealing copper arsenic bronze (Figure 7.3). Feature 8-'01, as a furnace charged and ready for combustion, was filled over and above its air intake hole when excavated, and we only recovered around 10% of the total volume of the feature in charcoal.

As excavated, Feature 8-'01 was less than half full at the time of excavation. Electron microscopy of the charcoal from this furnace indicated that there was little to no ash present in the samples. Wagner et al. (2001)

Figure 7.2 Feature 8-'01, Metalworking Furnace

suggest that this observation indicates that the materials in the furnace were never under combustion. The recovery of the remains of desiccated small branches from juvenile Fabaceae wood supports this theory. It is most likely that these small branches (0.77% by weight, 5.26% of the diversity) were added to the charcoal mix of the feature to aid in its ignition. The preservation of small, dry branches in the mix after 1000 years is unusual and relatively astounding. I suggest that their presence supports my hypothesis that these were additives for aiding combustion and ignition, and that this furnace was being prepared for use. The presence of dried small branches in other features of the Mound II complex likewise implicate preparation for use and the process of preheating, see discussion below regarding feature 9-'01a and b. In comparison with the contents of the other *porrón* related features recovered during the 2001 season (see below), Features 12-'01 and 14-'01 also contained substantially more fuel materials (Table 7.1). Of these, only Feature 14-'01 has uncarbonized juvenile Fabaceae twigs associated with it, which again may indicate that the presence of uncarbonized materials may be key representatives of features being prepared for use and not ignited prior to abandonment. Again, indicating that the presence of uncarbonized materials may be key representatives of features being prepared for use that were not ignited prior to abandonment.

Mound II: Explaining Carbonized Materials in Porrón Features

There are other features that indicate a mixing of carbonized and uncarbonized material in the Mound II area (Table 7.1). These are Features 21-'01 and 15-'01, both of which are in the feature class Old Porrón. The class of Old Porrón includes hearths where *porrones* were used to lend structure to the hearth feature. Potentially these are metal furnaces that fell into disrepair leaving only the lip of the ceramic vessel imbedded in the workshop floor. While no longer used as metal furnaces these reinforced features were used to contain other fires, some of which are discussed below. These features are found at Mounds I (n=1) and II at Huaca Sialupe, with the larger concentration of them at Mound II (n=6). They consist of large ceramic vessels, identical to the *porrones* used in the construction of the metal furnaces, but only the uppermost portion of the vessel imbedded in the floor, the inverted mouth of the vessel, is extant. This produces a resilient ceramic ring that can be used as a small, structurally sound and well-delimited locus for storage or fire making. Further clues to the use of both Features 12-'01 and 15-'01 come from a comparison of their contents. The densities for the materials recovered from Feature 21-'01 and their corresponding diversity rankings indicate that this feature has a composition similar to the charged furnace 8-'01.

Family	Determination	Feature #	11	14	16	21	Floor	Total Density g/ml	Taxon as Pct. of Total Assemblage
		Feature Type (Est. Vol. L)	Hearth (36.00)	Old Porron (20.70)	Hearth (12.60)	Old Porron (9.41)	(47.00)		
Unid	2mm Charcoal	Branchwood	1.68	1331.55	3.72	3.98	0.70	1341.63	47.46%
Fabaceae	*Prosopis* spp.	Branchwood	8.22	1136.92	15.57	3.28		1163.99	40.17%
Unid	4mm Charcoal	Branchwood	34.16	61.06	29.90	13.05	0.62	138.79	4.08%
Fabaceae	*Acacia macracantha*	Branchwood		60.76	38.91	34.43		134.10	4.07%
Capparidaceae	*Capparis ovalifolia*		1.17	47.45		18.04		66.67	1.99%
Capparidaceae	*Capparis angulata*	Branchwood		15.82		3.59		19.40	0.58%
Fabaceae	cf. Fabaceae	(Desiccated) Juv.	6.47	1.76	2.75	4.31		15.29	0.46%
Fabaceae	cf. Fabaceae	Juvenile		6.73	2.29	2.35		11.37	0.38%
Dung	*Llama* sp.	Dung	3.79	4.52	0.54	1.06		9.92	0.29%
Metal	Scoria	Amorphous		5.05				5.05	0.18%
Asteraceae	*Tessaria integrifolia*	Desiccated				4.99		4.99	0.16%
Fabaceae	*Prosopis* spp.	(Desiccated) Chip	3.15					3.15	0.08%
Asteraceae	*Tessaria integrifolia*	Mature	1.74			0.42		2.16	0.05%
Poaceae	*Zea mays*	Stem			0.46			0.46	0.01%
Poaceae	*Gynerium* sp.	Stem		0.42				0.42	0.01%
Fabaceae	*Acacia* sp.	Mature				0.42		0.42	0.01%
Unid	Wood Type CC	Branchwood		0.13				0.13	0.00%
Grand Total			60.39	2672.17	94.13	89.92	1.32	2917.94	100.00%

Table 7.2 Fuelwood Contents of Room 3, Mound II Features (Density of Weighed Charocal g/ml)

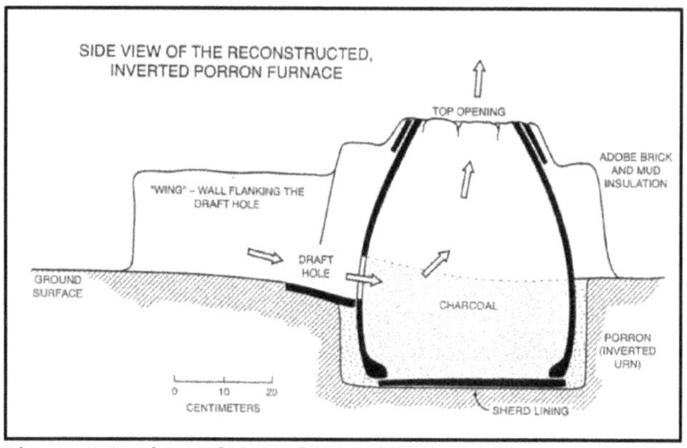

Figure 7.3 View of Experimental Furnace Showing Draft and Charcoal Level (Shimada and Wagner 2007)

The contents of Feature 15-'01, however, stand in stark contrast to the contents of 21-'01. Although both contain a substantial amount of desiccated remains, the overall contents lack a substantial presence, by density, of Fabaceae or Capparidaceae hardwood. Feature 15-'01 consists mostly *Gynerium* sp. (Poaceae) and *Tessaria integrifolia*, both plants that grow in river basins and along canals and that are commonly used as construction materials. In the case of Mound II these species appear to serve as expedient fuel additives. The presence of these species as desiccated, not carbonized, materials potentially indicates that these features were used for fuel storage. In the end it appears at least these two classes of fuels, desiccated and carbonized, were reserved and stored at Mound II for uses associated with metal production.

The similarities between 21-'01 and 8-'01, coupled with the differences with 15-'01, may indicate that 21-'01 provided a similar set of fire use potentials as 8-'01. Although it is possible that these phenomena demonstrate different taphonomic processes, it seems more likely that the similarities and differences in their fuel loads indicates salient qualities of these features. I suggest that 21-'01 was a place for collecting fuels to be used in a charged furnace, like that recovered as Feature 8-'01. This likelihood leads us to consider the role of storing and mixing of charcoal fuels used in the metalworks at Mound II.

From this data, it appears that not only was metal work production occurring at Mound II; the preparation of mixing of furnace charges by combining charcoal and wood fuels, and even charcoal production took place. The mixed materials in the charged furnace, consisting of desiccated and charcoal remains, demonstrates that fuel was mixed prior to its placement in the furnace. Based on these criteria, I have tentatively ascribed the use of charcoal mixing and/or charcoal production to the Feature 9-'01 parts A and B. Feature 9-'01A contained very little material (9.14 g/ml) from only three different taxa, *Llama* sp. dung, *T. integrifolia* wood, and some *Prosopis* sp. branchwood. Both parts of the feature exhibited areas of reddened, oxidized, compact earth, indicating that these areas were fired. The rest of Feature 9-'01A was mostly sand. Feature 9-'01B, on the other hand, was similarly sandy, but also contained a large rectangular pile of a variety of branchwood and a considerable amount of charcoal (1288.24 g/ml). Although ash does occur at the site in both Mounds I and II in features that are labeled ash deposits, there was a limited amount of ash in this feature; in addition, no desiccated material was removed from either part of Feature 9-'01.

In close proximity Feature 8-'01, I suggest that Feature 9-'01A and B both represent either small charcoal kilns or at least areas for pooling charcoal to charge the furnaces at Mound II. The A portion of the feature was mostly ephemeral at the time of excavation; the B portion, however, was partially enclosed by a low series of adobe blocks with the interior surface lined with clay, and contained no portions of a *porrón*. The more or less uniform-sized pieces of branchwood, 5-10 cm of *Prosopis* sp. and Capparidaceae, together with the oxidized surface, likely indicates its role as either an on-site charcoal kiln or, at the very least, a feature for preheating charcoal for use in the adjacent furnaces. Despite the morphological differences of Feature 9-'01 in comparison to Feature 8-'01, the similarities between their fuel remains highlights a related function. The role of additional fire-use features at Mound II that do not exhibit the *porrón* furnace characteristics can also be better understood through a thorough analysis of their fuel remains.

Mound II: Non-Porrón Features and Fine Metal Working

Unlike the charcoal production or storage features described above, some features classified as Old Porrones exhibit a distinct fuel assemblage and a closer examination is required to determine their relationship to metal production at the site. Features like 12-'01 and 14-'01 both contain higher densities of wood charcoal than any of the other features at Mound II. They also both contain combinations of taxa similar to the charged kiln, Feature 8-'01 and Feature 9-'01A. Feature 14-'01, however, contains over three times the quantity of *Prosopis* sp. branchwood than any other feature at Mound II (1136.32 g/ml). Its location in Mound II, Room 3, is shared with a number of other features, whose morphology also does not clearly match a *porrón* feature. Perhaps indicating a charge for a very hot, perhaps ingot melting or foundry fire.

Room 3, excavated in 2001, is the one clearly bounded area in Mound II, separated from the larger open area where the *porrones* are located; within it are features 14-'01, 16-'01, 11-'01, and 20-'01 (Figure 7.1). All of the features in this room, save for Feature 20-'01 and the Floor, contained desiccated materials within the charcoal matrix. Room 3 is also where the most of the fungal spores were recovered at this side of the site, indicating the presence of either dead wood or charcoal made from dead wood in this room. Feature 14-'01 contains non-carbonized materials, potentially indicating that the materials within were intended for combustion, either in that feature, or for use in another feature (Table 7.2). It is adjacent to two in-ground hearth features, features 11-'01 and 16-'01, the best preserved of which, in terms of the feature itself, is Feature 16-'01; Feature 11-'01 is more scattered and poorly preserved. These in-ground hearths

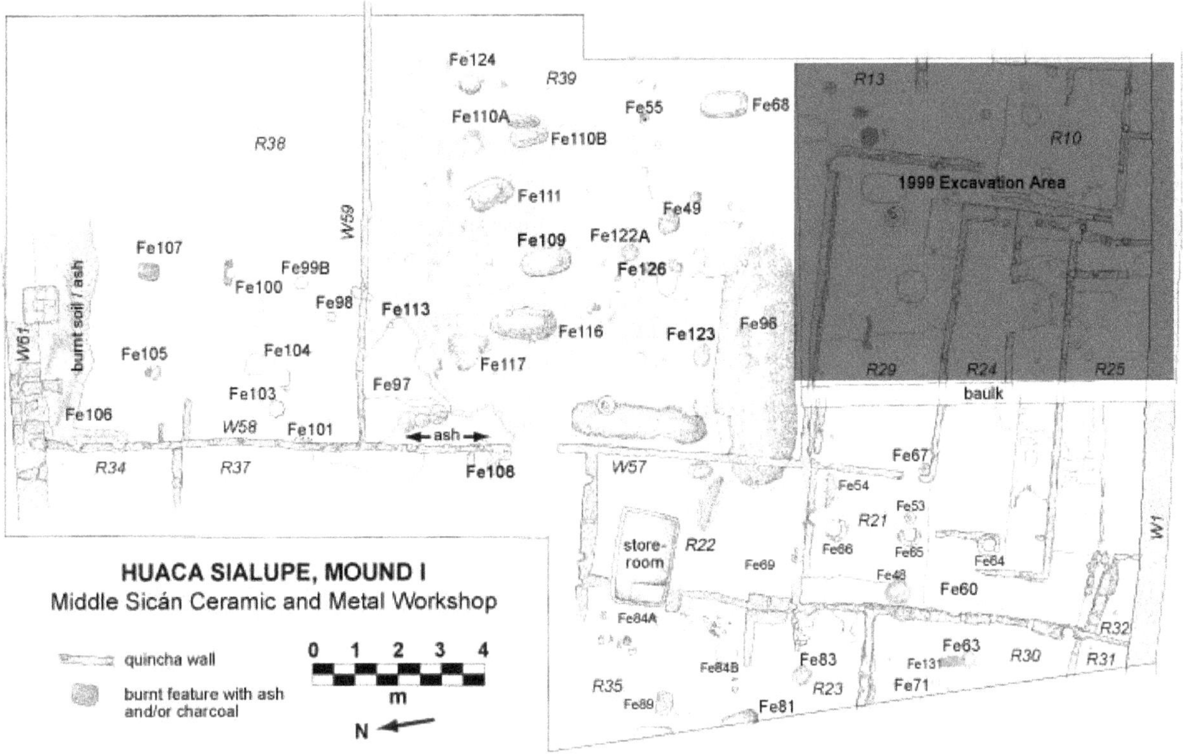

Figure 7.4 Mound I Ceramic Workshop Contexts

are small and compact, have a small pear-like shape, and are cut deeply into the floor to a depth of 10-20 cm. It is possible that these in-ground hearths represent small furnaces in a compact and shielded area that were used for small-scale production of soft metal (e.g., silver) goods. Sheet metal production of soft precious metals, e.g., silver and gold, require that areas be clear of small particulate matter, like dust and ash. It is likely that this area served as a means to sequester alloying and foundry processes necessary for the workshop to efficiently meet production demands without compromising quality.

Shimada and Merkel (Merkel et al. 1993; Shimada and Merkel 1991) describe for Huaca del Pueblo de Batán Grande and Cerro Cementarios/Cerro Huaringa pear-shaped smelting furnaces for ore refinement and ingot production. Descriptions for Inka period silver production also describe small-scale braziers and hearth features used for lead and silver metalworking in localized portions of patio groups in the Mantaro Valley (Howe and Petersen 1993). The full description of the furnaces, however, is not available, making comparison with the Sialupe contexts difficult (Howe and Petersen 1993). The desiccated materials within the charcoal matrix, however, indicate that the area was likely being prepared for use, prior to abandonment. These two distinctly-shaped small hearth features, and the more or less similar mix of fuels in this room, as compared with Furnace 8-'01 and the generalized *porrón* use area, points to a more specific application of metallurgy in this restricted area. The fact that feature 14-'01 was charged with such a high load of *Prosopis* sp. may lend further weight to the possibility that this room's use was restricted to foundry/casting purposes requiring exceptionally hotter fires than elsewhere at Mound II, 1200°C for arsenical-copper.

Metal Production at Mound II: Summary

About half of the charcoal recovered from Huaca Sialupe comes from the features at Mound II. Even so, although the floors and some of the non-*porrón* related features (11-'01,16-'01, and 9-'01) show signs of burning, they are relatively clean of charcoal; that is, the floors are not black and ashy, as we see in Mound I. Elsewhere, Shimada et al. (1999) have highlighted that cleanliness, especially keeping ash down, is key to working fine metal sheet, as it is easily deeply scratched, making later polishing and processing difficult. These conditions suggest that manufactured charcoal was either being brought to Mound II, its manufacture was limited to small and controlled production at Mound II, or, as was probably the case, a combination of both of these activities was occurring. This kept the floor at Mound II largely clean of ash and carbon debris. As demonstrated in the next section, the majority of this charcoal probably came from the production activities associated with reduced blackware ceramic production at Mound I.

A Qualitative Look at Fuel used in Mound I Ceramic Production

As mentioned in the previous chapters, Mound I is the portion of Huaca Sialupe that was the focus of ceramic production. Production features indicate that both reduced blackwares and oxidized redwares were produced at the site. The 2001 excavations and analyses,

Family	Determination	Mound I Room #	39							23	30	Total Density g/ml	Taxon as Pct. of All Mound I Feat.
		Feature #	68	109	110A	110B	111	117	96	81	131		
		Feature Type (Est. Vol L.)	Kiln (65.25)	Kiln (78.5)	Kiln (44.7)	Kiln (44.45)	Kiln (30.40)	Kiln (44.52)	Preheating Fire (90)	Kiln (27.61)	Kiln (67.50)		
Fabaceae	*Prosopis* spp.	Branchwood	1523.75	38.28		6.87	73.58	126.86	51.19	185.74		2006.27	23.96%
Fabaceae	*Prosopis* spp.	Mature	1147.13		27.97							1175.10	12.06%
Fabaceae	cf. Fabaceae	Juvenile		11.56			2.81		171.44	470.74	0.31	656.85	9.49%
Fabaceae	*Prosopis* spp.	Juvenile			151.05				242.02			393.07	7.51%
Fabaceae	*Acacia macracantha*	Branchwood	160.29	7.43			3.44	19.90		64.28	56.08	311.41	3.27%
Capparidaceae	*Capparis angulata*	Branchwood	24.73	8.02	1.22	98.72	3.54	4.48	0.49	84.35		225.55	2.23%
Dung	*Llama* sp.		5.61	20.48	24.54	14.41	0.56	0.69	4.09		2.46	72.85	1.01%
Fabaceae	*Acacia macracantha*	Juvenile							39.02			39.02	0.75%
Fabaceae	*Acacia macracantha*	Mature	69.33									69.33	0.70%
Unid	Wood Type CC	Branchwood						3.53	17.85			21.38	0.47%
Fabaceae	*Prosopis* spp.	Adze marks									109.15	109.15	0.45%
Ceramic	Burnt Clay						15.47				13.77	29.24	0.20%
Capparidaceae	*Capparis angulata*	Mature	17.26									17.26	0.17%
Capparidaceae	*Capparis ovalifolia*	Juvenile		0.38			16.56					16.95	0.16%
Unid	Wood Type AA	Mature	11.50									11.50	0.12%
Metal	Scoria	Amorphous							5.97			5.97	0.11%
Ceramic	Ceramic Redware					15.54						15.54	0.10%
Metal	Slag	Amorphous			4.81							4.81	0.09%
Poaceae	*Zea mays*				2.50				0.85			3.35	0.06%
Asteraceae	*Tessaria integrifolia*	Mature		1.19							10.38	11.58	0.06%
Unid	Wood Type BB	Mature	5.89									5.89	0.06%
Dung	Mammal							1.67				1.67	0.06%
Capparidaceae	*Capparis angulata*	Juvenile	1.44									1.44	0.01%
Unid	Wood Type AA	Branchwood				1.79						1.79	0.01%
Unid	Wood Type JJ	Branchwood		0.14								0.14	0.00%
Unid	2mm Charcoal	Branchwood	480.04	9.98		19.18	24.21	206.73	82.42	171.32	18.92	1012.80	16.18%
Unid	4mm Charcoal	Branchwood	1297.66	32.95		71.90	114.00	175.32			103.77	1795.59	20.70%
Grand Total			4744.63	130.42	212.08	228.41	254.18	539.17	615.34	976.43	314.85	8015.51	100.00%

Table 7.3 Fuel Contents All Mound I Kiln (Density of Weighed Charocal g/ml)

however, cover features only related to blackware production. These kilns, as mentioned in Chapters 3 and 5, are elongated cigar-shaped kilns, dug into the ground with a ceramic superstructure composed of vulva-like down curving walls. Kilns similar to those recovered at Huaca Sialupe appear to be part of a long-lived tradition of blackware production on the North Coast of Peru, and both our 1999 excavations and 2000 experiments indicate that we have a solid understanding of their technology (Shimada et al. 1998; Shimada and Wagner 2001; Wagner et al. 2001).

As noted in Chapter 6, the same suite of principal materials was recovered from Mound I as Mound II. Additionally, when it comes to the presence of wood fungal spores and signs of insect activity, the wood materials exhibit similar overall qualities with those from Mound II. Even with these similarities to the metalwork fuels, however, the contents of the features at Mound I appear qualitatively different. First, the majority of the remains from these features indicate a greater proportion of *Prosopis* sp. associated with kilns than with metal furnaces. Second, there is slightly more diversity, mostly consisting of unknown wood types, in the materials used for burning at Mound I. Finally, there is some evidence of large(r) pieces of hardwood in some of the kilns: in contrast to the furnace remains, we find a few, very limited, quantities of mature wood in some of the features representing the genera *Capparis*, *Acacia*, and *Prosopis* (Table 7.3). For the analysis performed here, 'mature' indicates that the juvenile or earliest growth phase of the wood, known as pith, was not present, completely subsumed in the innermost heart of the woody structure at the center of the branch. Additionally, the mature label indicates that only the diagnostic adult wood anatomical features for that particular species were present. Yet, while this material is labeled 'mature,' it does *not* indicate that these materials represent something other than branchwood; in most cases, the mature wood is, in fact, branchwood, indicated by the arc of the stem section. Although the charcoal recovered from the ceramic kilns largely consists of slightly smaller-sized branches than those seen in the metal furnaces, most representing branches around 0.5cm and no larger than

5.0cm in diameter, it also contains some examples of wood that came from mature stature trees.

There are a few instances, discussed below, indicating that larger pieces of mature wood, perhaps from large trunks or seasoned branches and measuring between 5-15 cm in diameter, were also used as fuel. Even so, the Mound I kiln areas are generally free from large pieces of charcoal and ash and contain limited remains of fuel use. The overall floor of the main kiln use area, Room 39, was largely blackened by the amount of combustion that occurred in the area; in terms of being 'dirty,' however, the floor and the areas between the kilns was largely free from debris. This may indicate that the kilns were cleaned prior to abandonment (See Below; Figure 7.4).

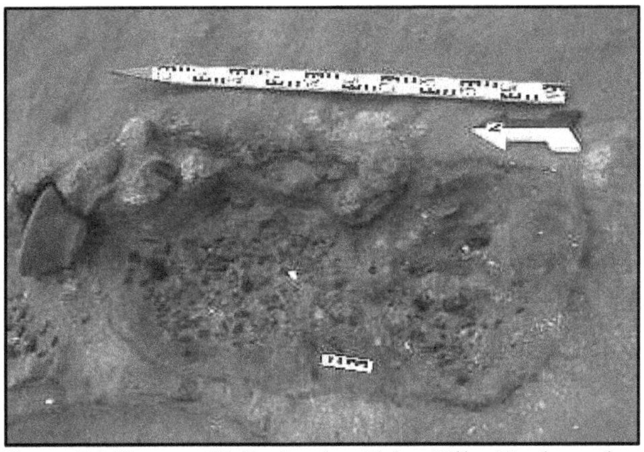

Figure 7.5 Feature 68, Reduction Firing Kiln, Uncleaned

Mound I: An Uncleaned Reduction Kiln, Feature 68

As is the case with the one presumably intact metalworking furnace in Mound II, we recovered one well-preserved kiln, Feature 68 (Figure 7.5), in the Room 39 floor (Figure 7.5). This feature was an exception to all of the other kiln features recovered from the Middle Sicán occupation at Mound I in that its post firing residual fuel load was present at excavation. This situation offers a unique insight into the qualities of the fuels used for ceramic firing. As noted above, the rest of the excavated kilns contained little residual fuel and looked more like Features 110A and 110B (Figure 7.5). Feature 68's fuel load, in comparison with the other features recovered in the kiln areas, mainly Room 39, likely indicate that the Sialupe artisans did not clean this kiln following firing. This feature alone contained a charcoal density of 4744.63 g/ml, accounting for 59% of all of the charcoal recovered from the nine kiln features at the site.

Feature 68 in Mound I, thus can serve as a type specimen against which we can compare the rest of the kiln contents. In addition to providing a 'standard' fuel assemblage, Feature 68 also provides a model for a ceramic kiln's overall shape and volume (circa 65 L). It is on the larger end of the spectrum of the kiln features recovered. Judging from one of the sidewalls of the kiln that was intact when excavated its shape conforms very closely to our experimental kiln, yet the experimental kiln's volume was much larger (137 L). When compared to the remains of other ceramic kilns at Mound I, apart from Feature 109, Feature 68 appears to have been more or less the standard size for the kilns at the site (Range 27.61-78.5 L; Mean 50.41 L).

The descriptive statistics of Feature 68 indicate that a high proportion of its contents consist of *Prosopis* sp. charcoal (57% of the feature). Of that, 1147.13 g/ml is accounted for by mature *Prosopis* sp. charcoal, about 43% of the total amount of *Prosopis* sp. and 25% of the overall feature. In this case, 'mature' signifies that the material probably came from larger branches (greater than 5 cm diameter), based on the visibility of the arc of the total diameter in the specimen. While no quantitative measure was made, no individual piece indicated that these mature specimens came from branches any larger than 10.0cm in diameter. One sample (S01-M159) that was removed from this kiln as a single piece of *Prosopis* sp. large branch charcoal weighed 191g. Similarly, other large pieces of charcoal or pieces that could be refit together indicate that a few, large, single branches were burned in the kiln. Along with these large branches, however, we also recovered wood fungal spores (n=85) and noted the presence of insect activity on many of the charcoal pieces recovered. Again, this indicates that dead or decaying branch wood were the main sources of fuel used at this part of the site, as seen at Mound II. These qualities are the case for both the mature and immature wood.

Another marked difference between the fuels used between Mounds I and II is that some pieces of larger branch wood, category mature, from Feature 131, demonstrate adze marks, potentially indicating that the wood was cut from a tree. Feature 131 is a kiln that is located in Room 30, not where the majority of the kilns are found. Instead it is a room attached to others, Rooms 22 and 23, where we know that clay preparation and mold use, if not production, was taking place. Additionally, this feature may pertain to a later phase of use at the site during the Middle Sicán, making it a potentially distinct fire use context. Regardless, adze marks are very rare in the assemblage. For instance, the only other place where this occurs is in the desiccated wood recovered from Mound II in Feature 11, the site of charcoal charge preparation or for small-scale soft metal ingot work in Room 3. The instance of finding chips or adze marks demonstrates that some tree cutting and woodworking was taking place; however, the fact that these remains are rare within the overall assemblage indicates that cutting was probably not the primary means of obtaining fuel wood.

Returning to our type kiln, Feature 68, in the southeast corner of Room 39, a suite of other fuels, apart from *Prosopis* sp., was recovered, including branchwood from *Acacia* sp., *Capparis ovalifolia*, and *Capparis angulata*, as well as more mature wood from *Capparis angulata*. Again, the designation mature wood in this case simply means that wood from branches that are larger than 5 cm are present in the assemblage. Feature 68 is distinct from the other kilns (Table 7.3), however, in its large proportion of *Prosopis* sp., the main constituent of the remains. If, as suspected, the kilns were being systematically cleaned of their residual charcoal, a natural by-product of reduction firing, and *Prosopis* sp. was a main constituent, then I expect that *Prosopis* sp. would be disproportionately underrepresented in the cleared kilns, as is the case at Mound I. In the kilns other than Feature 68, the main constituents are *Llama* sp. dung, *Acacia* sp., and *Capparis* spp. woods. All of these materials are also represented in the unclean Feature 68 assemblage; however, they are present in a lesser proportion to the *Prosopis* sp. wood.

In contrast to the various deposits of charcoal recovered in different parts of Mound II, the ceramic kilns are strikingly clean. In fact, it appears that in the area of Mound I, and in particular Rooms 38 and 39, the floors were particularly clear of charcoal. That said, Figure 7.5 illustrates the overall charcoal-black tinge to the soil, as well as an oxidized red hue to the floor, remnants of both substantial uses of fuel in these areas and the high temperature burns that occurred there. When taken together, the high proportion of charcoal at Mound II and its significantly lesser quantities at Mound I, appear to indicate that charcoal was removed from Mound I and was potentially used or stored elsewhere for use in Mound II or other activities.

Mound I: Preheating Fire Fuels

One of the other ceramic production features to consider in association with the kilns in Mound I is the large, elongated charcoal feature, Feature 97 (Figure 7.6). The feature is a low mound, 5-10 cm high, and consists of a mix of charcoal that spans an area 1.5 x 5 m running east to west in the northwest corner of Room 39. Shimada and Wagner (2002), Shimada and Montenegro (2002), and Goldstein et al. (i.p.) have all described this as a preheating fire for ceramics. Experimental (Shimada et al. 1998; Shimada and Wagner 2002) and ethnoarchaeological work (Arnold 1993; Cleland and Shimada 1998; Sillar 2000) have demonstrated the importance of preheating ceramics prior to placing them in the kiln for firing: this process drives off the final excess water within the clay matrix and reduces thermal shock to the ceramics as the temperature radically spikes within the kiln. Feature 96 corresponds with the characteristics associated with these features, which include elongated, shallow areas of charcoal or embers, where leather-hard, dried ceramics are placed prior to firing. Generally, as seen in Room 39, these features are placed in close proximity to the kilns themselves for easy transport and efficient use of heating. The efficient use includes the ability to place rows of vessels and turn them effectively in the space allowed. All of these are qualities observed during ethnoarchaeological fieldwork in the nearby town of Morropé (Cleland and Shimada 1998).

In Table 7.3, we see that its overall charcoal profile is similar to that of cleaned out ceramic kilns, with a marked absence of larger fuel remains, i.e. mature branch wood. Other fuel remains are more suggestive of a fire that was quickly thrown together from immediately available materials from the Sialupe locality: the major components are small branch wood and twigs of *Acacia* sp. and *Prosopis* sp., combined with *Llama* sp. dung. Type CC wood, possibly from a non-woody perennial herbaceous or euphorb plant, is also present and probably served as a fire starter. Wood anatomical features that include phloem and fiber bundles indicate its likely source as an herbaceous plant or some kind of liana or vine-like plant. As we have seen elsewhere, this constellation of materials probably represents some expedient source of collected fuel, or *poña*, used to help start or maintain the fire in the preheating area. Overall, the morphology, location, and fuel components of this feature points to its likely use as a ceramic preheating fire.

Production at Mound I: Summary

At the time of the site's abandonment, likely attributed to an ENSO event, it appears that the ceramic production season, or activities, was in a period of hiatus; while at Mound II, artisans were preparing the next set of metallurgical furnaces, none of the features recovered from the Mound I area contain uncarbonized material mixed in with plant charcoals. Additionally, the kiln features, save for one, were cleared out of their major charcoal constituents. Both of these indicate that there was no preparation occurring for imminent ceramic production. Finally, there were no super-floor kiln walls present at Mound I, indicating that the kilns were no longer in use. As we saw from our experiments and the case of Feature 68 the kiln walls extend above the floor surface. Feature 68, however, was the only kiln that still maintained significant portions of these walls; most of other features' super-floor walls were removed, with a few indicating failure under conditions of deluge, a potential ENSO event. Although our excavations only permit us to talk about the Mound I and II activities and not the buildings or activities in the surrounding areas, it appears that Mound I kilns were not being prepared for use even though the metal furnaces were actively in production. These ideas, in concert with the arguments for seasonal production of metal and ceramics will be discussed further in Chapter 8. A more detailed comparison of fuels between the kiln and metal working mounds provides additional insight into their integration with one another, and with other activities occurring at the workshops.

Comparison of Fuels in Kilns and Metal Working: Non-Wood Remains

Table 7.4 compares non-wood components of the major constituent features of the Mound I and II production areas and their associated floors. Identifying floor background 'noise' permits us to extract them from the feature analysis to locate particular plant remains and provides insight into the different production processes. Examining the materials recovered from the features relative to their attendant floor permits us to qualify the kinds of activities that took place at specific parts of the workshop. Although the floors from the Middle Sicán contexts were generally clean from major charcoal and most botanical or carbonized materials, a few taxa appear on both the floor and in individual features. As mentioned in Chapter 6, *Lippia* sp. seeds and fruits (Verbenaceae) are present in all of the contexts at the site, including the

Family	Determination	Feature Type	Mound I			Mound II				Total Density #/ml (Adjusted X Sample Size)	Taxon as Pct. of Total Feature Assemblage
		Organ/ Type	All Kilns	Preheating Fire (Fea. 96)	All Floors	Porron Contents (Fea. 8)	Charcoal Hearth (Fea. 9A&B)	All Old Porrons	All Floors		
Verbenaceae	*Lippia* sp.	Seed	7263.18	13176.67	443.85	33.17	22.74	104.17	62.00	21105.79	66.13%
Fabaceae	*Prosopis* spp.	Seed	550.56	9.49	75.19	294.08	12.57	252.78	360.02	1554.67	4.87%
Unid	Wood Fungi	Spore	411.03		26.23			909.96		1347.21	4.22%
Caviinae	*Cavia* sp.	Dung	100.22	15.00	2.70	126.95	22.48	467.21		734.56	2.30%
Shell	Shell 4mm	Shell	291.67	6.67		41.28	164.45	149.12	5.56	658.74	2.06%
Verbenaceae	*Lippia* sp.	Fruit	153.73	218.33	3.98	81.44		8.48	2.78	468.75	1.47%
Poaceae	*Zea mays*	Fruit	111.22		3.36	104.51	33.51	97.45	97.22	447.27	1.40%
Fabaceae	cf. Fabaceae	Seed	293.52	6.16	24.28	12.47	14.66	73.19	15.67	439.95	1.38%
Unid	Unid.	Seed	95.18	8.33	8.11	61.50	34.49	118.82	38.89	365.32	1.14%
Shell	*Donax* sp.	Shell	137.50		5.62	90.45	28.10	39.62	50.00	351.29	1.10%
Fabaceae	*Prosopis* spp.	Leaf	77.24	36.67	16.22	5.26	31.52	91.29		258.19	0.81%
Chenopodiaceae	*Chenopodium* sp.	Seed	150.75	16.67	29.81	13.71	1.65	17.31	27.78	257.68	0.81%
Boraginaceae	*Heliotropum* sp.	Seed	183.49	21.67	9.82	2.29		12.34	5.56	235.16	0.74%
Poaceae	*Zea mays*	Seed	104.63	3.33	15.82	16.69	15.87	14.62	48.11	219.08	0.69%
Malvaceae	*Gossypium* sp.	Seed	15.45			13.71	25.42	153.31	11.11	219.01	0.69%
Asteraceae	*Bidens* sp.	Seed						201.44		201.44	0.63%
Metal	Slag	Metal	56.16	103.33	12.53		4.19	22.85		199.06	0.62%
Fabaceae	*Acacia* sp.	Seed	88.34	6.16	13.51	50.98		39.12		198.11	0.62%
Fabaceae	*Prosopis* spp.	Fruit	26.28		1.64	25.14	3.31	131.64		188.00	0.59%
Fabaceae	cf. Fabaceae	Fruit	31.18	13.33				97.90	22.22	164.64	0.52%
Total			10141.33	13641.82	692.67	973.64	414.95	3002.60	746.92	29613.91	92.79%

Table 7.4 Some Non-Wood Components of Production Features (Density of Taxon Counts #/ml)

floors; a weed common to the human landscape, it is often used in households as an herb. Additionally, in experimental reconstructions of blackware firing performed by José Sosa (Shimada and Wagner 2000) and replication experiments of Formative Period kilns by the SAP by Victor Chang (Shimada et al. 1998) the potters used dry herbaceous plants and tree leaves, such as *Lippia* sp. as supplementary tinder, *poña,* to aid hardwood to catch fire.

At the site of Huaca Sialupe, the ubiquitous incidence of *Lippia* sp. as carbonized remains is most likely related to the expedient use of *poña* across Mounds I and II to start combustion. A common modern component of *poña*, *Lippia* sp. appears in disparate concentrations between Mounds I and II: much higher concentrations (i.e. densities) are found in the ceramic production of Mound I, than in the metal production of Mound II, where they occur in the metal working features almost as background noise. *Lippia* sp. composes up to 60% of the floor assemblage in Mound I and accounts for 12% of the diversity; at Mound II, it composes only up to 5% of the floors at Mound II, accounting for just over 5% of the diversity. This uneven distribution probably represents the distinct use of *poña* at the two mounds. Higher densities of *Lippia* sp. in both floor and feature contexts of Mound I point to a constant influx of this resource and its intensive use in ceramic production.

The relatively low levels of *Lippia* sp. at Mound II may reflect a distinct fuel strategy for metalworking. Lippia sp. seeds are small, recovered in the 0.425mm screen fraction, and their fruits are easily shattered and fractured when dry,especially when carbonized. As a result, the seeds and fruit do not transport well, especially as charcoal. Their presence in floor and features at Mound I likely demonstrates their use predominately in this area as a fuel starter. They end up on the floor during the post-firing disruption of fuel loads as charcoal is conserved and moved to Mound II. The low densities of *Lippia* sp. at Mound II indicates that only the larger charcoal pieces are surviving the trip across the site, which corresponds with the metalworkers' primary interest in energy-rich fuel. Additionally, the initiation of fire in the metal furnaces would not necessarily require the same kind of kindling initiation as used in starting the kilns, but rather the insertion of externally started hot coals to the charcoal load. Potentially, the movement of cleared charcoal from Mound I to Mound II accounts for the carbonized remains of *Lippia* sp. at Mound II. A second line of evidence corroborates this hypothesis: elevated densities of *Lippia* sp. fruits occur in the kilns, but are near non-existent on Mound II floors, appearing only occasionally in particular metal working features. If weed seeds are indicators of charcoal movement, then this may explain the similar pattern that is visible for seeds and fruits from other herbaceous plants in the ceramic workshop, e.g., *Heliotropum* sp. and *Chenopodium* sp. These plants, like *Lippia* sp. are disturbance-related taxa that are common in and around human settlement and herd animals, and most likely occur at the site as *poña.*

The patterning of other non-wood remains across the site, however, is not well explained in terms of the movement of charcoal residues from one mound to the other. A number of food remains all have distinct distributions, e.g., *Z. mays* seeds and fruits (cobs), seeds from *Capsicum* sp., seeds from *Lagenaria* sp. and other

Organ /Type	Item Description	Mound I					Mound II				Total Density #/ml (Adjusted X Sample Size)	Taxon as Pct. of Total Asmblg.
		All Communal Hearths	All Hearths	All Kilns	Old Porron Fea. 103	Preheating Fire Fea. 96	Charcoal Hearth Fea. 9-'01 A&B	All Hearth s	All Old Porrons	Porron Fea. 8-'01		
Fruit	**12-row**		3.13	11.77					2.50	2.63	20.03	2.28%
	16-row			10.49							10.49	1.19%
	Cob						33.51	43.48	78.07	2.29	157.34	17.91%
	Cupule		14.95	34.82					14.10	99.59	163.46	18.60%
	Glume		15.63	54.14							69.77	7.94%
Seed	**Embryo**	7.21	2.56	15.41		3.33	1.65	21.74		2.29	54.19	6.17%
	Flour			1.69	38.34				12.12	2.63	54.78	6.23%
	Mature	19.36		87.53			14.22	213.16	2.50	11.77	348.55	39.67%
Grand Total		26.56	36.26	215.85	38.34	3.33	49.38	278.38	109.29	121.20	878.60	100.00%

Table 7.5 *Z. mays* Remains by Mound and Feature Type (Density of Taxon Counts #/ml)

Cucurbitaceae, *Gossypium* sp. seeds, and *Erythoxylon coca* seeds. All of these materials and their presence within the fuel remains indicate that there is a connection between the broader domestic economy, i.e. subsistence activities associated with households probably established at Mounds III-V, and the production activities at Mounds I and II (Shimada and Montenegro 2002). Additionally, the close proximity of daily life processes of provisioning, consumption, and refuse deposition added to the potential for natural taphonomic processes, e.g., wind and rodent movement, to potentially bias the presence of these materials in the fuel assemblages.

Table 7.5 examines *Z. mays* remains from across the site, broken down by feature type. Here we see that the remains from maize cobs are present in a variety of contexts, and in particular, both the metalworking furnaces and ceramic kilns. Additionally, they also appear in contexts labeled as Hearths and as Old Porróns. In this case it appears to be significant that the one feature where they do not appear is in those called 'communal hearths,' discussed below. *Z. mays* seeds, on the other hand, appear in every context at the site, indicating that there is a general use of corn associated with the site. This ubiquitous distribution raises the question as to why so much food refuse, the product of domestic activity, is present across a site primarily associated with the workshop production of high status goods. The meanings and implications of these materials are discussed below.

Domestic Garbage and Fuel

There are some remains, e.g., *Z. mays*, *Capsicum* sp., and members of the Family Cucurbitaceae, that demonstrate deliberate domestic activity, including food serving and preparation garbage as fuel in several of the ceramic and metalworking contexts at Huaca Sialupe. The seeds of *Capsicum* sp. (chili peppers), for instance, are found in kilns, metal furnaces, old porrón features, and the floors at Mound II. Their presence indicates that there is likely a specific relationship between domestic garbage and fuel use. *Capsicum* sp. peppers contain high concentrations of capsaicin, the substance that makes peppers 'hot,' in their seeds, which are usually removed for discard not to be included in the served preparations. Their appearance, along with other potential food discard remains, e.g., *Cucurbita maxima*, *Cucurbita* sp. materials, and *Z. mays* cobs, indicates that food preparation discards were being used or reserved for fuel.

There are, among others, two potential alternate explanations. The first relates to the use of *Cavia* sp. and *Llama* sp. dung for fuel. Hastorf and Wright (1998) and Miller and Smart (1984) argue that animal dung can be a vector for seeds from fodder plants, e.g., Family Poaceae, Chenopodiaceae, Verbenaceae, and Boraginaceae; however, for remains of plants used for human consumption, their introduction to fuel use contexts is less certain. Camelids do not tend to eat human domestic refuse. *Cavia* sp. traditionally survives in modern and historic kitchens as a domestic animal that feed on waste from kitchen activities. It would be one potential vector for human-food-related material to enter the fuel contexts at the workshop. Corn cobs, seeds from Cucurbitaceae and *Capsicum* sp. seeds, however, are generally ground up by these animals and they are reduced to a fine pulpy mass resembling particle board. *Lippia* sp. seeds were the only recognizable materials recovered from dung during the analysis of the Huaca Sialupe remains; they were encountered very rarely, and may have contributed to the presence of *Lippia* sp. in the fuel contexts at the site. Domestic food production refuse, however, was not recorded.

A second possible vector for some of these domestic garbage materials relating to domestic kitchen refuse could be related to collecting *poña* from areas that may have been a common refuse dump for the overall settlement area, Sialupe Mounds I-V. Seed and fodder materials may enter the industrial contexts through either primary or secondary deposition of domestic garbage as fuel. While the presence of human domestic refuse is clear in the workshop fuel contexts, it is unclear, due to the probably movement of charcoal across the site between ceramic and metalwork areas, where they first entered the site. In the end, it is unclear what kinds of food production took place in association with the smaller hearths at the site, yet the presence of seeds from *Capsicum* sp. and *Lagenaria* sp., both indicate that people processing these materials for domestic use were

Family	Determination	Feature Location	Mound I					Mound II		Total Density g/ml	Taxon as Pct. of All Hearth Features
			Room 23	Room 30	Room 39			Room 3			
		Feature #	83	63	49	122	123	11	16		
Fabaceae	*Prosopis* spp.	Branchwood		4.77	413.74	353.38	275.94	8.22	15.57	1071.62	37.44%
Fabaceae	*Acacia macracantha*	Branchwood				405.47	311.59		38.91	755.98	24.73%
Ceramic	Ceramic Redware					74.03				74.03	2.34%
Capparidaceae	*Capparis angulata*	Branchwood					42.32			42.32	1.44%
Fabaceae	cf. Fabaceae	Juvenile			4.36	17.59			2.29	24.24	0.80%
Fabaceae	cf. Fabaceae	(Desic.) Juv.						6.47	2.75	9.22	0.31%
Apocynaceae	*Vallesia glabra*	Juvenile			0.79	7.13				7.92	0.26%
Dung	*Llama* sp.				3.13			3.79	0.54	7.46	0.25%
Ceramic	Ceramic Blackware				3.38					3.38	0.13%
Fabaceae	*Prosopis* spp.	(Desic.) Chip						3.15		3.15	0.09%
Asteraceae	*Tessaria integrifolia*	Mature			0.87			1.74		2.61	0.07%
Poaceae	Wood Type FF	Stem				2.31				2.31	0.07%
Shell	*Donax* sp.				1.44					1.44	0.06%
Capparidaceae	*Capparis ovalifolia*	Branchwood						1.17		1.17	0.03%
Poaceae	Zea mays	Stem							0.46	0.46	0.02%
Unid	4mm Charcoal	Branchwood			33.36	495.16	274.78	34.16	29.90	867.35	28.33%
Unid	2mm Charcoal	Branchwood	0.24	1.04	5.64	35.63	61.16	1.68	3.72	109.11	3.64%
Grand Total			0.24	5.80	466.72	1390.69	965.80	60.39	94.13	2983.78	100.00%

Table 7.6 Fuel Contents of Hearth Features Mounds I and II (Density of Weighed Charocal g/ml)

either performing those tasks at Sialupe Mounds I and II, or in close association with it at Sialupe Mounds III-V.

Other Forms of Production at Huaca Sialupe

Shimada (2000) and Shimada and Montenegro's (2002) site reports make it clear that food consumption was taking place at the site: animal remains reviewed by Melody Shimada, indicate that terrestrial resources, e.g., *Canis* sp., *Llama* sp., and *Cavia* sp, and marine resources *Donax* sp., as well as some fish species were all consumed in close proximity or at the workshops during the Middle Sicán period, especially in Rooms 38 and 39. The *Llama* sp. bones are mostly from the distal ends of the long bones and feet; dog (*Canis* sp.) bones have gnaw marks and were broken open for marrow extraction. Based on the cuts of meat represented, Shimada interprets these data to indicate that meat protein sources were not of the highest quality at Sialupe. The extraction of marrow was interpreted by Melody Shimada as an indicator of taking maximum advantage of scarce protein resources.

Clearly the inhabitants of Huaca Sialupe consumed meat at the site, and what was consumed was not the best quality available. The presence of a restricted set of camelid and dog remains at Mound I strengthens the idea that cooked food was consumed at the Mound I workshop. Animal parts were divided by cut off-site, and brought to the site cooked or were cooked and eaten at Sialupe. Below I examine other fire-use features present in the workshops that I suggest were used in food production activities at the site. With the plant and animal remains from the site considered together, i.e. the consumption of cuts of meat, not whole animals, and the remains of non-consumed plant parts, it appears that cooking took place in close proximity to the Huaca Sialupe workshop. Ultimately, as only insignificant bone remains were recovered from the fire-use features at the site, it appears that the organic garbage was separated from animal bone waste. The organic materials, separate from bone were used as a fuel resource. Garbage that contained animal wastes, bone, was generally not used as a fuel resource. This observation is important as I try to reconstruct fuel resources. It demonstrates that garbage, houschold or kitchen waste, was potentially set aside and divided in ways that prevented animal bone from entering fuel contexts.

Hearths in the Huaca Sialupe Workshop

Hearths constitute another set of fire-use features systematically sampled at Sialupe. These are mostly ephemeral features without solid enclosing structures. The location and morphology of these features leads us to believe that they were ancillary to industrial production and related to the domestic economy of the site. Hearths are small, ca. 15 cm in diameter, and rarely larger than 30 cm. Burning tended to be complete, leaving a shallow

Family	Determination	Feature Loc.	Room 38			Room 39			Room 23		Room 30		Room 35		Total Density g/ml	Taxon as Pc of All Comm Hearth
		Feature #	106	98B	100	97	99B	Floor	83	Floor	63	Floor	84A	84B		
		Feature Type	Comm. Hearth	Ash Pit	Ash Pit	Comm. Hearth	Ash Pit		Hearth		Hearth		Ash Pit	Ash Pit		
Fabaceae	*Prosopis* spp.	Branchwood	104.26	137.15	16.50	5.72	29.54	7.76		5.27	4.77	28.25	1.92	2.24	1386.45	30.87
Fabaceae	*Acacia macracantha*	Branchwood	36.90		1.83		10.54	3.54		1.03		15.31		12.46	798.68	17.59
Asteraceae	*Tessaria integrifolia*	Mature	126.44	0.72	5.49	0.90		0.68		1.00		0.27		0.62	136.99	4.00%
Dung	*Llama* sp.		1.73	0.68	41.07	14.74		5.86				2.31	0.96	20.62	91.11	2.39%
Ceramic	Ceramic Redware														74.03	1.52%
Apocynaceae	*Vallesia glabra*	Juvenile	45.36									0.57			53.86	1.19%
Fabaceae	cf. Fabaceae	Juvenile	13.22			0.63		11.14					2.03	2.68	51.63	1.13%
Capparidaceae	*Capparis angulata*	Branchwood	6.47												48.79	1.13%
Ceramic	Burnt Clay			12.09											12.09	0.18%
Capparidaceae	*Capparis ovalifolia*		8.41									3.33			11.74	0.31%
Ceramic	Ceramic Blackware				2.31							1.05			6.74	0.22%
Metal	Scoria	Amorphous	5.95												5.95	0.13%
Fabaceae	cf. Fabaceae	(Desic.) Juv.	1.14				1.32							1.46	3.93	0.09%
Unid	Wood Type FF														2.31	0.05%
Fabaceae	cf. Fabaceae	Branchwood			1.85										1.85	0.05%
Shell	*Donax* sp.														1.44	0.04%
Poaceae	*Zea mays*				0.39										0.39	0.01%
Dung	Mammal											0.15			0.15	0.01%
Poaceae	*Gynerium* sp.													0.11	0.11	0.00%
Unid	2mm Charcoal	Branchwood	197.18	33.15	28.82		14.01	1.16	0.24	2.40	1.04	14.94			395.37	10.77
Unid	4mm Charcoal	Branchwood	366.18	51.06				5.24		9.73		18.67			1254.19	28.32
Grand Total			913.25	234.85	98.26	21.99	55.42	35.38	0.24	19.43	5.80	84.85	4.91	40.19	4337.79	100.0

Table 7.7 Cross Feature Comparison of Potential Cooking Fuel Contents (Density of Weighed Charocal g/ml)

ashy lens and a visibly smaller quantity of charcoal when compared with the production features. Most hearths are ashy, 'dirty,' with ochre colored earth around them, demonstrating that they have not been cleared; as a result, some of these features were classified as Ash Pits in the field. Based on these characteristics, their use is not immediately clear. It is likely that craftspersons constructed them for other uses like those associated with the production process, e.g., small preheating fires, or others that are not yet clearly defined at this time.

In general the paleoethnobotanical data indicate that the materials used for burning in the hearth features depended less on specified combinations of wood fuel components, and are more diverse overall in terms of the taxa used than in the craft production contexts: pretty much any combustible was used as fuel. This pattern may indicate that organic garbage was being burned as an expedient way to deal with garbage disposal. Table 7.6 demonstrates that the overall diversity of the hearth fuel components draws from the same universe of materials as the craft production features; their overall fuel assemblages, however, are more limited in what taxa are combined, with only one or two making up the bulk of the fuel used. As is the case with most of the site, *Prosopis* sp. and *Acacia* sp. branch wood compose the majority of the fuels in these hearths. Also the overall quantities of fuel materials are relatively lower in these features than in the others at the site. These two trends indicate that burning or fire use in these hearths was limited in comparison to the other workshop features that were both more substantial in fuel load recovered and the associated feature superstructure.

The hearth features that occur on the Mound I side of the site are predominantly associated with Room 38 (Figure 7.4). Recall that Room 38 is separated from Room 39, where the majority of the kilns are located, by a low partition wall, probably the remnants of a quincha separator (Wall 59). The partition divided this exterior sector of the site in terms of activity, with kilns on one side and a series of small hearths on the other. The contents of some of these hearths included materials that indicate the area was a locus for food preparation. For instance, Feature 105, Room 38, contained the remains of a deposit of dried seaweed and a large ashy residue. The seaweed in this case is similar to the *cochayuyo* that is commonly gathered and sold in the modern markets of Peru as a food additive (Masuda 1985). I suggest that some of these Ash Pit and small Hearth features may relate to expedient food production for the people working at the site. Unfortunately, the patterns within the fuels used and the diversity in morphology makes further distinctions about hearths, in general, and their use, difficult. In particular, as I illustrate below, the larger of these Hearth features, where hearth features are

Family	Determination	Organ/ Type	Feature 106 Room 38 Communal Hearth est. Vol. 39.9L	Feature 97 Room 39 Communal Hearth est. Vol. 27.00L	Total Density #/ml (Adjusted X Sample Size)	Taxon as Pct. of Total Feature Assemblage
Shell	Shell 4mm	Shell	780.37	Missing Data	780.37	37.69%
Verbenaceae	*Lippia* sp.	Seed	38.60	250.68	289.28	13.97%
Fabaceae	*Acacia* sp.	Seed	59.30	158.04	217.34	10.50%
Fabaceae	cf. Fabaceae	Seed	65.36	95.37	160.73	7.76%
Fabaceae	*Prosopis* spp.	Seed	147.91		147.91	7.14%
Unid	Wood Fungi	Spore	117.94	8.17	126.11	6.09%
Chenopodiaceae	*Chenopodium* sp.	Seed	23.95	27.25	51.20	2.47%
Unid	Unid.	Fruit	23.30	19.07	42.37	2.05%
Solanaceae	*Nicotiana* sp.	Seed	34.38		34.38	1.66%
Poaceae	*Zea mays*	Seed	10.21	16.35	26.56	1.28%
Caviinae	*Cavia* sp.	Dung	9.86	10.90	20.76	1.00%
Unid	Unid.	Seed	8.64	10.90	19.54	0.94%
Fabaceae	UKN#64-120.1	Fruit	17.39		17.39	0.84%
Unid	Unid.	Dung	15.00		15.00	0.72%
Boraginaceae	*Heliotropum* sp.	Seed	6.81	5.45	12.26	0.59%
Malvaceae	*Gossypium* sp.	Seed	12.04		12.04	0.58%
Fabaceae	*Acacia* sp.	Flower	8.70	2.72	11.42	0.55%
Euphorbiaceae	*Euphorbia* sp.	Seed	9.62		9.62	0.46%
Bone	Mammal	Bone	9.02		9.02	0.44%
Poaceae	*Bromus* sp.	Seed	8.70		8.70	0.42%
Ceramic	Ceramic Redware	Ceramic	8.61		8.61	0.42%
Erythoxylaceae	*Erythoxylon coca*	Seed	5.91		5.91	0.29%
Fabaceae	UKN#12-177	Seed	5.00		5.00	0.24%
Oxalidaceae	*Oxalis* sp.	Seed	1.41	2.72	4.13	0.20%
Solanaceae	cf. Solanaceae	Seed	3.01		3.01	0.15%
Lithic	Lithic		3.01		3.01	0.15%
Malvaceae	cf. Malvaceae	Seed	2.90		2.90	0.14%
Cyperaceae	*Scirpus* sp.	Seed	2.90		2.90	0.14%
Metal	Slag	Metal	2.90		2.90	0.14%
Fabaceae	*Phaseolus* sp.	Seed	2.82		2.82	0.14%
Shell	*Donax* sp.	Shell		2.72	2.72	0.13%
Plantaginaceae	*Plantago* sp.	Seed		2.72	2.72	0.13%
Malvaceae	*Sida* sp.	Seed		2.72	2.72	0.13%
Poaceae	*Panicum* sp.	Seed	2.50		2.50	0.12%
Unid	UKN#13-177	Seed	2.50		2.50	0.12%
Amaranthaceae	*Amaranthus* sp.	Seed	1.41		1.41	0.07%
Bone	Fish	Bone	1.41		1.41	0.07%
Verbenaceae	*Lippia* sp.	Fruit	1.41		1.41	0.07%
Grand Total			1454.79	615.80	2070.59	100.00%

Table 7.8 Non-Wood Contents of Communal Hearth Features (Density of Taxon Counts #/ml)

aggregated in certain rooms, provide clearer indications of the reasons for and manner of food preparation that took place at Mound I.

Communal Hearth Features at Mound I

Regardless of the array of small Hearth or Ash Pit features, the largest and most significant cooking features that are real indicators of the integration of the different activities at the site is the large area of burning situated along the northwestern corners of Room 38 and 39, Features 106 and 97, respectively. These fires often burned to ash that resulted in the accumulation of an ash lens in these two areas. The formation of these large features contrast to the scattered single use or single occurrence individual hearths that occur throughout Rooms 38 and 39 (e.g. 100, 103, 105, 107) and likely represent the remains associated with communal cooking. Based on ethnographic analogies of fermented beverage production on the North Coast of Peru in the modern (Camino 1987; Schaedel 1988) and Prehispanic periods (Goldstein et al. i.p.; Moore 1989; Shimada 1994a), I argue here that these areas are loci of the production of corn-based fermented beverages.

Features 106 and 97 represent Communal Hearth cooking activities that resulted from a complex formation process of the aggregation of 10-30 cm diameter cooking fires (Figure 7.6). Their fuel composition is similar to the fuels used in the kilns of Room 39. Macrobotanically, they share some production features described above, but also differ in significant ways (Table 7.7). Unlike in other hearths and fire-use features at Mound I, *Prosopis* sp., although present, does not dominate the assemblage of either Feature 106 or 97; instead, each contains a diverse composition of taxa. In Feature 106, *Acacia macracanatha* (36.90 g/ml: 15.63% of the total taxa present) and *Tessaria integrifolia* (126.44 g/ml: 9.38% of the total taxa present) are the dominant identified materials, Whereas in Feature 97, *Llama* sp. dung (14.74 g/ml: 40% of the total taxa present) is the dominant fuel feature. *Z. mays* cobs are absent (Table 7.5), which may be due to the reservation of *Z. mays* cobs for use in the kilns and organic material hoarding areas of Mound II.

Qualitatively, Feature 97 is more ephemeral than Feature 106 and clean of residual fuel materials, possibly indicating that its charcoal was removed along with other left over kiln fuels from the same room. Yet, Feature 97 did have a large number of gastropods recovered within it, indicating, perhaps, repeated use for production. With no kilns in Room 38, it may have been that 'cleaning' the fuel materials from Feature 106 was less of a priority, there being no similar trend to remove fuel from this area; thus, Feature 106 had a larger overall fuel load remaining when excavated. Additionally, the base of the adobe wall was discolored from firing, indicating that it was used later in time than Feature 97. This pattern appears to indicate that the role of these Communal Hearths at the site shifted during the use of the Mound I as locus for reduction firing. At abandonment, ceramic reduction kilns (Features 113 and 117) appear superimposed over Feature 97 in Room 39. Feature 106, however, with a full fuel load, was probably used up to the end of occupation.

Figure 7.6 Feature 97, Communal Hearth Feature

Wall 56 was likely constructed as a barrier between Rooms 38 and 39, which may reflect that the intensification of reduction kilns required the separation of food production from the workshop contexts, leaving Room 39 exclusively for ceramic firing. An analysis of non-wood materials sheds more light on the integrated role these features play in the craft production workshop.

Communal Hearths: Non-Wood Materials

Non-wood materials present in Features 106 and 97 (Table 7.8) highlight activities related to domestic production occurring at Mound I. *Donax* sp. shell composes the major element of non-wood material recovered, particularly in Feature 106. In the case ofFeature 97, a single portion of this feature containing the *Donax* sp. shell was removed separately for analysis; different from the case of Feature 106 where the entire feature was sampled. Despite a difference in sampling the separately removed portion of Feature 97 included several hundred *Donax* sp. shells, the quantity and sample are still under evaluation. These shells are found littering most of the archaeological sites in the area, and appear to have been a common food source in antiquity. *Donax* sp. are bivalves that are gathered along the coastline and are steamed or boiled to be eaten as communal/social food during group gatherings. The pronounced presence of the *Donax* sp. shells clearly points to food consumption: the shells were largely intact, neither calcined nor crushed, which might otherwise implicate their use in ceramic production as temper.

Several of the non-wood materials recovered from Features 97 and 106 (Table 7.4) indicate that apart from fuel resources, some non-ceramic and non-metalworking economic activities were also being shared at these locations. Their presence gives insight into the probability that cooking was associated with activities like spinning and cleaning cotton. Cotton seeds, *Gossypium* sp., representing the possible cleaning residues of cotton boles are present in the Communal Hearths and kiln features at Mound I, as well as the metal furnaces, and charcoal mixing and production features at Mound II. This indicates that detritus associated with other activities, e.g., the cleaning, spinning and production of cotton textiles, were finding their way, not just into the workshop as a whole, but more specifically into the fuel load itself. This occurrence may simply be due to a generalized pattern of these activities across the site. The fact that these seeds are heavy, and large, and not often wind dispersed indicates that their deposition at the site was likely due to activities taking place at the site or in very close proximity.

Erythoxylon coca seeds appear in the ceramic and metalworking portions of the site. They only occur at Mound II, however, in the charged furnace (n=2). They occur with higher density and frequency in the Mound I kilns (n=3), and hearths (n=5) associated with that side of the site. In particular, they appear far more frequently and with greater density in the Communal Hearth (n=2), Ash Pit (n=2) and, Hearth (n=1) features. Overall, there are not that many *coca* seeds associated, normally, with chewing *Erythoxylon coca* as only the leaves are cut from the tree. Only occasionally, in the modern context, does someone come across seeds when chewing a bag of *coca*. So in fact, to find them at all at the site is quite remarkable, yet they do demonstrate that *coca* chewing was a part of domestic and production activities in the workshop.

Ethnographically, spinning and coca chewing are activities that occur within a variety of contexts related to domestic activities that would be easily recognized by social scientists like Folbre (1986), Halperin (1994), and Moore (1992) as household related (Allen 2002; Schaedel 1988). The appearance of cotton and *coca* seeds may serve as indicators for the presence of other kinds of activities on site at Huaca Sialupe, that are attached to the actual metal and ceramic crafting activities to some degree. The recovery of a complete backstrap loom and associated spinning equipment from a burial placed near a corner of Room 39, the central ceramic kiln area, provides an additional line of evidence supporting the practice of weaving at Mound I (Shimada and Montenegro 2003). Both *coca* chewing and spinning are associated with household activities and, ethnographically, are performed in both private and public social spaces, kitchens and workshops being both (Allen 2002; Gillin 1947; Schaedel 1988).

In the Communal Hearth features, cotton and *coca* both indicate that there are continuities between household activities and workshop space at Mound I (Table 7.8). Yet, we can see in both cases that the features are very diverse in terms of their contents with a variety of herbaceous, euphorb, and grass vegetation seeds, some food elements, dung from *Cavia* sp. and *Llama* sp., as well as *E. coca* and *Gossypium* sp. seeds. As mentioned above, large quantities of *Donax* sp. shells, a potential indicator of communal food consumption, were recovered in association with the Communal Hearth features. When taken together with the cotton and *coca* seeds, I suggest that they implicate processes of food production that include long periods of eating and potential collective cooking and consumption activities. Goldstein and Shimada (i.p.) describe the cleaning of cotton and the chewing of *coca* as 'waiting' activities associated with long-term cooking and ceramic firing. One aspect of production that both fermented beverage production and the reduction firing of ceramics share is 17 to 24 hours of continuous firing; a fire that needs to be tended giving people time to multi-task. It provides a degree of 'downtime' to spend with other people and work on communal projects, e.g., cleaning of cotton, and perform tasks that are both ritual in nature and foster social solidarity, e.g., chewing *coca* (Allen 2002). Goldstein and Shimada (i.p.) have also argued, based on feature morphology and ethnographic analogy, that these are remains of suprahousehold food production that included the production of fermented-corn beverages. The implications of the presence of these kinds of processes in a workshop context are discussed below.

Large Communal Cooking Features: An Explanation

Ethnographically, the two corner areas, Features 106 and 97 of Rooms 38 and 39, respectively, that we are associating here with suprahousehold production have some precedence. Brüning's (Schaedel 1987) photographs and notes and Gillin's (1973) ethnography captured 19th and 20th century North coastal life quite vividly. In a number of these documents we see commensual activities and other communal affairs, including the placement of large fires in the corners of large rooms, occurring in fenced in or quincha walled outdoor patios attached to adobe block structures, mostly houses. One facet of suprahousehold production is the manufacture of corn beer, previously explored by Moore (1989) and Shimada (1994). Moore (1989) establishes the archaeological correlates of what he calls *chicha* production as grinding stones, straining baskets and cloths, large ceramic vessels, *mate* or gourd bowls, stirring implements, and production residues, *Z. mays* cobs and *alfrecho*, the cooked sediment strained after boiling and before fermenting. Shimada (1994) cites that the most likely evidence for *chicha* making, noted by Bernabe Cobo and Cieza de Leon noted in the colonial era, is a set of complementary, differently shaped ceramic vessels. We know from the ceramic analysis at Huaca Sialupe that the artisans were producing an array of different shaped small bowls and plates, as well as the blackware bottles, and *porrones*. The array of ceramics produced relied on small oxidation and reduction kilns for the production of certain products as well as large pit kilns for firing larger vessels and utilitarian ceramics. Based on the production process of *chicha*, these correlates are reasonable and are likely to be preserved on the arid North Coast of Peru. At Huaca Sialupe we have several inorganic and organic correlates of fermented beverage production during the Middle Sicán: room corner repeated use hearth fires, maize remains, cotton bags and textile production, large porrones, gourd bowls remains, and the complementary differently shaped ceramic vessels.

Gillin and Brüning both discuss the presence of *tavernas* in most extended households. These are separate buildings or attached salons where *chicha* is produced for suprahousehold consumption and some of the photos included their works capture these phenomena. In some cases these *tavernas* are part of or attached to a living space, or they can be buildings or fenced in areas held communally between a series of houses expressly for the use of brewing and drinking (Camino 1987). Both observers indicate that these are typical suprahousehold phenomena, as is the communal cooking of food between households and across neighborhoods. Interesting at Sialupe is not the implication of fermented-beverage production, often used in Andean Archaeology to demonstrate reciprocal labor exchange for alcohol at a state level. Instead, I highlight its cultural association of communal food production and fire use in conjunction with craft production and other technological activities that have complementary fuel needs.

As I consider the residents at Sialupe as a suprahousehold of both craftspersons and their support. The ethnographic reality of this kind of household production is one where the roles, e.g., apprentice, master, cook, spinner, fuel gatherer, may have changed fluidly throughout their lives. From the paleoethnobotanical analysis, the expenditure of labor, fuel, and other resources for food production at a suprahousehold level was similar to that of the other types of craft production associated with the site. Most particularly, the production of blackware bottles appears to have been closely associated with these suprahousehold repeated-use cooking contexts in Rooms 38 and 39. People involved in ceramic production, for instance, used the spent resources, e.g., maize cobs, cotton, and *coca* seeds, from suprahousehold cooking and other activities, and employed the debris of cooking to fire kilns. At least one of the food production processes supplied drink, and probably roast meat, to those involved with ceramic production, ancestor veneration, or metalwork. Similar household contexts with attendant areas of mundane and festal suprahousehold food production areas, including ancestor veneration rites, are reported for the Moche Period site of El Brujo (Gummerman i. p.).

Combining Features and Site Wide Trends

This analysis of the role that food production played at Huaca Sialupe serves to bring the overall analysis to a conclusion and begin the interpretations regarding fuel use at the site. In the next and final chapter we will see that these fuel remains permit not only an ecological, but a holistic cultural understanding of how craft production was a part of the lives of the people working there. Once established, I then reflect on its meaning to the main consumers of the products manufactured at the site.

Although imported resources and symbols of power were held dear by Middle Sicán elites, this study demonstrates that some portion of the actual development of products like blackware bottles and exquisite metalwork was borne on the shoulders of small-scale craft producers. These individuals' actions created the residues of activity areas that, in effect, are palimpsests representative of the technological choices that these artisans made in the Middle Sicán. Both their cultural preferences and their available forest resources circumscribed these actions and their resultant archaeological contexts. Moreover, they were intimately tied to a domestic world that played an active physical presence in the workshop contexts. Drawing from these data, I provide a sketch of the ecological dynamics of Huaca Sialupe during the Middle Sicán; we can also demonstrate how its craftspeople apportioned fuel resources and integrated craft production into the domestic economy at the site.

CHAPTER 8
DRY TROPICAL FOREST USE IN THE MIDDLE SICÁN: CONCLUSIONS AND FUTURE AVENUES

The archaeobotanical materials recovered from excavations at Huaca Sialupe indicate the presence of integrated and allied activities of crafting and domestic economy across the site. Rather than seeing economic activities being controlled from outside, it seems that crafters at Sialupe made rather independent decisions based on their available means along lines theorized by Wolf (1997) and Halperin (1994). The most substantive result of my analysis graphically illustrates that paleoethnobotanical analysis permits ecological and economic reconstruction of many different kinds of specialized activities, beyond crafting alone. Agriculture, food production, social solidarity activities, textile production, and the management of living and renewable resources around the site are all tightly interconnected at Huaca Sialupe. This review of the ecological and social implications of the archaeobotanical materials provides a holistic model in which to interpret previously established datasets, e.g. ceramic production (Shimada and Wagner 2001; Rospigliosi 2007; Taylor 2002), metal production (Bezúr 2003; Shimada et al. i.p.; Shimada and Wagner 2007), and human remains (Klaus 2003), as well as the historical, technological, and regional framework (Shimada and Wagner 2007). From my own endeavors, I have provided a social context for production, developed from Wolf's (1997) kin-based (household) production model; most importantly, I have established an ecological framework for interpreting the different production contexts at the site, developed from Newsom's (1993) ecological and functional approach to evaluating resource quality and availability.

I revisit the two fundamental questions initially driving this research, addressing labor and activity organization and the ecological implications of production. As with all research, addressing questions of production from a more holistic approach has amplified the number and depth of queries that can further the interpretation of data, including issues of seasonality and human perception of the environment, as well as raising certain methodological concerns of ecological reconstruction. I will address these at the end of this chapter, as well as discuss the need to develop complementary datasets that to aid future investigations, paleoethnobotanical and otherwise, in the La Leche River Valley and the Andes. The holistic approach of this study, carried out over multiple years, yielded results that very nicely illustrate the intricacies and intimacies linking subsistence, domestic production, and crafting activities in the archaeological record. Moreover, my work, as presented and executed, demonstrates the critical need for anthropological research of production contexts to include a heavy field component centered on extracting ecological information of the research area. Until the relationships between plants, humans, and their environments are better understood, production contexts will remain only loosely situated within their historical, economic and, most importantly, social contexts.

Question One: Organization and Competition

In the introduction I posed the question, 'From looking at the organic remains from fuel-use areas, what can we discern about how labor was organized around these two industries?' From there, I broke this query into two parts: To what extent can we reconstruct a holistic understanding of organic resource use across the site, and to what extent did ceramic production and metalworking complement or compete with each other in regard to the critical fuel supply and use? In this section I begin with the implications of the archaeobotanical data on labor and activity relationships and finish with an assessment of fuel use competition between the different industries at the site. Although some previous studies have evaluated the political economy of Sicán from consumption-based contexts, the investigation of crafting at Huaca Sialupe provides a window into its production aspects, as well as demonstrates how it articulates with local domestic economies.

The archaeobotanical evidence demonstrates that ceramic and metal production drew upon the same pool of resources at Huaca Sialupe. Artisans employed a variety of taxa to meet their needs, looking to both the dry forest, as well as local industries, such as camelid herding and agriculture, to supplement wood fuels. Juvenile branchwood of *Prosopis* sp. and *Acacia* sp. dominate the assemblages of both. The presence of fruits, flowers, leaves and stems from dry tropical forest species, as well as agricultural remains, however, indicates that the workshop shared fuels with activities tied to the site's wider economy. The presence of garbage and herbaceous disturbance taxa in the fire-use features further accentuates that fuel resource pooling, from domestic to production contexts, was critical to production in general at the site. A general evaluation of the fuel resources points to a catchment area that included dry tropical forest, agricultural fields, and household production contexts.

The close examination of the archaeobotanical materials has provided a more complex picture of metal and ceramic production loci at Huaca Sialupe. For example, the specialized nature of fuels and features in Room 3 of Mound II must be revisited to evaluate the potential role of fine metalworking and foundry work at the site. Additionally, the location of potential brewing and cooking facilities, adjacent to the ceramic production areas, has firmly identified domestic activity within the architectural footprint of the workshop. Establishing these activities reinforces the multi-family or household basis of production, perhaps along the lines of the kin-based production seen regionally in the Early Colonial period by Netherly (1977) and Ramírez (1996). Lastly, although camelid and *Cavia* sp. dung was noted as present during excavation, closer scrutiny of the fuel remains indicates that herding, or at least camelid raising, played a critical role to the productive capacities of the site: the prevalence of *Cavia* sp. and camelid dung across the site reflects a tight interconnection between the domestic economy and workshop production in the

Middle Sicán. The interrelatedness of different economic activities at the site reinforces the patterns of interconnected resource use that are reflected in the archaeological record at Sialupe. Clearly, the intensive investigation of the fuel remains provides insight for economic reconstruction that goes beyond an evaluation of ceramic and metal production alone.

The interplay between domestic and political economy, apparent in the fuel remains at Sialupe, reinforces a model of autonomous production at the site. There is very little to indicate that much overdetermination from an Earle-styled (1997) wealth or staple finance economy was functioning at Sialupe (Shimada and Wagner 2001, 2007; Taylor 2002). Instead, fuel use remains suggest that residents had a sophisticated system at work for using and reusing organic materials. In particular, the domestic economy was centered on renewable resources, camelid dung, managed tree species, disturbance weeds, and agricultural production. All of these materials, once identified in the archaeological record, permit a detailed analysis and reconstruction of household economy and local human environment interaction. From this perspective Wolf's (1997) ideas regarding household based kin structured production models better illustrate the normative processes that took place at the site. For instance, we see that almost all aspects of food production, from seed to stalk to table, are represented in these remains. We also see that people took full advantage of the resources around them to satisfy their needs, exploiting some wild foods, like *Cordia* sp. fruit. Although previous holistic analyses of the production contexts at Huaca Sialupe implicated the presence of household and domestic processes at the site, analysis of the archaeobotanical remains illustrates their high degree of integration into a wide variety of economic processes, elucidating the ways that different activities articulate. Yet, it remains to be see how, in fact, the elite political economy, radiating from its center at the site of Sicán, influenced production activities at the site.

Metal and ceramic production fuel remains from Huaca Sialupe support Matson's (1989) understanding of ceramic production as the hub of pyrotechnology. At Mound I, where ceramic production occurred, we see the highest concentration of *Prosopis* sp.; the dominant tree species of the dry tropical forest, it was also judged to be the highest-quality fuel resource based on its long-term combustion stability and caloric output (Harker et al 1982; Shimada and Merkel 1991; Shimada and Wagner 2001). The analysis indicates that ceramic production, along with other household/farming processes, largely supported metal production, supplying charcoal fuels for furnaces. Matson's (1989) estimation of ceramic production as primary is further reinforced by the presence of *taverna* styled brewing and cooking facilities at Mound I. Although cooking was at one time a part of the kiln complex in Mound I, Room 39, it was either expanded or moved to a more independent space in Room 38. Overall, the Huaca Sialupe data solidly supports Matson's (1989) model.

With ceramic production at the base of all other production at Huaca Sialupe, it appears that fuel sharing, rather than fuel competition, characterized production dynamics. Apart from crafting, activities such as food preparation may also have drawn on fuel sources seen as critical to artisans at the site. Evidence for the seasonality of workshop activities, however, allows us to understand how these activities were maintained based on a limited and shared fuel resource base. Seasonality indicators demonstrate that ceramic and metal production occurred at different times of the year, permitting fuel sharing. In particular, high quality charcoal, created in the dry ceramic production season, could be redistributed to metal working furnaces. In addition, *Llama* sp. dung, the product of maintaining a population of animals at the site, provided a supplementary source of year-round fuel, suitable to all fire needs at the site. This renewable source of fuel at Sialupe, in conjunction with *poña* or domestic garbage, served to extend the limited high-quality fuel resources, like hardwood charcoal. Either consciously or unconsciously, their use of camelid dung and recycling of domestic garbage created less dependency on technologically superior charcoal resources and less of a chance to undermine local fuel supplies via deforestation.

As Matson (1989; 1966) predicted, metalworking fuel requirements at Huaca Sialupe appears to be more flexible: unlike in ceramic firing, other wood species, e.g. *Capparis* spp. fuelwood or charcoal, were integrated into the fuel load for metal production. The production of charcoal within the confines of the metal workshop at Mound II also served to generate high quality fuels for their furnaces on an *ad hoc* basis. In the end, it appears that metalwork and ceramic production, apart from being on their own seasonal timeline, do not appear to have been in competition. Instead, their use of fuels was apparently well integrated, both seasonally and ecologically.

The framework outlined of 'technological choice' (Lemonnier 1992) presents a good theoretical context to explain the integration of fuel use and production at Sialupe. The 'choices' made by artisans in this case are negotiations and actions that mediate how artifacts are culturally conceptualized and physically produced. Artisans at Sialupe used their culturally informed view of the available resources with the culturally acceptable actions of artifact production in the context of their setting, e.g., gesture (Mauss 1979), skill (Ingold 2000), distinction (Bourdieu 1994), operation (Leroi-Gourhan 1964). In this case, the artifacts are fires; the technological choices artisans were making was directly related to how fuels were being apportioned across the site. The gross example of this is the removal of charcoal from blackware kilns for loading and preparing metalwork furnaces at Mound II. Charcoal was potentially moved and stored in the Mound II furnaces as the concentration of industrial efforts moved from Mound I (ceramics) to Mound II (metalwork). Additionally, their idea to mix fuels, perhaps, in adjacent features at Mound II, with desiccated or dried branches indicates that a system of renewable resource use was in force at Sialupe.

This kind of complementarity between different social and economic processes is illustrated in ethnographic situations about farm smallholding and articulation with the political economy (Reed 1995) and in instances of craft production more particularly (Schmidt 1997; Sillar 2000; Winterhalder et al. 1974). In the case of Reed (1995) we see that domestic or household processes, like those described by Wolf (1997), Folbre (1986), and Halperin (1994) all take precedence over pressures from the external political economy. Even when confronting dire economic and political circumstances, the allocation of labor and food held in common and the familial social forces together with ecological intuition and experience prevent people from overexploiting their resources in spite of needs for capital (Reed 1995). In the instance of work by Sillar (2000), the ecological realities of exploiting dung are critical for maintaining ceramic production, the household's main source of capital and connection with the political economy. The option to use dung being selected over overexploiting wood resources and overextending household labor to gather fuelwood. So, in fact, even in light of dire circumstances of placing indigenous identity against capital interests of the modern political economy, Reed (1995) and Sillar (2000) each indicate that household producers are reluctant to overexploit resources culturally interpreted as vulnerable. This case is even more evident when other options, fencepost production in the case of Reed (1995) or dung use in Sillar (2000), are visible, salient, and readily available.

This last point raises questions about evaluating of the overall ecological footprint of Middle Sicán society, a topic that is current in modern political discourse, as well as a major topic in interpreting the prehistory of Lowland South America (Balée and Erickson 2006). Again, the topic here cannot be fully addressed based on the datasets available. It does appear, however, further research is now in a position to hypothesize and potentially illustrate a model for some form of low intensity or sustainable resource use among Middle Sicán small-holders, farmers and/or craft producers. For the region, and slightly later periods, survey and excavation data of the site of Galindo (Bawden 1982) might serve as a potential model for developing the kind of settlement reconnaisance necessary to begin assessing Middle Sicán sustainability and household level provisioning. Yet, even in Bawden's (1982) case organic remains were ignored, and to the kind of research I propose here, looking at landscape use from archaeobotanical data, requires not only settlement data but also environmental proxy data, and a well developed and analyzed archaeobotanical dataset.

In the end, one aspect that remains to be discussed about the wood fuel used at the site is its advent at the site itself. How are the fuel demands made and how are they met at a site like Sialupe? As Adegbehin and Omijeh (1994) and Bogach (1985) indicate, the collection of fuelwood is commonly the job of small children, women, and even geriatric portions of the population in the modern non-industrial contexts. Allen (2002), Arnold (1985), Sillar (2000), Weismantel (1998), and Winterhalder (1974) all similarly indicate that this is the case in the Andean context for the modern setting. Given the context of the vital domestic economy present at Huaca Sialupe, as determined through the archaeobotanical remains, firewood collection by these same members of the family is the path of least resistance. Additionally, this may support, at least in rationale, why collected materials favor small branches and not large stem or branchwood that might be the object of more intensive labor investment. We know that harvesting of large timbers was a part of monumental construction technology at the site of Sicán (Shimada 1995). These constructions required large boles of trees, as well as *poña* materials with potential ritual significance (Shimada 1986). Additionally, the more mundane structures from Sialupe, as well as those seen elsewhere in the valley, also required large tree parts as the long lasting structural components of building, as seen today (Correa Álamo 2000). How these were harvested is uncertain. Clearly management techniques as described above, and those like girdling (Reed 1995), would have been possible ways to take down large trees.

From ethnography we know that households often preserve old timbers and place them in new buildings (Schaedel 1988), this kind of reuse was certainly possible and probable in the past. Yet, again no real studies of timber use in the modern sense or in antiquity exist for the area. The indications at Sialupe, based on a very limited visible presence of elite overdetermination of production, are that the collection of wood fuel, and its movement and other renewable fuel sources across the site was probably done as a part of domestic labor apportioning that considered resource investments across the five mound complex at Sialupe. Until considerable time is devoted to looking at timber and wood technology in the Middle Sicán, as has been done for metal production (Bézur 2003) and ceramic production (Cleland and Shimada 1994; Shimada and Wagner 2001;2007), detailing the intricacies of forest management in the Middle Sicán remains elusive outside of the indications from this analysis.

In terms of complementarity, the initial point for the conception of this research, it appears that wood fuel was part of a series of cycles of resource use at the site. These were divided up between a series of different production considerations, as well as some considerations that I believe here were cultural constructs for how fuel best served the needs of the individuals at the site. While ceramic production may have served as a hub from a technological standpoint, from an economic standpoint agriculture, provisioning, subsistence, and herding all played significant roles, leaving a marked footprint on the dataset. Ideas about ritual use of trees, cultural views of sustainability, and the impact of construction and agricultural clearance at Huaca Sialupe and at the main site of Sicán remain topics for future research.

Question Two: Ecological Setting and Potential Impacts of Production

The second question posed at the outset of this dissertation was, 'what can we discern from the botanical remains regarding the local ecology and did craft production in the Middle Sicán impact local forests?' To address this issue in a substantive way requires that a dataset is established demonstrating the historical change of the forest, or resources from the forest, over time. This research establishes a baseline for the Middle Sicán, but future research with a similar approach to archaeological production contexts in the region is necessary to understand long-term ecological impacts. In addition, this study can only discuss the materials that were actually consumed/used, which leaves only speculation regarding the overall forest composition and structure relative to the stability of the ecological communities in and around Huaca Sialupe.

Ecological Setting

As Craig and Shimada (1986) outline, a sediment core analytical data are necessary to determine the actual components of the dry forest in the La Leche Valley and their historical change; the archaeobotanical data of this study are limited in their capacity to reconstruct the past environment. Sediment cores are the primary means for paleoenvironmental reconstruction and future work in coastal swamps called *patanos*, along the Lambayecano coastline, could permit the development of this type of dataset. These cores when analyzed for sedimentation, trace elements, diatoms, and pollen provide an excellent source for environmental proxies for reconstruction ancient environments. The fact that the bulk of the Huaca Sialupe wood data consists of more than 90% juvenile branchwood also limits the utility of this dataset for ecological reconstruction. As many of charcoal remains demonstrate anatomy indicative of very young growth, I have only a very limited sense regarding the role that mature trees played in the local ecological community. The data reflects a forest that was growing; I cannot, however, see a baseline for species richness. With only the products of human selection at my disposal, I cannot reconstruct the entire universe of ecological resources from which this selection was originally made. For instance, I do not know if the branchwood selected was from long-lived mature trees of a healthy forest or from compromised forests that only consisted of immature woody vegetation or primary growth that were trying to reestablish themselves, as they are today. A more concrete understanding of the ecological communities of the dry tropical forests of the 10th century would highlight how the branches selected for fire use articulate with the corresponding ecological community. Only a good microbotanical and geochemical record establishes this type of ecological background.

Despite these shortcomings, this study demonstrates that routine collection of distal growth, young branches or fallen deadwood, was a standard practice amongst species like *Prosopis* and *Acacia* and it was the essential wood fuel used for the domestic economy and production overall. Additionally, in their collection of immature branchwood, either directly culled from trees or gathered from the forest floor, artisans at Sialupe recognized that these branches would regenerate, indicating that they likely perceived trees as a renewable resource and understood their limitations. This branchwood occurs in all fire use applications at Huaca Sialupe, which demonstrates its primacy for fire use, their fossil fuel. Most importantly, this idea of resource perception fits into a broader framework of understanding renewable resources to suit their energy needs. This knowledge is applied throughout the fuel choices made at the site, e.g., using camelid dung, *Cavia* sp. dung, and domestic garbage as vital complements to the quality branchwood. On the one hand I cannot determine that the pattern for resource use and the fuel choices represented at Sialupe were determined by an overarching fear or conception of resource strain. On the other hand, I can say that renewable resources, those that dependably regenerate so that production can continue in subsequent seasons, characterized the fuels collected by artisans at Huaca Sialupe during the Middle Sicán. These resources were characterized as either seasonally regenerated branches through pruning practices, e.g., coppicing or pollarding, or strict collection of deadwood naturally abscised. In either case, the considerable amount of insect and fungal activity indicates that the wood used was either stockpiled or spent some time on the forest floor where decompositional processes dominated.

Human disturbance in and around Huaca Sialupe generated an ecology of herbaceous taxa very similar to what exists alongside traditional farming practices in the area today. Instead of creating an ecology that reflected the demands of elite aspects of the political economy, *ala* Earle (1997), this landscape was the materiality of the sum of their domestic and other production activities. It was the dominance of the domestic economy that sculpted the Huaca Sialupe landscape instead of the extractive processes of status item production, e.g., blackware ceramics or precious metalwork, or the influence of the additive elements of production, e.g., the import of gold. Whereas an overdetermined political economy might have dominated the extractive processes of the materiality of landscape, as seen in Earle's three examples (1997), from the residues of production this kind of elite dominance is not clearly visible. Instead more local extractive forces, however influenced by the political economy at large, appear visible in this dataset. This reality is evident in the presence of species like *Lippia* sp. and *Heliotropum* sp. in quantity and ubiquity at the site. Surprisingly, there is a marked absence of grass taxa: although seeds from the family Poaceae are present in the dataset, they are scarcer than seeds and fruits from other herbaceous or weedy taxa. A high presence of grass taxa is generally tied to open spaces with limited forest cover (Jones 1988). *Heliotropum* sp., and *Trianthema* sp., two disturbance taxa in the area can grow to be quite large, 2-5 m in height; these are normally weedy herbs. In areas surveyed with forest cover, however, it seems that they grow quite well and tall when sheltered from the tropical sun. The depressed signature of grasses in the Huaca Sialupe assemblage may indicate that forest cover in the surrounding areas, save those utilized for agricultural production, was the norm. With the presence of *Heliotropum* sp. and *Trianthema* sp. seeds we might be able to say that forest cover, in and around, or at least adjacent to Sialupe's domestic and agricultural features, was wooded.

Additionally, the presence of Camelids and *Cavia* sp. at the site cannot be discounted and the lack of grasses may indicate the overburden of grazing and feed supply. These kinds of processes may be what we see in areas along the slopes of the lower La Leche where overgrazing in the modern era has lead to a state of deflected succession. To evaluate this situation of the Middle Sicán, however, we would need a better idea of what the landscape looked like in terms of overall vegetation to better evaluate if there was another dominant herbaceous species that outcompeted grasses locally, which can be the case between euphorbs and grasses in arid climates. Again, only a comprehensive sediment core analysis can confirm or deny this hypothesis.

Finally, even though it is clear that the people working and living around Huaca Sialupe had clear and distinct articulations with the elite Middle Sicán economy, the importation of precious metal and the export of blackware ceramics, these interactions do not dominate the signatures of the archaeobotanical remains at the site. Instead, generalized domestic practices, fuel gathering, herding, agriculture and food production, all dominate the palimpsest of the Middle Sicán occupation at the site. In return, the landscape that we have a glimpse of is one that looks like those of household groups in small scale societies in some parts of present day South America (Balée and Erickson 2006; Reed 1995) and ethnohistorically in Lambayeque (Netherly 1977, 1984; Ramírez 1996; Schaedel 1988).

Impact of Craft Production

From the outset of this research design, I was aware of the problem of assessing potential overharvesting of fuel during the Middle Sicán: lack of systematic pollen and geochemical core data and systematic excavation of historically earlier production sites in the region fail to set up an appropriate botanical baseline to assess the impact of production on regional ecologies. Based on my analysis of technological choice and domestic production; it is clear from the archaeobotanical data that there was no apparent conflict of fuel use between Mounds I and II at the Sialupe workshop; as noted above, there was a general sharing of the same resource pool. At the most refined level of analysis, feature-to-feature, certain fuel preferences represent *ad hoc* decisions based on production needs at a specific locus of production, i.e. ceramic, metal, or food producing fire. In terms of the organic resources being utilized, all are renewable, e.g. dung, weeds, branchwood, garbage; had we seen large boles of trees or great quantities of mature wood, these would have served as indicators that large portions of the forest were being cleared and cut, severely impacting forest regeneration and the stability of the existing ecosystem. Adze marks on some pieces of charcoal and wood chips, as well as the diameter of postholes in the roofed section of Mound I (circa 25 cm dia.) indicate that the artisans were capable of chopping down large trees. The lack of recovery of Middle Sicán adzes, axes, or pruning tools, combined with the overwhelming preference of branchwood over mature wood at Huaca Sialupe further demonstrates a focus on renewable resources. The felling of trees was not apparently necessary to fulfill the needs of production at the site or was seen as unwise due to the weight of other cultural factors such as an understanding of unsustainable land use. Given the importance of *Prosopis* sp. in the modern context (Chapter 2), and through the many cross cultural examples of indigenous societies today that understand trees as important historical and cultural markers of sacred and human landscapes (Basso 1996; Coles and Coles 1986; Fowler 1996; Búrquez and de los Angeles Q. 1994; Reed 1995; Rival 1998), tree use at Huaca Sialupe may have been based on such considerations. Yet, in the end, whether mature trees existed at Huaca Sialupe in the 10th century remains up for debate for the reasons stated previously.

Based on experimental data, estimates of Middle Sicán copper alloy production posited by Merkel and Shimada (1991) and Goldstein et al. (2007) may have underestimated the ingenuity of artisans processing ingots at Huaca Sialupe and overestimated their impacts on local forest. Estimates of fuel consumption in metal and ceramic production were based largely on the technological constraints and inorganic remains (i.e. furnace and kiln types) of these processes. This study however, demonstrates that fuel resources for craft production must be considered as part of broader domestic economies and regional ecological processes. The systematic recovery of well-preserved organic remains from consistently well-excavated production contexts is key to developing a holistic understanding of production processes generally, and both ecological impact and fuel use more specifically.

Lastly, a clearer representation of how Middle Sicán elites viewed, articulated, and even used a site like Huaca Sialupe may give broader insight into how production demands may have forced a more aggressive harvesting of fuels in the area. Given Earle's (1997) conception of how political power can be brought to bear on small scale producers like those at Huaca Sialupe, we might expect that increased demand or influence by elites at the site could dramatically alter the way that the present analysis reconstructs the domestic economy at Huaca Sialupe. Instead of a generalized pattern of fuel use across the site we may see more intensive dedication of certain resources in certain areas, or the wholesale abandonment of certain technologies in favor of others. In particular, it would be good to see how forest resources were used at the main site of Sicán, the smelting centers, and a few other domestic settings during the Middle Sicán for comparison.

Without a broader dataset, it is difficult at this time to see how a political economy model would adequately explain the patterns of fuel use visible at the site. That said, certainly elites were visible to people at Huaca Sialupe to some extent. Some of the Sialupe products ended up with elites, and they sent precious materials to the site for refinement and production. From looking at the fuel contexts alone at this one site, the influence of the political economy remains unclear, to the extent that demand, or any interaction, may have influenced an increased demand on environmental resources in the form of fuel wood.

Seasonality and Human Perception of the Environment

As mentioned above, one of the most interesting aspects of this research has been the ability to interpret workshop production-based seasonal constraints. Only through a direct analysis of organic fuel remains can we address seasonality. Previous ethnoarchaeological studies (Cleland and Shimada 1998) have confirmed observations made by Arnold (1985) and Sillar (2000) that ceramic production necessitates periods of predictably dry weather. The recovery of flowers and fruits in the dataset of this study identifies a more defined seasonality of production. Additionally, the multi-crafting production at Sialupe indicates that sharing space does not necessarily imply its simultaneous use. Ceramic and metal production at Sialupe were probably occurred at different times of the year; the results of this analysis indicate, with some surety, that ceramic production and firing took place during the summer months and metal working was a wintertime activity. It is only from the organic remains that we can evaluate the timing of these different activities, creating a compelling argument for the comprehensive recovery of botanical datasets.

The analysis of organic remains from Huaca Sialupe, as noted above, has also opened a new set of discussions regarding the integration of craft production with other domestic activities at the site. Craft production was part of a broader set of cultural and ecological processes occurring at the site. This integration raises the question as to how human perceptions of the ecological processes around the site impacted and in turn were impacted by the instituted domestic activities within Huaca Sialupe. The idea of a Middle Sicán recognition or internalization of understanding resource renewability has some other implications regarding resource perception at Huaca Sialupe.

One question that has arisen during the course of my research relates to the unique nature of the dry tropical forests of North Coastal Peru. Block and Richter (2000) argue that part of the key regenerative cycle of these forests is fire. These ecosystems tend to be pyrogenic and fires are critical to forest regeneration and other growth processes. In clearing out underbrush, they allow new trees to germinate. For this reason, most plants native to dry tropical forests are pyrophytic and adapted to their periodic occurrence, e.g. deep tap roots, succulent and waxy stems/leaves, dense heartwoods, dense barks, and epidermal nectaries exude sap. In the historical record, however, there is no mention of forest fires in and around Huaca Sialupe (Delvaud 1984; Kaulicke 1993; Mabres Torrello 1993). As mentioned in Chapter 2, prolonged droughts between ENSO events put pressure on trees, causing them to drop leaves and young branches, and threatens their general well being; this state makes forests susceptible to fires sparked by lightening strikes or human accident, much like the situation in the American West today. Fowler, (1996) in her example of the Shoshone, demonstrates the role humans played in preventing forest fires by reducing the fuel load on forest floors. She argues that human populations scavenging for fuel wood significantly and knowingly diminished the potential for forest fires. Fowler's informants were clearly aware of the fact that their predation on the fuel load of the forests of the California Great Basin and Death Valley region kept the area 'clean' and maintained a cultural sense of beauty and landscape in and around their dwelling places (1996). Informants argued that once moved off the land, fires became more prevalent and severe. Additionally, the inability to return to seasonal fishing holes or cultivation plots allowed weeds and successional species previously controlled by Shoshone practice to return. Species like *Salix* sp., were allowed to grow back, clogging water holes, and mesquites when unmaintained permitted the establishment of sand dunes that blighted the landscape. Moreover, the removal of Shoshone practices significantly compromised land resources that were once productive to these people who depended on seasonal watercourses and cultivation plots.

Although the dry forests of the North America are in a geographically distinct location from the forests of North Coast Peru, they are ecologically very similar and, as dry forests, are subject to the same natural pressures. I suggest that one reason there is no record of forest fires in the pre-Hispanic/Colonial transition period on the North Coast of Peru is due to fuel scavenging of local populations from the forest floor. Essentially playing the role of forest fires, human populations cleared the underbrush, allowing new growth to occur and simultaneously prevented catastrophic fires. Due to fuel collection strategies that relied on renewable resources of immature or abscised branchwood, the artisans at Huaca Sialupe likely played an integral role in the regeneration and conservation of these dry forests. This hypothesis leads us back again to the need to develop a better understanding of cultural practice in Peru's coastal dry tropical forests today, and field more ethnohistorical and archaeological material relating for fuel use practice in the past. Again, confirmation of this hypothesis requires a broader signature of the historical change in the regional botanical composition, a dataset that does not yet exist for the region.

The dataset generated in this study permits both ecological and cultural reconstructions of forest resource use, as well as a holistic interpretation of craft production. My analysis of ancient fuel remains has located the taxa important to Middle Sicán artisans and at the same time, demonstrated the roles that they played in their ecological community. Implied in their use is a reconstruction of how the residents of Huaca Sialupe used and perceived the variety of choices open to them for serving their energy needs. Additionally, this dataset developed arguments about the seasonal organization and the domestic implications of production at the site. Without an in depth survey of the ecological communities of Lambayeque, this research would have been impossible. Likewise, without the systematic recovery and analysis of fuel remains from the discreet features reconstructed through Shimada and Wagner's (2007) holistic perspective, understanding domestic economy and the potential impacts of production would not be possible. In the end, the entrance of a paleoethnobotanical approach into the holistic framework employed by the SAP has more than demonstrated its potential benefits for reconstructing ancient production, the cultural contexts for production, and the ecological realities that circumscribed Middle Sicán economic activities.

Concluding Remarks

For the past 29 years the SAP's approach, under Dr. Izumi Shimada, has aimed to reconstruct the totality of Middle Sicán craft production, centered on understanding the materials and their ecological and cultural values held within that society. The research presented here is a direct result of his understanding of holism (Shimada and Wagner 2007), where all production remains are examined and interpreted. In a region like the North Coast of Peru, where exceptionally preserved archaeobotanical remains are the norm, it is a testament to Shimada's persistent outlook that this dissertation is unique among research generated from this region. Since the broad availability of Towle's (1961) *Ethnobotany of Pre-Colombian Peru,* it is difficult to understand why more archaeobotanical remains are not described in site reports or investigated as datasets in their own right. Towle, herself, never even visited Peru, and her work continues to be a standard reference guide some 46 years after it was published and nearly 60 years since her dissertation research at Harvard was completed. The work completed, on all accounts, by the SAP at Huaca Sialupe covers an ostensibly brief period of time; yet with its ethnoarchaeological, ceramic technology, archaeometric, bioarchaeological, and now paleoethnobotanical reconstructions, the site stands as the most intensively reconstructed 850 square meters in Peru. Despite this precedent, the level of intensity, dedication to a region, and overall holistic vision to interpreting ancient society remains rare in Andean Archaeology today.

For me Huaca Sialupe presented itself as a test case scenario for applying field excavation, wood anatomy, seed determination, and spatial analysis to a dataset whose overall ecological systems were unknown to me at the time. The majority of this work was focused on developing an intimate understanding of the local ecology and the potential reality that it presented for artisans in the 10th century C.E. The surface of the resultant dataset, explored in the pages above, has only begun to be scratched; there are many more avenues of inquiry that require follow up. The goals of the research, however, were met and served to create a document that demonstrates the power of patience and skill in the field when combined with an ecological outlook for data interpretation. Although focus on Huaca Sialupe was primarily to understand the production contexts in light of Middle Sicán needs and desires, the final analyses have reached into the homes of people who lived some time ago. Within this dataset we see people making decisions about what they needed and when, and then follow them as they apply their knowledge to create artifacts, make food, construct walls, grow corn, and feed their herds. From outside the site itself, we see flowers blooming, trees growing, weeds finding their niche next to human space, and people bearing coca to chew in the company of their friends and family. The elegance of the features studied and the composite activities that created them is barely touched upon in this analysis. Yet, knowing the situation in Peru regarding the archaeological resources available for study reminds me that there are several small sites like Huaca Sialupe out there both excavated, in storage units somewhere, and waiting to be excavated. At some point, those features will be examined, their contents determined, and their ecological and social realities reconstructed in a similar, hopefully more thorough manner.

In conclusion, even with a holistic outlook, like that proposed by Shimada and Wagner (2007), it seems that production, especially of artifacts requiring sophisticated pyrotechnology, is often discussed without considering fuel. In the case of this study, fuel took center stage and illustrated very well the important intersections of local ecology, domestic activity, agricultural practice, and artisanal production. While in many ways these different aspects of the site were hypothesized prior to reviewing the fuel remains data, the analyses served to make more concrete the interdependence of all aspects of the site and its surroundings, as well as to provide a more complex picture of the daily realities facing Middle Sicán artisans. Ideally, this work demonstrates the effective nature of paleoethnobotanical research that draws together ecology, plant biology, ethnoarchaeology, and archaeometry to develop a broad and thorough model for archaeological fieldwork. Although this point has often been made by many of my forebears in anthropology, my hope here is to make this point in an Andean context, where only a few are advancing this research agenda.

APPENDICES

APPENDIX A

List of All Modern Plant Specimens Collected During the 2002 Field Season

Specimen Number	**1**	**3**	**4**	**5**	**6**	**7**	**8**	**9**
Genus, Species	*Loxopterygium huarango*	*Bursera graveolens*	*Bursera graveolens*	*Capparis avicennifolia*	*Cercidium praecox*	*Capparis angulata*	*Cordia lutea*	*Cercidium praecox*
Botanical Family	**Anacardiaceae**	**Burseraceae**	**Burseraceae**	**Capparidaceae**	**Fabaceae (Caesalpinoidiae)**	**Capparidaceae**	**Boraginaceae**	**Fabaceae (Caesalpinoidiae)**
Surrounding Vegetation	Trees. Herbs	Trees. Herbs	Trees. Herbs	Trees. Herbs	Trees. Herbs	Trees. Herbs	Trees. Herbs	Trees, Shrubs
Terrain	Inclined	Inclined	Inclined	Inclined	Inclined	Inclined	Inclined	Inclined
Soil	Sand/ Pebbles	Sand/ Pebbles	Sand/ Pebbles	Sand/ Pebbles	Sand/ Pebbles	Sand/ Pebbles	Sand/ Pebbles	Sand/ Pebbles
Habitat	Hilly Dry Forest	Hilly Dry Forest	Hilly Dry Forest	Hilly Dry Forest	Hilly Dry Forest	Hilly Dry Forest	Hilly Dry Forest	Hilly Dry Forest
Branching Form	Globular	Scanty	Scanty	Shrubby	Scanty	Spiked	Globular	Globular
DBH	12 cm	16.4cm	15.7cm	20.5 cm	17.2 cm	7 cm	13 cm	20.5 cm
Height (m)	5.62	2.49	6.13	5.68	10	2.75	3.6	5.7
Form	Liana	Tree (Juvenile)	Tree	Tree	Tree	Tree (Juvenile)	Shrub	Tree
Sample Type	Wood from a mature tree includes Trunk, Primary, and Secondary branches.	Wood from Trunk, primary, and secondary branches	Wood from Trunk, primary, and secondary branches, Leaves and fruit for voucher specimen	Wood from Trunk, primary, and secondary branches, Leaves and fruit for voucher specimen	Wood from Trunk, primary, and secondary branches, Leaves and fruit for voucher specimen	Wood from Trunk, primary, and secondary branches.	Wood from Trunk, primary, and secondary branches. Flower, fruts, and branches	Wood from Trunk, primary, and secondary branches.
UTM N	92,87,701	92,87,701	92,87,701	92,87,701	92,87,701	92,87,701	92,87,701	92,87,821
UTM E	6,55,287	6,55,287	6,55,287	6,55,287	6,55,287	6,55,287	6,55,287	6,55,494
IGM Quad	Jayanca	Jayanca	Jayanca	Jayanca	Jayanca	Jayanca	Jayanca	Jayanca
Place	Cerro Motupillo, NW Slope	Cerro Motupillo, NW Slope	Cerro Motupillo, NW Slope	Cerro Motupillo, NW Slope	Cerro Motupillo, NW Slope	Cerro Motupillo, NW Slope	Cerro Motupillo, NW Slope	Cerro Motupillo, NW Slope
Date Collected	5/9/2002	5/9/2002	5/9/2002	5/9/2002	5/9/2002	5/9/2002	5/9/2002	5/10/2002

Specimen Number	10	11	13	14	15	16	17	19	21
Genus, Species	*Bursera graveolens*	*Capparis angulata*	*Cordia lutea*	*Acacia macarantha*	*Prosopis pallida*	*Capparis avicennifolia*	*Cordia lutea*	*Valesia glabra*	*Acacia macarantha*
Botanical Family	**Burseraceae**	**Capparidaceae**	**Boraginaceae**	**Fabaceae (Mimosoidiae)**	**Fabaceae (Caesalpinoidiae)**	**Capparidaceae**	**Boraginaceae**	**Apocynaceae**	**Fabaceae (Mimosoidiae)**
Surrounding Vegetation	Trees, Shrubs	Trees, Shrubs	Trees, Shrubs	Trees, Shrubs	Trees, Shrubs	Trees, Shrubs	Shrubs, Herbs	Shrubs, Herbs	Shrubs, Herbs
Terrain	Inclined	Inclined	Inclined	Inclined	Inclined	Inclined	Falda de Canal	Flat	Inclined
Soil	Sand/ Pebbles	Sand/ Pebbles	Sand/ Pebbles	Sand/ Pebbles	Sand/ Pebbles	Sand/ Pebbles	Sand/ Clay	Clay	Clay
Habitat	Hilly Dry Forest	Hilly Dry Forest	Hilly Dry Forest	Hilly Dry Forest	Hilly Dry Forest	Hilly Dry Forest	River	Human Disturbed	Succesional Dry Forest
Branching Form	Scanty	Globular	Shrubby	Spiked	Scanty	Shrubby	Shrubby	Parallel	Scanty
DBH	19 cm	23 cm	11.5cm	2.7 cm	9.5 cm	6 cm	Cluster of stalks	15 cm	8
Height (m)	6.9m	4.2	3.6	2.2 m	3.6	2.1	5	3	4.9
Form	Tree	Tree	Tree	Tree (Juvenile)	Tree (Juvenile)	Shrub	Shrub	Shrub	Tree (Juvenile)
Sample Type	Wood from Trunk, primary, and secondary branches.	Wood from Trunk, primary, and secondary branches. Taken from a tree that has been heavily pruned for wood.	Trunk, primary and secondary branch wood.	Trunk, primary and secondary branch wood.	Trunk, primary and secondary branch wood.	Trunk, primary and secondary branch wood.	Large shrubby plant, tall and in flower. Voucher samples of flowers leaves and branches, as well as woody trunk and branches.	Wood from primary, secondary branches and trunk, as well as leaves and fruits for vouchering. All of the primary and trunk wood should show coppiced growth as they are eminating from a central stem that has been cut often.	Trunk, primary and secondary branches.
UTM N	92,87,821	92,87,821	92,87,821	92,87,821	92,87,821	92,87,821	92,87,118	92,86,777	92,86,758
UTM E	6,55,494	6,55,494	6,55,494	6,55,494	6,55,494	6,55,494	6,53,678	6,53,284	6,53,290
IGM Quad	Jayanca	Jayanca	Jayanca	Jayanca	Jayanca	Jayanca	Jayanca	Jayanca	Jayanca
Place	Cerro Motupillo, NW Slope	Cerro Motupillo, NW Slope	Cerro Motupillo, NW Slope	Cerro Motupillo, NW Slope	Cerro Motupillo, NW Slope	Cerro Motupillo, NW Slope	Canal de Regadillo (W of Motupillo)	W of Motupillo	Cuenca Sanjon
Date Collected	5/10/2002	5/10/2002	5/10/2002	5/10/2002	5/10/2002	5/10/2002	5/10/2002	5/10/2002	5/11/2002

Specimen Number	22	23	24	25	27	28	29	30	31
Genus, Species	*Mutingia calabura*	*Rumex* sp.	*Acacia macarantha*	*Tessaria integrifolia*	*Trianthema portulacastrum*	*Cassia bicapsularis*	*Scoparia dulcis*	*Scoparia dulcis*	*Caesalpinia spinosa*
Botanical Family	**Elaeocarpaceae**	**Portulacaceae**	**Fabaceae (Mimosoidiae)**	**Asteraceae**	**Azioaceae**	**Fabaceae (Caesalpinoidiae)**	**Scrophuliaceae**	**Scrophuliaceae**	**Fabaceae (Caesalpinoidiae)**
Surrounding Vegetation	Shrubs, Herbs	Shrubs, Herbs	Shrubs, Herbs	Shrubs, Herbs	Shrubs, Herbs	Shrubs, Herbs	Shrubs, Herbs	Shrubs, Herbs	Trees, Shrubs, Herbs
Terrain	Inclined	Flat	Flat	Flat	Inclined	Flat	Flat	Flat	Flat
Soil	Clay	Clay	Clay	Clay	Clay	Clay	Clay	Clay	Clay
Habitat	Standing Water	Standing Water	Standing Water	Standing Water	River	River	River	River	Succesional Dry Forest
Branching Form	Scanty	Scanty	Scanty	Spiked	Shrubby	Shrubby	Spiked	Crawling	Shrubby
DBH	26	8.5	31.5	10	vaRiverus	1-3 cm	1-2	2 cm	2-3
Height (m)	10.9	10.37	11.34	6.5	1.5	2.1	0.4	1	2
Form	Tree	Tree (Juvenile)	Tree	Shrub	Shrub	Shrub	Herb	Herb	Shrub
Sample Type	Trunk, primary and secondary branches. Voucher specimens including flowers, fruit, and leaves	Trunk, primary and secondary branches. Voucher specimens includes leaves.	Trunk, primary and secondary branches.	Trunk, primary and secondary branches. Flowers, leaves and branches.	Trunk, primary and secondary branches. Shrubby growth, and creeping along ground and climbing along tree branches.	Trunk, primary and secondary branches. Shrubby growth, and creeping along ground and climbing along tree branches.	Trunk, primary and secondary branches. Flowers, fruits and leaves for voucher. Flowers white and minute.	Trunk, primary and secondary branches.	Wood from trunks and branches, also flowers and leaves for voucher specimens. Flowers yellow.
UTM N	92,86,758	92,86,758	92,86,758	92,86,758	92,86,758	92,86,758	92,86,758	92,86,758	92,82,190
UTM E	6,53,290	6,53,290	6,53,290	6,53,290	6,53,290	6,53,290	6,53,290	6,53,290	6,52,095
IGM Quad	Jayanca	Jayanca	Jayanca	Jayanca	Jayanca	Jayanca	Jayanca	Jayanca	Jayanca
Place	Cuenca Sanjon	Cuenca Sanjon	Cuenca Sanjon	Cuenca Sanjon	Cuenca Sanjon	Cuenca Sanjon	Cuenca Sanjon	Cuenca Sanjon	Guevara Forest
Date Collected	5/11/2002	5/11/2002	5/11/2002	5/11/2002	5/11/2002	5/11/2002	5/11/2002	5/11/2002	5/11/2002

Specimen Number	32	33	34	35	37	38	39	40	41
Genus, Species	*Capparis crotonoides*	*Capparis crotonoides*	*Encelia* sp.	*Encelia* sp.	*Cordia lutea*	*Capparis angulata*	*Cercidium praecox*	*Capparis avicennifolia*	*Prosopis pallida*
Botanical Family	**Capparidaceae**	**Capparidaceae**	**Asteraceae**	**Asteraceae**	**Boraginaceae**	**Capparidaceae**	**Fabaceae (Caesalpinoidiae)**	**Capparidaceae**	**Fabaceae (Caesalpinoidiae)**
Surrounding Vegetation	Shrubs, Herbs	Shrubs, Herbs	Shrubs, Herbs	Shrubs, Herbs	Shrubs, Herbs	Shrubs, Herbs	Shrubs, Herbs	Shrubs, Herbs	Shrubs, Herbs
Terrain	Inclined	Inclined	Inclined	Inclined	Inclined	Inclined	Inclined	Inclined	Inclined
Soil	Sand/ Clay	Sand/ Clay	Sand/ Clay	Sand/ Clay	Sand/ Clay	Sand/ Clay	Sand/ Clay	Sand/ Clay	Sand/ Clay
Habitat	Hilly Dry Forest	Hilly Dry Forest	Hilly Dry Forest	Hilly Dry Forest	Hilly Dry Forest	Hilly Dry Forest	Hilly Dry Forest	Hilly Dry Forest	Hilly Dry Forest
Branching Form	Globular	Shrubby	Shrubby	Scanty	Scanty	Globular	Scanty	Shrubby	Scanty
DBH	4	6	1	1	8	3	4	8	12
Height (m)	1.9	2.5	1.14	1.75	1.9	1.47	2.20	2.62	3.1
Form	Shrub	Shrub	Shrub	Shrub	Shrub	Shrub	Tree (Juvenile)	Shrub	Tree (Juvenile)
Sample Type	Trunk and primary and secondary branch wood. Plant show signs of having been coppiced at the base of the trunk.	Trunk and primary and secondary branch wood. Plant show signs of having been coppiced at the base of the trunk.	Trunk and primary and secondary branch wood. Flowers, leaves, and stem for voucher specimens. Resin smells like oranges, flowers yellow.	Trunk and primary and secondary branch wood. Resin smells like oranges, flowers yellow.	Trunk and primary and secondary branch wood. From a coppiced trunk.	Trunk and primary and secondary branch wood.	Trunk and primary and secondary branch wood. Heavuly overgrazed by goats below 1.5 m	Trunk and primary and secondary branch wood. Coppiced from a central stem.	Trunk and primary and secondary branch wood. Voucher Specimen of Flowers and leaves.
UTM N	92,88,101	92,88,101	92,88,101	92,88,101	92,88,101	92,88,101	92,88,101	92,88,101	92,88,101
UTM E	6,55,210	6,55,210	6,55,210	6,55,210	6,55,210	6,55,210	6,55,210	6,55,210	6,55,210
IGM Quad	Jayanca	Jayanca	Jayanca	Jayanca	Jayanca	Jayanca	Jayanca	Jayanca	Jayanca
Place	Cerro Motupillo	Cerro Motupillo	Cerro Motupillo	Cerro Motupillo	Cerro Motupillo	Cerro Motupillo	Cerro Motupillo	Cerro Motupillo	Cerro Motupillo
Date Collected	5/13/2002	5/13/2002	5/13/2002	5/13/2002	5/13/2002	5/13/2002	5/13/2002	5/13/2002	5/13/2002

Specimen Number	**42**	**43**	**45**	**46**	**48**	**49**	**50**	**51**	**52**
Genus, Species	*Acacia macarantha*	*Cercidium praecox*	*Valesia glabra*	*Prosopis pallida*	*Capparis angulata*	*Heliotropum arborescens*	*Cordia lutea*	*Capparis angulata*	*Acacia macarantha*
Botanical Family	**Fabaceae (Mimosoidiae)**	**Fabaceae (Caesalpinoidiae)**	**Apocynaceae**	**Fabaceae (Caesalpinoidiae)**	**Capparidaceae**	**Boraginaceae**	**Boraginaceae**	**Capparidaceae**	**Fabaceae (Mimosoidiae)**
Surrounding Vegetation	Shrubs, Herbs	Shrubs, Herbs	Shrubs, Herbs	Shrubs, Herbs	Shrubs, Herbs	Shrubs, Herbs	Trees, Shrubs, Herbs	Trees, Shrubs, Herbs	Trees, Shrubs, Herbs
Terrain	Inclined	Inclined	Inclined	Inclined	Inclined	Inclined	Flat	Flat	Flat
Soil	Sand/ Clay	Sand/ Clay	Sand/ Clay	Sand/ Clay	Sand/ Clay	Sand/ Clay	Clay	Clay	Clay
Habitat	Hilly Dry Forest	Hilly Dry Forest	Hilly Dry Forest	Hilly Dry Forest	Hilly Dry Forest	Hilly Dry Forest	Succesional Dry Forest	Succesional Dry Forest	Succesional Dry Forest
Branching Form	Scanty	Spiked	Parallel	Scanty	Scanty	Scanty	Scanty	Scanty	Scanty
DBH	3	3	5	3	24.9	8	7	3	7.5
Height (m)	1.95	1.45	2	2.15	4.1	2	2.55	1	6.46
Form	Tree (Juvenile)	Tree (Juvenile)	Shrub	Tree (Juvenile)	Tree	Shrub	Shrub	Tree (Juvenile)	Tree
Sample Type	Trunk and primary and secondary branch wood.	Trunk and primary and secondary branch wood. Heavily eaten by goats and other grazing animals.	Wood specimen includes trunk, primary and secondary branches. Coppiced at base and overgrazed by goats.	Wood specimen includes trunk, primary and secondary branches. Coppiced at base and overgrazed by goats.	Wood sample from trunk to quaternary branches. Voucher Specimen Flowers and branches	Wood sample of trunk as well as flowers, leaves and branches as voucher specimens. Flowers are white and minute.	Trunk and primary and secondary branch wood.	Trunk and primary and secondary branch wood.	Trunk and primary and secondary branch wood.
UTM N	92,88,101	92,88,101	92,88,101	92,88,101	92,88,101	92,88,101	92,83,552	92,83,552	92,83,552
UTM E	6,55,210	6,55,210	6,55,210	6,55,210	6,55,210	6,55,210	6,54,462	6,54,462	6,54,462
IGM Quad	Jayanca	Jayanca	Jayanca	Jayanca	Jayanca	Jayanca	Jayanca	Jayanca	Jayanca
Place	Cerro Motupillo	Cerro Motupillo	Cerro Motupillo	Cerro Motupillo	Cerro Motupillo	Cerro Motupillo	Bosque Patapon	Bosque Patapon	Bosque Patapon
Date Collected	5/13/2002	5/13/2002	5/13/2002	5/13/2002	5/13/2002	5/13/2002	5/15/2002	5/15/2002	5/15/2002

Specimen Number	53	54	55	56	57	58	59	60	61
Genus, Species	*Acacia macarantha*	*Capparis avicennifolia*	*Capparis angulata*	*Prosopis pallida*	*Cordia lutea*	*Prosopis pallida*	*Capparis avicennifolia*	*Cercidium praecox*	*Prosopis pallida*
Botanical Family	**Fabaceae (Mimosoidiae)**	**Capparidaceae**	**Capparidaceae**	**Fabaceae (Caesalpinoidiae)**	**Boraginaceae**	**Fabaceae (Caesalpinoidiae)**	**Capparidaceae**	**Fabaceae (Caesalpinoidiae)**	**Fabaceae (Caesalpinoidiae)**
Surrounding Vegetation	Trees, Shrubs, Herbs	Trees, Shrubs, Herbs	Trees, Shrubs, Herbs	Trees, Shrubs, Herbs	Trees, Shrubs, Herbs	Trees, Shrubs, Herbs	Trees, Shrubs, Herbs	Trees, Shrubs, Herbs	Trees, Shrubs, Herbs
Terrain	Flat	Flat	Flat	Flat	Flat	Flat	Flat	Flat	Flat
Soil	Clay	Clay	Clay	Clay	Clay	Clay	Clay	Clay	Clay
Habitat	Succesional Dry Forest	Succesional Dry Forest	Succesional Dry Forest	Succesional Dry Forest	Succesional Dry Forest	Succesional Dry Forest	Succesional Dry Forest	Succesional Dry Forest	Succesional Dry Forest
Branching Form	Scanty	Shrubby	Scanty	Scanty	Shrubby	Scanty	Scanty	Scanty	Scanty
DBH	10.5	18.5	13.5	3	16	10	11.5	4.5	23
Height (m)	6.02	4.05	3.95	3.1	2.08	5.15	4.06	1.35	6.89
Form	Tree	Shrub	Tree (Juvenile)	Tree (Juvenile)	Shrub	Tree	Shrub	Tree (Juvenile)	Tree
Sample Type	Trunk and primary and secondary branch wood.	Trunk and primary and secondary branch wood.	Trunk and primary and secondary branch wood.	Trunk and primary and secondary branch wood.	Trunk and primary and secondary branch wood.	Trunk and primary and secondary branch wood.	Trunk and primary and secondary branch wood. Coppiced material probably or nearby fence or wall quincha.	Trunk and primary and secondary branch wood. One of very few of its species visible in the area.	Trunk to last branch wood.
UTM N	92,83,552	92,83,552	92,83,552	92,83,552	92,83,552	92,83,552	92,83,552	92,83,552	92,83,552
UTM E	6,54,462	6,54,462	6,54,462	6,54,462	6,54,462	6,54,462	6,54,462	6,54,462	6,54,462
IGM Quad	Jayanca	Jayanca	Jayanca	Jayanca	Jayanca	Jayanca	Jayanca	Jayanca	Jayanca
Place	Bosque Patapon	Bosque Patapon	Bosque Patapon	Bosque Patapon	Bosque Patapon	Bosque Patapon	Bosque Patapon	Bosque Patapon	Bosque Patapon
Date Collected	5/15/2002	5/15/2002	5/15/2002	5/15/2002	5/15/2002	5/15/2002	5/15/2002	5/15/2002	5/15/2002

Specimen Number	**62**	**63**	**65**	**66**	**67**	**68**	**69**	**70**	**71**
Genus, Species	*Prosopis pallida*	*Lippia* sp.	*Cordia lutea*	*Acacia macarantha*	*Capparis avicennifolia*	*Prosopis pallida*	*Prosopis pallida*	*Capparis angulata*	*Prosopis pallida*
Botanical Family	**Fabaceae (Caesalpinoidiae)**	**Verbenaceae**	**Boraginaceae**	**Fabaceae (Mimosoidiae)**	**Capparidaceae**	**Fabaceae (Caesalpinoidiae)**	**Fabaceae (Caesalpinoidiae)**	**Capparidaceae**	**Fabaceae (Caesalpinoidiae)**
Surrounding Vegetation	Trees, Shrubs, Herbs	Herbs	Trees, Shrubs	Trees, Shrubs	Trees, Shrubs	Trees, Shrubs	Trees, Shrubs	Trees, Shrubs	Trees, Shrubs
Terrain	Flat	Flat	Flat	Flat	Flat	Flat	Flat	Flat	Flat
Soil	Clay	Clay	Clay	Clay	Clay	Clay	Clay	Clay	Clay
Habitat	Succesional Dry Forest	Chapparal	Succesional Dry Forest	Succesional Dry Forest	Succesional Dry Forest	Succesional Dry Forest	Succesional Dry Forest	Succesional Dry Forest	Succesional Dry Forest
Branching Form	Scanty	Shrubby	Shrubby	Scanty	Shrubby	Scanty	Scanty	Globular	Globular
DBH	22	1.4	22.5	7.5	33.5	20	2.5	24.5	9.5
Height (m)	8.42	1.6	4.71	5.8	3.95	7.98	2.9	2.854	5.58
Form	Tree	Herb	Shrub	Tree (Juvenile)	Shrub	Tree	Tree (Juvenile)	Tree	Tree (Juvenile)
Sample Type	Trunk to last branch wood. Thought to be from the '83 El Nino.	Shrubby herb. Woody stem and branch as well as flowers and leaves collected for voucher specimen.	Trunk, primary and secondary branchwood.	Trunk, primary and secondary branchwood.	Trunk, primary and secondary branchwood. Coppiced growth from base of central trunk.	Trunk, primary and secondary branchwood.	Trunk, primary and secondary branchwood.	Primary and secondary branchwood. Primary branch is from a coppiced trunk and branch is about 8cm dia.	Trunk, primary and secondary branchwood. Sampled from drain of present day canal system.
UTM N	92,83,552	92,83,552	92,83,151	92,83,151	92,83,151	92,83,151	92,83,151	92,83,151	92,83,151
UTM E	6,54,462	6,54,462	6,54,515	6,54,515	6,54,515	6,54,515	6,54,515	6,54,515	6,54,515
IGM Quad	Jayanca	Jayanca	Jayanca	Jayanca	Jayanca	Jayanca	Jayanca	Jayanca	Jayanca
Place	Bosque Patapon	Bosque Patapon	Bosque Patapon	Bosque Patapon	Bosque Patapon	Bosque Patapon	Bosque Patapon	Bosque Patapon	Bosque Patapon
Date Collected	5/15/2002	5/15/2002	5/17/2002	5/17/2002	5/17/2002	5/17/2002	5/17/2002	5/17/2002	5/17/2002

Specimen Number	72	73	74	75	76	77	78	80	81
Genus, Species	*Capparis angulata*	*Acacia macarantha*	*Capparis avicennifolia*	*Capparis angulata*	*Cordia lutea*	*Prosopis pallida*	*Cordia lutea*	*Cercidium praecox*	*Valesia glabra*
Botanical Family	**Capparidaceae**	**Fabaceae (Mimosoidiae)**	**Capparidaceae**	**Capparidaceae**	**Boraginaceae**	**Fabaceae (Caesalpinoidiae)**	**Boraginaceae**	**Fabaceae (Caesalpinoidiae)**	**Apocynaceae**
Surrounding Vegetation	Trees, Shrubs	Trees, Shrubs	Trees, Shrubs	Trees, Shrubs	Trees, Shrubs	Trees, Shrubs	Trees, Shrubs	Shrubs, Herbs	Shrubs, Herbs
Terrain	Flat	Flat	Flat	Flat	Flat	Flat	Flat	Flat	Flat
Soil	Clay	Clay	Clay	Clay	Clay	Clay	Clay	Sand/ Clay	Sand/ Pebbles
Habitat	Succesional Dry Forest	Succesional Dry Forest	Succesional Dry Forest	Succesional Dry Forest	Succesional Dry Forest	Succesional Dry Forest	Succesional Dry Forest	Chapparal	Scanty Dry Forest
Branching Form	Scanty	Scanty	Shrubby	Globular	Scanty	Spiked	Shrubby	Parallel	Shrubby
DBH	18.5	8	20.6	11	2.5	2	11	0.10	27.5
Height (m)	3.84	6.89	3.62	2.3	3	2.8	3.18	5.58	2.65
Form	Tree	Tree	Shrub	Tree	Shrub	Tree (Juvenile)	Shrub	Tree	Shrub
Sample Type	Trunk, primary and secondary branchwood. Big old tree that has been coppiced several times as well has having major branches pruned.	Trunk, primary and secondary branchwood.	Trunk, primary and secondary branchwood.	Trunk, primary and secondary branchwood.	Trunk, primary and secondary branchwood. Heavily browsed by grazing animals.	Trunk, primary and secondary branchwood.	Trunk, primary and secondary branchwood.	Trunk, primary, and secondary branch wood.	Primary, and secondary branch wood.
UTM N	92,83,151	92,83,151	92,83,151	92,83,151	92,83,151	92,83,151	92,83,151	92,83,151	92,87,088
UTM E	6,54,515	6,54,515	6,54,515	6,54,515	6,54,515	6,54,515	6,54,515	6,54,515	6,46,212
IGM Quad	Jayanca	Jayanca	Jayanca	Jayanca	Jayanca	Jayanca	Jayanca	Jayanca	Jayanca
Place	Bosque Patapon	Bosque Patapon	Bosque Patapon	Bosque Patapon	Bosque Patapon	Bosque Patapon	Bosque Patapon	Bosque Patapon	Nueva Pacora (Pativilca)
Date Collected	5/17/2002	5/17/2002	5/17/2002	5/17/2002	5/17/2002	5/17/2002	5/17/2002	5/17/2002	5/18/2002

Specimen Number	82	83	84	85	86	87	88	89	90
Genus, Species	*Capparis avicennifolia*	*Valesia glabra*	*Cercidium praecox*	*Capparis avicennifolia*	*Valesia glabra*	*Prosopis pallida*	*Prosopis pallida*	*Prosopis pallida*	*Prosopis pallida*
Botanical Family	**Capparidaceae**	**Apocynaceae**	**Fabaceae (Caesalpinoidiae)**	**Capparidaceae**	**Apocynaceae**	**Fabaceae (Caesalpinoidiae)**	**Fabaceae (Caesalpinoidiae)**	**Fabaceae (Caesalpinoidiae)**	**Fabaceae (Caesalpinoidiae)**
Surrounding Vegetation	Shrubs, Herbs	Trees, Shrubs, Herbs	Trees, Shrubs, Herbs	Trees, Shrubs, Herbs	Trees, Shrubs, Herbs	Trees, Shrubs, Herbs	Trees, Shrubs, Herbs	Trees, Shrubs, Herbs	Trees, Shrubs, Herbs
Terrain	Flat	Flat	Flat	Flat	Flat	Flat	Flat	Flat	Flat
Soil	Sand/ Pebbles	Sand/ Pebbles	Sand/ Pebbles	Sand/ Pebbles	Sand/ Pebbles	Sand/ Pebbles	Sand/ Pebbles	Sand/ Pebbles	Sand/ Pebbles
Habitat	River	Scanty Dry Forest	River	Scanty Dry Forest	Scanty Dry Forest	Scanty Dry Forest	Scanty Dry Forest	Scanty Dry Forest	Scanty Dry Forest
Branching Form	Spiked	Shrubby	Crawling	Shrubby	Shrubby	Scanty	Scanty	Scanty	Scanty
DBH	6.5	14.5	5 cm	14.0	40	15	14.5	27	25
Height (m)	4.71	4.06	3.7	5.58	6.02	NA	Tallada	Tallada	Tallada
Form	Shrub	Shrub	Tree	Shrub	Shrub	Tree	Tree	Tree	Tree
Sample Type	Trunk, primary, and secondary branch wood.	Trunk, primary, and secondary branch wood.	Trunk, primary, and secondary branch wood, includes root. Plant is growing out along ground towards and above canal bottom.	Trunk, primary, and secondary branch wood.	Trunk, primary, and secondary branch wood. Shows evidence of having been coppiced at base, all branches from shoot demonstrate coppiced growth.	root and stump wood. Know from wood and bark to be algarrobo.	root and stump wood. Know from wood and bark to be algarrobo.	root and stump wood. Know from wood and bark to be algarrobo. Sample is of a root buttress.	root and stump wood. Know from wood and bark to be algarrobo.
UTM N	92,87,088	92,87,088	92,87,088	92,87,088	92,87,088	92,87,088	92,87,088	92,87,088	92,87,088
UTM E	6,46,212	6,46,212	6,46,212	6,46,212	6,46,212	6,46,212	6,46,212	6,46,212	6,46,212
IGM Quad	Jayanca	Jayanca	Jayanca	Jayanca	Jayanca	Jayanca	Jayanca	Jayanca	Jayanca
Place	Nueva Pacora (Pativilca)	Nueva Pacora (Pativilca)	Nueva Pacora (Pativilca)	Nueva Pacora (Pativilca)	Nueva Pacora (Pativilca)	Nueva Pacora (Pativilca)	Nueva Pacora (Pativilca)	Nueva Pacora (Pativilca)	Nueva Pacora (Pativilca)
Date Collected	5/18/2002	5/18/2002	5/18/2002	5/18/2002	5/18/2002	5/18/2002	5/18/2002	5/18/2002	5/18/2002

Specimen Number	91	92	93	94	95	98	99	100	101
Genus, Species	*Prosopis pallida*	*Prosopis pallida*	*Prosopis pallida*	*Prosopis pallida*	*Prosopis pallida*	*Prosopis pallida*	*Prosopis pallida*	*Acacia macarantha*	*Prosopis pallida*
Botanical Family	**Fabaceae (Caesalpinoidiae)**	**Fabaceae (Caesalpinoidiae)**	**Fabaceae (Caesalpinoidiae)**	**Fabaceae (Caesalpinoidiae)**	**Fabaceae (Caesalpinoidiae)**	**Fabaceae (Caesalpinoidiae)**	**Fabaceae (Caesalpinoidiae)**	**Fabaceae (Mimosoidiae)**	**Fabaceae (Caesalpinoidiae)**
Surrounding Vegetation	Trees, Shrubs, Herbs	Trees, Shrubs, Herbs	Trees, Shrubs, Herbs	Trees, Shrubs, Herbs	Trees, Shrubs, Herbs	Trees, Shrubs, Herbs	Trees, Shrubs, Herbs	Trees, Shrubs, Herbs	Trees, Shrubs, Herbs
Terrain	Flat	Flat	Flat	Flat	Flat	Flat	Flat	Flat	Flat
Soil	Sand/ Pebbles	Sand/ Pebbles	Sand/ Pebbles	Sand/ Pebbles	Sand/ Pebbles	Clay	Clay	Clay	Clay
Habitat	Scanty Dry Forest	Scanty Dry Forest	Scanty Dry Forest	Scanty Dry Forest	Scanty Dry Forest	Semi Dense Dry Forest	Semi Dense Dry Forest	Semi Dense Dry Forest	Semi Dense Dry Forest
Branching Form	Scanty	Scanty	Scanty	Scanty	Scanty	Scanty	Scanty	Globular	Globular
DBH	11	8.5	10.5 cm	20.5	20	57	49 cm	52.5	39 cm
Height (m)	Tallada	Tallada	Tallada	9.29	9.73	Tallada	Tallada	Tallada	Tallada
Form	Tree	Tree	Tree	Tree	Tree	Tree	Tree	Tree	Tree
Sample Type	root and stump wood. Know from wood and bark to be algarrobo.	root and stump wood. Know from wood and bark to be algarrobo.	root and stump wood. Know from wood and bark to be algarrobo.	Wedge taken from trunk of mature tree.	Wedge taken from trunk of mature tree, primary to quaternary branches.	Trunk of old and recently felled mature tree. 98a- prim. Branch, 98b-h templadores(butresses), 98i- sec. Branch, 98j- Tunkwood.	Trunk of old and recently felled mature tree. 99a-b templadores, 99c-d secondary roots, 99e Trunk.	Trunk of old and recently felled mature tree. 100a trunk, 100b secondary root, 100c secondary root, 100d secondary and tertiary root.	Trunk of old and recently felled mature tree. 101a primary branch, 101b secondary root, 101c tap root, 101d secondary tap root
UTM N	92,87,088	92,87,088	92,87,088	92,87,088	92,87,088	92,83,684	92,83,684	92,83,684	92,83,684
UTM E	6,46,212	6,46,212	6,46,212	6,46,212	6,46,212	6,52,665	6,52,665	6,52,665	6,52,665
IGM Quad	Jayanca	Jayanca	Jayanca	Jayanca	Jayanca	Jayanca	Jayanca	Jayanca	Jayanca
Place	Nueva Pacora (Pativilca)	Nueva Pacora (Pativilca)	Nueva Pacora (Pativilca)	Nueva Pacora (Pativilca)	Nueva Pacora (Pativilca)	Roadside to Patapon	Roadside to Patapon	Roadside to Patapon	Roadside to Patapon
Date Collected	5/18/2002	5/18/2002	5/18/2002	5/18/2002	5/18/2002	5/20/2002	5/20/2002	5/20/2002	5/20/2002

Specimen Number	102	103	104	106	107	108	109	110	111
Genus, Species	*Prosopis pallida*	*Valesia glabra*	*Prosopis pallida*	*Cassia bicapsularis*	*Pithecellobium multiflorum*	*Scoparia dulcis*	*Acacia macarantha*	*Cassia bicapsularis*	*Scoparia dulcis*
Botanical Family	**Fabaceae (Caesalpinoidiae)**	**Apocynaceae**	**Fabaceae (Caesalpinoidiae)**	**Fabaceae (Caesalpinoidiae)**	**Fabaceae (Mimosoidiae)**	**Scrophuliaceae**	**Fabaceae (Mimosoidiae)**	**Fabaceae (Caesalpinoidiae)**	**Scrophuliaceae**
Surrounding Vegetation	Trees, Shrubs, Herbs	Trees, Shrubs, Herbs	Trees, Shrubs, Herbs	Trees, Shrubs, Herbs	Trees, Shrubs, Herbs	Trees, Shrubs, Herbs	Trees, Shrubs, Herbs	Trees, Shrubs, Herbs	Trees, Shrubs, Herbs
Terrain	Flat	Flat	Flat	Flat	Flat	Flat	Flat	Flat	Flat
Soil	Clay	Sand/ Clay	Clay	Clay	Clay	Clay	Clay	Clay	Clay
Habitat	Semi Dense Dry Forest	Semi Dense Dry Forest	Dense Dry Forest	Dense Dry Forest	Dense Dry Forest	Dense Dry Forest	Dense Dry Forest	Dense Dry Forest	Dense Dry Forest
Branching Form	Globular	Shrubby	Scanty	Shrubby	Scanty	Scanty	Scanty	Shrubby	Scanty
DBH	44.7	0.21	38.5	2.5	16	1	9	3	4
Height (m)	Tallada	4.3	12.67	2.51	10.60	1	10.17	3.5	0.8
Form	Tree	Shrub	Tree	Shrub	Tree	Herb	Tree (Juvenile)	Shrub	Herb
Sample Type	Trunk of old and recently felled mature tree. 102a primary root templador, 102b secondary root, 102c trunk, 102d secondary branch, 102e primary branch	Wood from trunk, primary, and secondary branches.	Wood from trunk, primary, and secondary branches.	Wood from trunk to tertiary branches.	Trunk wood	Stem wood.	Trunk, primary, and secondary branchwood.	Trunk, primary, and secondary branchwood. Flower, leaves and stems for voucher specimens. Liana type growth taking over cereso, algarrobo, and faiques growing adjacent.	Woody stem and branches.
UTM N	92,83,684	92,82, 661	92,82, 661	92,82, 661	92,82, 661	92,82, 661	92,82, 661	92,82, 661	92,82, 661
UTM E	6,52,665	6,52,016	6,52,016	6,52,016	6,52,016	6,52,016	6,52,016	6,52,016	6,52,016
IGM Quad	Jayanca	Jayanca	Jayanca	Jayanca	Jayanca	Jayanca	Jayanca	Jayanca	Jayanca
Place	Roadside to Patapon	Guevara Forest	Guevara Forest	Guevara Forest	Guevara Forest	Guevara Forest	Guevara Forest	Guevara Forest	Guevara Forest
Date Collected	5/20/2002	5/20/2002	5/20/2002	5/20/2002	5/20/2002	5/20/2002	5/20/2002	5/20/2002	5/20/2002

Specimen Number	112	113	114	115	116	117	118	119	120
Genus, Species	*Prosopis pallida*	*Cestrum auriculatum*	*Grabowskia boerhaviaefolia*	*Grabowskia boerhaviaefolia*	*Cryptocarpus pyriformus*	*Prosopis pallida*	*Cassia bicapsularis*	*Mutingia calabura*	*Prosopis pallida*
Botanical Family	**Fabaceae (Caesalpinoidiae)**	**Solanaceae**	**Solanaceae**	**Solanaceae**	**Nyctaginaceae**	**Fabaceae (Caesalpinoidiae)**	**Fabaceae (Caesalpinoidiae)**	**Elaeocarpaceae**	**Fabaceae (Caesalpinoidiae)**
Surrounding Vegetation	Trees, Shrubs, Herbs	Trees, Shrubs, Herbs	Trees, Shrubs, Herbs	Trees, Shrubs, Herbs	Trees, Shrubs, Herbs	Trees, Shrubs, Herbs	Trees, Shrubs, Herbs	Trees, Shrubs, Herbs	Trees, Shrubs, Herbs
Terrain	Flat	Flat	Flat	Flat	Flat	Flat	Flat	Flat	Flat
Soil	Clay	Clay	Clay	Clay	Clay	Clay	Clay	Clay	Clay
Habitat	Dense Dry Forest	Dense Dry Forest	Dense Dry Forest	Dense Dry Forest	Dense Dry Forest	Dense Dry Forest	Dense Dry Forest	Dense Dry Forest	Dense Dry Forest
Branching Form	Scanty	Spiked	Shrubby	Crawling	Escalando	Scanty	Shrubby	Globular	Globular
DBH	22.5	3	9.5	10.5	6.5	76	3	0.12	44.5
Height (m)	10.27	1.8	2	3	10.05	17.15	2.2	8.63	11.69
Form	Tree	Herb	Shrub	Shrub	Liana	Tree	Shrub	Tree	Tree
Sample Type	Two shoots growing upwards out of bases of tree along the upward surface of tunk. Tall erect shoots sampled for primary and secondary branchwood. Mimics coppiced growth. 112a from main trunk. 112b taken from a tall shoot of 9 cm dbh.	Woody stem, and branches. Feeding off roots of specimen 112.	Woody stem, and branches.	Woody stem, and branches.	Woody stem, and branches. Covering an algarrobo.	Woody stem, and branches.	Woody stem, and branches. Growing in plain, next to others and under shade of big algarrobo.	Woody stem, and branches.	Trunk wood and branches.
UTM N	92,82, 844	92,82, 844	92,82, 844	92,82, 844	92,82, 844	92,82, 844	92,82, 844	92,82, 844	92,82, 844
UTM E	6,52,020	6,52,020	6,52,020	6,52,020	6,52,020	6,52,020	6,52,020	6,52,020	6,52,020
IGM Quad	Jayanca	Jayanca	Jayanca	Jayanca	Jayanca	Jayanca	Jayanca	Jayanca	Jayanca
Place	Guevara Forest	Guevara Forest	Guevara Forest	Guevara Forest	Guevara Forest	Guevara Forest	Guevara Forest	Guevara Forest	Guevara Forest
Date Collected	5/21/2002	5/21/2002	5/21/2002	5/21/2002	5/21/2002	5/21/2002	5/21/2002	5/21/2002	5/21/2002

Specimen Number	121	122	123	124	125	126	127	128	129
Genus, Species	*Prosopis* sp.	*Cryptocarpus pyriformus*	*Heliotropum arborescens*	*Prosopis pallida*	*Heliotropum arborescens*	*Valesia glabra*	*Parkinsonia aeuelata*	*Laurel* sp.	*Galvesia* sp.
Botanical Family	**Fabaceae (Caesalpinoidiae)**	**Nyctaginaceae**	**Boraginaceae**	**Fabaceae (Caesalpinoidiae)**	**Boraginaceae**	**Apocynaceae**	**Fabaceae (Caesalpinoidiae)**	**Lauraceae**	**Scrophuliaceae**
Surrounding Vegetation	Trees, Shrubs, Herbs	Trees, Shrubs, Herbs	Trees, Shrubs, Herbs	Trees, Shrubs, Herbs	Trees, Shrubs, Herbs	Trees, Shrubs, Herbs	Trees, Shrubs, Herbs	Trees, Shrubs, Herbs	Trees, Shrubs, Herbs
Terrain	Flat	Flat	Flat	Flat	Flat	Flat	Fluvial	Fluvial	Fluvial
Soil	Clay	Clay	Clay	Clay	Clay	Clay	Sand/ Pebbles	Sand/ Pebbles	Sand/ Pebbles
Habitat	Dense Dry Forest	Dense Dry Forest	Dense Dry Forest	Dense Dry Forest	Dense Dry Forest	Dense Dry Forest	River	Standing Water	Standing Water
Branching Form	Parallel	Crawling	Globular	Globular	Shrubby	Shrubby	Scanty	Shrubby	Shrubby
DBH	11	6	1	21.5	13.5	11.5	7	9	41
Height (m)	8.87	8.56	0.65	12.56	5.58	5.47	6.98	6.36	2.2
Form	Tree	Liana	Herb	Tree	Shrub	Liana	Tree	Shrub	Shrub
Sample Type	Trunk wood and branches.	Trunk wood and branches. Growing below a fiaque and together with a cereso.	Leaves, flowers, and stems for voucher. Trunk, branch and root wood. Flowers are composite and white.	Trunk and branch wood. Tall and erect tree with little branching.	Trunk and branch wood.	Trunk and branch wood.	Wood from trunk to quaternary branches. Voucher specimens include flowers (Yellow) and leaves.	Wood from trunk to quaternary branches. Voucher specimens include flowers (fuchsia and showy) and leaves.	Wood from trunk to quaternary branches. Voucher specimens include flowers (red and dainty) and leaves.
UTM N	92,82, 844	92,82, 844	92,82, 844	92,82, 844	92,82, 844	92,82, 844	92,86,758	92,86,758	92,86,758
UTM E	6,52,020	6,52,020	6,52,020	6,52,020	6,52,020	6,52,020	6,53,290	6,53,290	6,53,290
IGM Quad	Jayanca	Jayanca	Jayanca	Jayanca	Jayanca	Jayanca	Jayanca	Jayanca	Jayanca
Place	Guevara Forest	Guevara Forest	Guevara Forest	Guevara Forest	Guevara Forest	Guevara Forest	Cuenca Sanjon	Cuenca Sanjon	Cuenca Sanjon
Date Collected	5/21/2002	5/21/2002	5/21/2002	5/21/2002	5/21/2002	5/21/2002	5/22/2002	5/22/2002	5/22/2002

Specimen Number	130	132	133	134	135	136	137	138	139
Genus, Species	*Ipomea* sp.	*Tessaria integrifolia*	*Prosopis pallida*	*Parkinsonia aeuelata*	*Cryptocarpus pyriformus*	*Amaranthus spinosus*	*Porophyllium elipticum*	*Prosopis* sp.	*Prosopis* sp.
Botanical Family	**Convulvucaceae**	**Asteraceae**	**Fabaceae (Caesalpinoidiae)**	**Fabaceae (Caesalpinoidiae)**	**Nyctaginaceae**	**Amaranthiaceae**	**Asteraceae**	**Fabaceae (Caesalpinoidiae)**	**Fabaceae (Caesalpinoidiae)**
Surrounding Vegetation	Trees, Shrubs, Herbs	Trees, Shrubs, Herbs	Trees, Shrubs, Herbs	Trees, Shrubs, Herbs	Trees, Shrubs, Herbs	Trees, Shrubs, Herbs	Trees, Shrubs, Herbs	Trees, Shrubs, Herbs	Trees, Shrubs, Herbs
Terrain	Fluvial	Fluvial	Fluvial	Fluvial	Fluvial	Fluvial	Fluvial	Fluvial	Fluvial
Soil	Sand/ Pebbles	Sand/ Pebbles	Sand/ Pebbles	Sand/ Pebbles	Sand/ Pebbles	Sand/ Pebbles	Sand/ Pebbles	Sand/ Pebbles	Sand/ Pebbles
Habitat	River	Standing Water	Standing Water	River	River	Standing Water	Standing Water	River	River
Branching Form	Shrubby	Spiked	Scanty	Scanty	Crawling	Scanty	Scanty	Globular	Parallel
DBH	10.5	11	3	7	7.5	1.2	1.3	52	11
Height (m)	3.6	10.16	1.7	10.07	0.30	0.60	1	18.10	6.33
Form	Liana	Shrub	Tree (Juvenile)	Tree	Liana	Herb	Herb	Tree	Tree (Juvenile)
Sample Type	Woody stem collected. Voucher specimens include flowers (purple and bellshaped) and leaves.	Wood from trunk and branches collected.	Wood from trunk and branches collected.	Wood from trunk and branches collected.	Wood from trunk and branches collected, sections of spreading roots and suckers also collected. Some suckers are up to 20m long. Old plant repeatedly growing from single stalk.	Woody herb. Branches and stalk materials collected with flowers and leaves for voucher specimens.	Woody herb. Branches and stalk materials collected with flowers and leaves for voucher specimens. Particularly squared stem and cruciform branching.	Fallen, but still living tree visible in the cut of the river bank. Roots and trunk sampled. 138 a taproot, 138b temblador, 138c trunk sample.	Shoot growing out from sample 138. Shoots probably started growing with the 1998 El Nino event. Branch, and trunk wood sampled.
UTM N	92,86,758	92,86,758	92,86,758	92,86,758	92,86,758	92,86,758	92,86,758	92,82,009	92,82,009
UTM E	6,53,290	6,53,290	6,53,290	6,53,290	6,53,290	6,53,290	6,53,290	6,52,271	6,52,271
IGM Quad	Jayanca	Jayanca	Jayanca	Jayanca	Jayanca	Jayanca	Jayanca	Jayanca	Jayanca
Place	Cuenca Sanjon	Cuenca Sanjon	Cuenca Sanjon	Cuenca Sanjon	Cuenca Sanjon	Cuenca Sanjon	Cuenca Sanjon	Sanjon River II	Sanjon River II
Date Collected	5/22/2002	5/22/2002	5/22/2002	5/22/2002	5/22/2002	5/22/2002	5/22/2002	5/23/2002	5/23/2002

Specimen Number	140	141	142	143	144	145	146	147	148
Genus, Species	*Prosopis* sp.	*Heliotropum arborescens*	*Acacia macarantha*	*Capparis avicennifolia*	*Parkinsonia aeuelata*	*Grabowskia boerhaviaefolia*	*Valesia glabra*	*Cordia lutea*	*Caesalpinia spinosa*
Botanical Family	**Fabaceae (Caesalpinoidiae)**	**Boraginaceae**	**Fabaceae (Mimosoidiae)**	**Capparidaceae**	**Fabaceae (Caesalpinoidiae)**	**Solanaceae**	**Apocynaceae**	**Boraginaceae**	**Fabaceae (Caesalpinoidiae)**
Surrounding Vegetation	Trees, Shrubs, Herbs	Trees, Shrubs, Herbs	Trees, Shrubs, Herbs	Trees, Shrubs, Herbs	Trees, Shrubs, Herbs	Trees, Shrubs, Herbs	Trees, Shrubs, Herbs	Trees, Shrubs, Herbs	Trees, Shrubs, Herbs
Terrain	Fluvial	Fluvial	Fluvial	Fluvial	Fluvial	Fluvial	Fluvial	Fluvial	Fluvial
Soil	Sand/ Pebbles	Sand/ Pebbles	Sand/ Pebbles	Sand/ Pebbles	Sand/ Pebbles	Sand/ Pebbles	Sand/ Pebbles	Sand/ Pebbles	Sand/ Pebbles
Habitat	River	River	River	River	River	River	River	River	River
Branching Form	Globular	Parallel	Scanty	Shrubby	Scanty	Scanty	Shrubby	Shrubby	Shrubby
DBH	0.65	3	29.5	10.5	6	1.5	27	7.5	3
Height (m)	13.16	2.8	9.73	4.49	7.55	2.1	4.05	4.27	2.3
Form	Tree	Shrub	Tree	Shrub	Tree (Juvenile)	Shrub	Shrub	Shrub	Shrub
Sample Type	Branch, and trunk wood sampled.	Branch, and trunk wood sampled.	Branch, and trunk wood sampled.	Branch, and trunk wood sampled.	Branch, and trunk wood sampled.	Branch, and trunk wood sampled. Big trunk growing over old and fallen trees.	Branch, and trunk wood sampled. Mature shrub.	Branch, and trunk wood sampled. Mature shrub.	Branch, and trunk wood sampled. Mature shrub.
UTM N	92,82,009	92,82,009	92,82,009	92,82,009	92,82,009	92,82,009	92,82,009	92,82,009	92,82,009
UTM E	6,52,271	6,52,271	6,52,271	6,52,271	6,52,271	6,52,271	6,52,271	6,52,271	6,52,271
IGM Quad	Jayanca	Jayanca	Jayanca	Jayanca	Jayanca	Jayanca	Jayanca	Jayanca	Jayanca
Place	Sanjon River II	Sanjon River II	Sanjon River II	Sanjon River II	Sanjon River II	Sanjon River II	Sanjon River II	Sanjon River II	Sanjon River II
Date Collected	5/23/2002	5/23/2002	5/23/2002	5/23/2002	5/23/2002	5/23/2002	5/23/2002	5/23/2002	5/23/2002

Specimen Number	149	150	151	153	154	155	156	157	158
Genus, Species	*Pithecellobium multiflorum*	*Mutingia calabura*	*Capparis angulata*	*Parkinsonia aeuelata*	*Capparis avicennifolia*	*Prosopis pallida*	*Pithecellobium multiflorum*	*Inga feuillei*	*Waltheria ovata*
Botanical Family	**Fabaceae (Mimosoidiae)**	**Elaeocarpaceae**	**Capparidaceae**	**Fabaceae (Caesalpinoidiae)**	**Capparidaceae**	**Fabaceae (Caesalpinoidiae)**	**Fabaceae (Mimosoidiae)**	**Fabaceae (Mimosoidiae)**	**Sterculiaceae**
Surrounding Vegetation	Trees, Shrubs, Herbs	Trees, Shrubs, Herbs	Shrubs, Herbs	Shrubs, Herbs	Shrubs, Herbs	Shrubs, Herbs	Shrubs, Herbs	Shrubs, Herbs	Shrubs, Herbs
Terrain	Fluvial	Flat	Flat	Flat	Flat	Flat	Flat	Flat	Flat
Soil	Sand/ Pebbles	Sand/ Clay	Sand/ Clay	Clay	Clay	Clay	Clay	Clay	Clay
Habitat	River	River	River	Ag. Field	Ag. Field	Ag. Field	Ag. Field	Ag. Field	Ag. Field
Branching Form	Scanty	Globular	Globular	Scanty	Shrubby	Scanty	Globular	Parallel	Shrubby
DBH	13.5	18	33	6.5	13	18.5	45.5	18	3
Height (m)	9.51	15.62	trunk height1.9	4.27	5.36	7.11	10.60	9.64	2
Form	Tree	Tree	Tree	Tree (Juvenile)	Shrub	Tree (Juvenile)	Tree	Tree (Juvenile)	Shrub
Sample Type	Wood from trunk to quaternary branches.	Trunk and primary-tertiary branch wood. Growing together with a group of four other large trees.	Primary branch, Trunk wood, and primary and secondary roots. Runk fallen and in river bottom from recent El Nino event.	Trunk, and primary-tertiary branchwood.	Trunk, and primary-tertiary branchwood.	Trunk, and primary-tertiary branchwood. Four trunks coming from a single trunk base, coppiced growth. Growing in a Ag. Field, reclaimed land. Growth is since the 1998 ENSO.	Trunk, and primary-tertiary branchwood.	Trunk, and primary-tertiary branchwood. Flowers leaves and branches for voucher specimens. Flower is white and yellow. Showy anthers are yellow. Only four vouchers taken.	Woody stems and branches. Flowers and leaves for voucher specimens.
UTM N	92,82,009	92,82,009	92,82,009	92,82,190	92,82,190	92,82,190	92,82,190	92,82,190	92,82,190
UTM E	6,52,271	6,52,271	6,52,271	6,52,095	6,52,095	6,52,095	6,52,095	6,52,095	6,52,095
IGM Quad	Jayanca	Jayanca	Jayanca	Jayanca	Jayanca	Jayanca	Jayanca	Jayanca	Jayanca
Place	Sanjon River II	Sanjon II	Sanjon II	Guevara Forest south (agfields)	Guevara Forest south (agfields)	Guevara Forest south (agfields)	Guevara Forest south (agfields)	Guevara Forest south (agfields)	Guevara Forest south (agfields)
Date Collected	5/23/2002	5/23/2002	5/23/2002	5/23/2002	5/23/2002	5/23/2002	5/24/2002	5/24/2002	5/24/2002

Specimen Number	159	160	161	162	163	164	165	166	167
Genus, Species	*Prosopis pallida*	*Prosopis* sp.	*Prosopis* sp.	*Capparis angulata*	*Capparis angulata*	*Acacia macarantha*	*Prosopis* sp.	*Pithecellobium multiflorum*	*Inga feuillei*
Botanical Family	**Fabaceae (Caesalpinoidiae)**	**Fabaceae (Caesalpinoidiae)**	**Fabaceae (Caesalpinoidiae)**	**Capparidaceae**	**Capparidaceae**	**Fabaceae (Mimosoidiae)**	**Fabaceae (Caesalpinoidiae)**	**Fabaceae (Mimosoidiae)**	**Fabaceae (Mimosoidiae)**
Surrounding Vegetation	Trees, Shrubs, Herbs	Trees, Shrubs, Herbs	Trees, Shrubs, Herbs	Trees, Shrubs, Herbs	Trees, Shrubs, Herbs	Trees, Shrubs, Herbs	Trees, Shrubs, Herbs	Trees, Shrubs, Herbs	Shrubs, Herbs
Terrain	Flat	Flat	Flat	Flat	Flat	Flat	Flat	Flat	Flat
Soil	Clay	Clay	Clay	Clay	Clay	Clay	Clay	Clay	Clay
Habitat	Succesional Dry Forest	Succesional Dry Forest	Succesional Dry Forest	Succesional Dry Forest	Succesional Dry Forest	Succesional Dry Forest	Dense Dry Forest	Semi Dense Dry Forest	Ag. Field
Branching Form	Scanty	Globular	Globular	Globular	Globular	Spiked	Globular	Globular	Globular
DBH	4.5	20	50	23.5	18	1.5	62	10 cm	30
Height (m)	7.98	14.09	10.23	5.15	10.17	2.5	12.67	5	9.60
Form	Tree (Juvenile)	Tree	Tree	Tree	Tree	Tree (Juvenile)	Tree	Tree	Tree
Sample Type	Woody stems and branches. From 1998 El Nino.	Woody stems and branches. From 1983 El Nino.	Woody stems, branches, and young shoots . From 1972 El Nino. 161 a-c are from shoot in nE facing side of tree over an ag.field.	Woody stems, branches, and young shoots. Old and mature Sapote that once once in open field, but now completely covered by succesional forest growth, probably from the 1983 El Nino.	Woody stems, branches, and young shoots. Old and mature Sapote that once once in open field, but now completely covered by succesional forest growth, probably from the 1983 El Nino. Excellent shade tree, nothing gets through.	Woody stems, branches, and young shoots. Among a group of 20 faiques growing up through a hole in the succesional forest cover. An example of succesional growth anatomy.	Trunk to the quaternary branches.	Trunk to the quaternary branches.	Trunk to the quaternary branches. Growing along side a canal sluce and in the limit of the living fence. One of seven trees same species growing in a line
UTM N	92,82,190	92,82,190	92,82,190	92,82,190	92,82,190	92,82,190	92,82,538	92,82,538	92,85, 955
UTM E	6,52,095	6,52,095	6,52,095	6,52,095	6,52,095	6,52,095	6,52,140	6,52,140	6,50,334
IGM Quad	Jayanca	Jayanca	Jayanca	Jayanca	Jayanca	Jayanca	Jayanca	Jayanca	Jayanca
Place	Guevara Forest south (successional area)	Guevara Forest south (successional area)	Guevara Forest south (successional area)	Guevara Forest south (successional area)	Guevara Forest south (successional area)	Guevara Forest south (successional area)	Guevara Forest	Guevara Forest	Navarro Fields
Date Collected	5/24/2002	5/24/2002	5/24/2002	5/24/2002	5/24/2002	5/24/2002	5/24/2002	5/24/2002	5/27/2002

Specimen Number	168	169	172	173	174	176	177	178	179
Genus, Species	*Inga feuillei*	*Alternathera halmifolia*	*Annona* sp.	*Persea americana*	*Waltheria ovata*	*Salix* sp.	*Salix* sp.	*Lippia* sp.	*Acacia macarantha*
Botanical Family	**Fabaceae (Mimosoidiae)**	**Amaranthiaceae**	**Annonaceae**	**Lauraceae**	**Sterculiaceae**	**Salicaceae**	**Salicaceae**	**Verbenaceae**	**Fabaceae (Mimosoidiae)**
Surrounding Vegetation	Shrubs, Herbs	Shrubs, Herbs	Shrubs, Herbs	Shrubs, Herbs	Shrubs, Herbs	Trees. Herbs	Trees. Herbs	Trees. Herbs	Trees. Herbs
Terrain	Flat	Flat	Flat	Flat	Flat	Inclined	Inclined	Inclined	Inclined
Soil	Clay	Clay	Clay	Clay	Clay	Clay	Clay	Clay	Clay
Habitat	Ag. Field	Ag. Field	Ag. Field	Ag. Field	Ag. Field	River	River	River	River
Branching Form	Globular	Crawling	Globular	Parallel	Parallel	Parallel	Parallel	Shrubby	Globular
DBH	3.5	0.8	34	37	18	23.5	19.5	1.5	4
Height (m)	3.1	0.40	7.11	7.64	6.11	10.85	11.69	1.2	3.7
Form	Tree (Juvenile)	Herb	Tree	Tree	Shrub	Tree	Tree	Shrub	Tree (Juvenile)
Sample Type	Shoot from root of mature tree, same tree as 167. Use as an example of rapid growth	Voucher specimens of flowers, stems, and leaves. Samples of woody stems.	Trunk, branch, and shoot wood. 25 year old tree.	Trunk, branch, and shoot wood. 35 year old tree.	Shrubby tree in field. Wood from trunk and branches.	Branch and trunk wood.	Branch and trunk wood.	Woody stem and branches.	Wood from trunk and branches.
UTM N	92,85, 955	92,85, 955	92,85, 955	92,85, 955	92,85, 955	92,85,972	92,85,972	92,85,972	92,85,972
UTM E	6,50,334	6,50,334	6,50,334	6,50,334	6,50,334	6,50, 918	6,50, 918	6,50, 918	6,50, 918
IGM Quad	Jayanca	Jayanca	Jayanca	Jayanca	Jayanca	Jayanca	Jayanca	Jayanca	Jayanca
Place	Navarro Fields	Navarro Fields	Navarro Fields	Navarro Fields	Navarro Fields	Acequia Grande (San Luis)	Acequia Grande (San Luis)	Acequia Grande (San Luis)	Acequia Grande (San Luis)
Date Collected	5/27/2002	5/27/2002	5/27/2002	5/27/2002	5/27/2002	5/27/2002	5/27/2002	5/27/2002	5/27/2002

Specimen Number	**180**	**181**	**182**	**183**	**184**	**185**	**186**	**187**	**188**
Genus, Species	*Amaranthus spinosus*	*Ludwingia peruvana*	*Salix* sp.	*Ricinus* sp.	*Prosopis pallida*	*Prosopis* sp.	*Prosopis* sp.	*Acacia macarantha*	*Prosopis pallida*
Botanical Family	**Amaranthiaceae**	**Onagraceae**	**Salicaceae**	**Euphorbiaceae**	**Fabaceae (Caesalpinoidiae)**	**Fabaceae (Caesalpinoidiae)**	**Fabaceae (Caesalpinoidiae)**	**Fabaceae (Mimosoidiae)**	**Fabaceae (Caesalpinoidiae)**
Surrounding Vegetation	Trees. Herbs	Trees. Herbs	Trees. Herbs	Trees. Herbs	Trees, Shrubs, Herbs	Trees, Shrubs, Herbs	Trees, Shrubs, Herbs	Trees, Shrubs, Herbs	Trees, Shrubs, Herbs
Terrain	Inclined	Inclined	Inclined	Inclined	Flat	Flat	Flat	Flat	Flat
Soil	Clay	Clay	Clay	Clay	Clay	Clay	Clay	Clay	Clay
Habitat	River	River	River	River	Dense Dry Forest	Dense Dry Forest	Dense Dry Forest	Dense Dry Forest	Dense Dry Forest
Branching Form	Globular	Globular	Parallel	Shrubby	Globular	Scanty	Globular	Spiked	Scanty
DBH	1.5	3.5	33	18	44	34.5	38	3	1.5
Height (m)	2	2.2	16.43	6.30	10.17	14.52	13.00	4.48	1.43
Form	Herb	Herb	Tree	Tree	Tree	Tree	Tree	Tree (Juvenile)	Tree (Juvenile)
Sample Type	Wood from stem and branches.	Wood from stem and branches.	Wood from trunk and branches.	Wood from trunk and branches.	Trunk to quaternary branchwood.	Trunk to quaternary branchwood.	Trunk to quaternary branchwood.	Trunk to quaternary branchwood.	Trunk to quaternary branchwood.
UTM N	92,85,972	92,85,972	92,85,972	92,85,972	92,82,502	92,82,502	92,82,502	92,82,502	92,82,502
UTM E	6,50, 918	6,50, 918	6,50, 918	6,50, 918	6,52,078	6,52,078	6,52,078	6,52,078	6,52,078
IGM Quad	Jayanca	Jayanca	Jayanca	Jayanca	Jayanca	Jayanca	Jayanca	Jayanca	Jayanca
Place	Acequia Grande (San Luis)	Acequia Grande (San Luis)	Acequia Grande (San Luis)	Acequia Grande (San Luis)	Forest Guevarra	Forest Guevarra	Forest Guevarra	Forest Guevarra	Forest Guevarra
Date Collected	5/27/2002	5/27/2002	5/27/2002	5/27/2002	5/28/2002	5/28/2002	5/28/2002	5/28/2002	5/28/2002

Specimen Number	189	190	191	195	196	197	200	202	203
Genus, Species	*Cestrum auriculatum*	*Lippia* sp.	*Valesia glabra*	*Sapindus saponifera*	*Sapindus saponifera*	*Ficus nymphaeifolia*	*Polygonum hydropipertoides*	*Mimosa pigra*	*Piper* sp.
Botanical Family	**Solanaceae**	**Verbenaceae**	**Apocynaceae**	**Anacardiaceae**	**Anacardiaceae**	**Moraceae**	**Polygonaceae**	**Fabaceae (Mimosoidiae)**	**Piperaceae**
Surrounding Vegetation	Trees, Shrubs, Herbs	Trees, Shrubs, Herbs	Trees, Shrubs, Herbs	Shrubs, Herbs	Shrubs, Herbs	Shrubs, Herbs	Shrubs, Herbs	Shrubs, Herbs	Shrubs, Herbs
Terrain	Flat	Flat	Flat	Flat	Flat	Inclined	Inclined	Inclined	Inclined
Soil	Clay	Clay	Clay	Clay	Clay	Clay	Clay	Clay	Clay
Habitat	Dense Dry Forest	Scanty Dry Forest	Semi Dense Dry Forest	Ag. Field	Ag. Field	River	River	River	Ag. Field
Branching Form	Scanty	Scanty	Shrubby	Globular	Spiked	Globular	Parallel	Parallel	Parallel
DBH	10.5	1.2	21.5	38.5	0.25	106	14.5	3	18.5
Height (m)	7.98	1.35	3.77	14.09	0.30	15.98	5.96	2.1	6.52
Form	Shrub	Shrub	Shrub	Tree	Tree (Juvenile)	Tree	Tree	Tree (Juvenile)	Shrub
Sample Type	Wood sampled from trunk to the last branch.	Wood sampled from trunk to the last branch.	Trunk and primary branch. Old plant, has been growing in a living fence for some time. Leaves minute and obviously stressed. (other species???)	Trunk and branchwood. Voucher specimens include flowers, leaves, and branches.	Trunk and branchwood. Shoot propogating from root system of larger mature tree.	Wood from main trunk and root wood. The tree is very old. Composed of mutliple trunks, living in base of deep and old canal on edge of cane field.	Voucher specimen of flowers, leaves and stems. Slight flowers with sweet perfume, light purple internally with purple petals. Trunk and branch specimens. May not native. Tree has been battered several times by innundations.	Trunk and branch wood.	Shoot from base of big trunk was sampled for wood, branches, and root. Voucher specimens of flower, leaf, and stems.
UTM N	92,82,502	92,82,502	92,82,502	92,86,113	92,86,113	92,86,607	92, 85, 972	92, 85, 972	92, 85, 972
UTM E	6,52,078	6,52,078	6,52,078	6,50,184	6,50,184	6,50,370	6,50, 918	6,50, 918	6,50, 918
IGM Quad	Jayanca	Jayanca	Jayanca	Jayanca	Jayanca	Jayanca	Jayanca	Jayanca	Jayanca
Place	Forest Guevarra	Forest Guevarra	Forest Guevarra	San Luis	San Luis	Acequia Tres Puentes	Acequia Grande (San Luis)	Acequia Grande (San Luis)	Casa de Alfonso di Avala, b/n San Luis and Motupillo
Date Collected	5/28/2002	5/28/2002	5/28/2002	5/29/2002	5/29/2002	5/29/2002	5/29/2002	5/29/2002	5/29/2002

Specimen Number	206	207	208	209	210	212	213	214	216
Genus, Species	*Psidium guayava*	*Bixa orellana*	*Baccharis salicifolia*	*Mimosa pigra*	*Typha angustifolia*	*Tessaria integrifolia*	*Psidium guayava*	*Inga feuillei*	*Spondias* sp.
Botanical Family	**Myrtaceae**	**Bixaceae**	**Asteraceae**	**Fabaceae (Mimosoidiae)**	**Typhaceae**	**Asteraceae**	**Myrtaceae**	**Fabaceae (Mimosoidiae)**	**Anacardiaceae**
Surrounding Vegetation	Trees. Herbs	Trees. Herbs	Trees. Herbs	Shrubs	Shrubs	Shrubs	Trees, Shrubs, Herbs	Trees, Shrubs, Herbs	Herbs
Terrain	Flat	Inclined	Flat	Flat	Flat	Flat	Inclined	Inclined	Inclined
Soil	Clay	Clay	Clay	Sand/ Pebbles	Sand/ Pebbles	Sand/ Pebbles	Clay	Clay	Clay
Habitat	Ag. Field	Ag. Field	River	River	River	River	Human Disturbed	Human Disturbed	Ag. Field
Branching Form	Parallel	Globular	Crawling	Globular	Parallel	Parallel	Globular	Globular	Encorvida
DBH	11.5	21.5	3	7	5	4.5	41	19.5	36.5cm
Height (m)	6.61	2.96	1.79	4.93	5.80	6.30	10.95	9.13	6.11
Form	Shrub	Tree	Shrub	Tree (Juvenile)	Shrub	Shrub	Tree	Tree	Tree
Sample Type	Wood from trunk and branches. 7 year old tree	4 year old tree/shrub.	Low shrubby herb. Trunk and branch wood. Several trunks eminating from series of clumped stalks.	Trunk and branch wood.	Trunk and branch wood.	Trunk and branch wood.	Trunk and branch wood. Shoot was sampled that is coming from associated root system. Growing out of canal side.	Trunk and branch wood. Growing above canal side.53	Palos Tertiary Branch, in flower on branches. Flowers acaulescent
UTM N	92,85, 955	92,85, 955	92,87,088	92,85,955	92,85,955	92,85,955	92,87,088	92,87,088	92,83,552
UTM E	6, 50, 334	6, 50, 334	6,46,212	6,49,462	6,49,462	6,49,462	6,46,212	6,46,212	6,54,462
IGM Quad	Jayanca	Jayanca	Jayanca	Jayanca	Jayanca	Jayanca	Jayanca	Jayanca	Jayanca
Place	Ag. Field Navarro	Ag. Field Navarro	Tres Puentes (Ancient Pacora)	River La Leche	River La Leche	River La Leche	Before Pativilca Crossing	Before Pativilca Crossing	Bosque Patapon (Ag. Field en route to Guevaras)
Date Collected	5/30/2002	5/30/2002	5/30/2002	5/30/2002	5/30/2002	5/30/2002	5/30/2002	5/30/2002	9/17/2002

Specimen Number	217	218	219	220
Genus, Species	*Salix* sp.	*Caesalpinia paipai*	*Capparis mollis*	*Prosopis pallida var. pallida*
Botanical Family	**Salicaceae**	**Fabaceae (Caesalpinoidiae)**	**Capparidaceae**	**Fabaceae (Caesalpinoidiae)**
Surrounding Vegetation	Trees. Herbs	Shrubs, Herbs	Trees, Shrubs, Herbs	Herbs
Terrain	Inclined	Flat	Flat	Flat
Soil	Sand/ Clay	Sand	Clay	Clay
Habitat	River	River	Succesional Dry Forest	Ag. Field
Branching Form	Parallel	Globular	Globular	Globular
DBH	10.5 cm	64 cm	24 cm	1.26m
Height (m)	5.26	7.42 m	8.42	19.46
Form	Tree	Tree	Tree	Tree
Sample Type	Ramas	Wood, 2-5 branches, flowers and branches for vouchers.	Wood, leaves, no flowers, but voucher taken. Primary to 5 branches taken.	Excellent tree. In great shape, never moved or bent its form. Heavily coppiced from, but a grand old tree that stands erect and proud. The last of a generation.
UTM N	92,89,550	92,89,551	92,87,323	92,87,324
UTM E	6,52,003	6,52,003	6,50,125	6,50,125
IGM Quad	Jayanca	Jayanca	Jayanca	Jayanaca
Place	Mancuhria	Mancuhria	Tres Puentes 'B'	Tres Puentes 'B'
Date Collected	9/18/2002	9/18/2002	9/18/2002	9/18/2002

APPENDIX B

Field Recording Form and Form Criteria for 2002 Ecological Survey

Reconocimiento de Plantas Combustibles en el Valle Rio La Leche
Survey of Fuel Plants in the La Leche River Valley

Specimen Number | **Date of Collection** | **Investigators**

Botanical Family | **Scientific Name**

Common Names

Location | **Duplicates**

IGM Quadrant | **UTM E** | **UTM N** | **Weather**

Location Description

Reconnaisance Marker

Sample Type

Sample Description

Plant Form: | **Height** | **Breast Height** | **Diameter at Breast Height**

Height of First Branch: | **Form of Branching:** | **Width of Branching E-W** | **Width of Branching N-S**

Inclination of Trunk

Inclination of Branches

Habitat Type: | **Ground Slope/Slope Direction**

Soil Type: | **Terrain:** | **Strata of Vegetation:**

General Description of the Ecology | **Photograph** ☐

Weather-Tiempo:
A. Plena Sol- Sunny-
B. Semi-Nublado- Partly Cloudy-
C. Nublado- Cloudy-
D. Lluvia- Raining-

Plant Form- Forma de Planta:
E. AM. Árbol (maduro)- Tree (mature)-
F. Árbol (juvenil)- Tree (Juvenile)-
G. Arbusto- Shrub-
H. Parra/Vid- Vine/Liana-
I. Monte de Hoja- Herb-
J. Grama- Grass-
K. Cactus- Cactus-

Form of Branching-Ramificación:
L. Sin Tallo- Acaulescent (no stem)-
M. Ramas se Encorvan- Ascending (obliquely upward, curved)-
N. Escalando- Climbing-
O. Se Arrastrando- Creeping-
P. Copa de Globo- Deliquescent (irregular branching/open)-
Q. Copa Rala- Divergent (spreading branches/scarce)-
R. Ramas Paralelas- Fastigiate (broomlike/parallel)-
S. Arbustivo- Frutescent (shrubby)-
T. Como Espiga- Orthotropic (Vertical growth)-
U. Como Palma- Palmate (palm-like)-
V. Aplastado- Prostrate (Flat on the ground)-
W. Como Almohadón- Pulvinate (Cushion-like)-
X. Como Cactus (Tuna)- Succulent (Cactus-like)-

Habitat Type- Tipo de Habitat:
SD. Bosque Seco(Semi-Denso/llano)- Dry Forest (Semi-dense/plains)-
RL. Bosque Seco(Ralo/llano)- Dry Forest (Sparse/plains)-
CO. Bosque Seco (Colina)- Dry Forest (Hill)-
SA. Bosque Seco (Sabana)- Dry Forest (Savanah)-
ES. Bosque Seco (En Establecimiento)- Dry Forest (Successional)-
CH. Chaparral Chaparral-
MA. Matorral- Brush/Scrub-
AG. Agricultura- Agricultural Field-
DE. Desierto-Desert-
ZP. Zona de Población- Populated Area-
RI. Zona del Rio- River Bottom-
LB. Lago/Bajo- Standing Water-

Soil Type-Tipo de Suelo:
1. Arena/Piedras (del Rio)- Sand/Pebbles (fluvial)-
2. Arena- Sand-
3. Arena/Arcilla- Sand/Clay-
4. Arcilla- Clay-

Terrain-Tipo de Terreno

5. **Plano-** Flat-
6. **Poco de Inclinación-** Slight Rise-
7. **Semi-Colina-** Moderate Hill-
8. **Colina Fuerte-** Steep Slope-
9. **Bajo-** Swamp-
10. **Rio-** Fluvial-

Strata of Vegetation-Estratos de Vegetación:

11. **Árboles, Arbustos, Malezas-** Arboreal, Shrubs, Herbs
12. **Árboles, Malezas-** Arboreal, Shrubs-
13. **Arbustos, Malezas-** Shrubs, Herbs-
14. **Árboles, Malezas-** Arboreal, Herbs-
15. **Arbustos-** Shrubs-
16. **Malezas/Hierbas-** Herbs-

APPENDIX C

Huaca Sialupe 2001 Excavations Field Feature and Floor Recording Forms

2001 HUACA SIALUPE FLOOR RECORDING FORM

Floor Number:

Area/Sector

Unit Number(floor):

Unit Dimensions(

Room Number(

Floor Provenience:

Date of Excavation:

Date of Record:

Excavators:

Floor Photo Number

Floor Drawing Number:

Depth Below Datum: **cm**

Floor Description

Association:

Floor Matrix:

2001 HUACA SIALUPE FLOOR RECORDING FORM

Floor Number:

Area/Sector

Unit Number(floor): **Unit Dimensions**

Room Number

Floor Provenience:

Date of Excavation: **Date of Record:**

Excavators:

Floor Photo Number **Floor Drawing Number:**

Depth Below Datum: **cm**

Floor Description

Association:

Floor Matrix:

APPENDIX D

Visual Atlas of Unknown Seeds Recovered from the Huaca Sialupe Assemblage
(Each Background Black and White Square Measures 1cm)

cf. Asteraceae-UKN#25-37

cf. Asteraceae-UKN#50-142

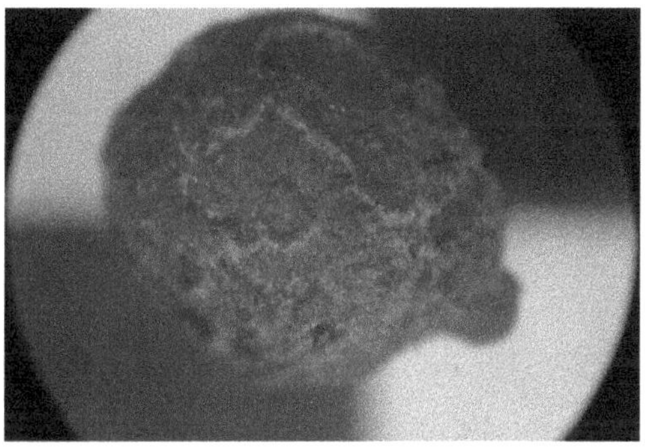

cf. Fabaceae-UKN#12-177

cf. Fabaceae-UKN#43-121.1

cf. Fabaceae-UKN#64-120.1

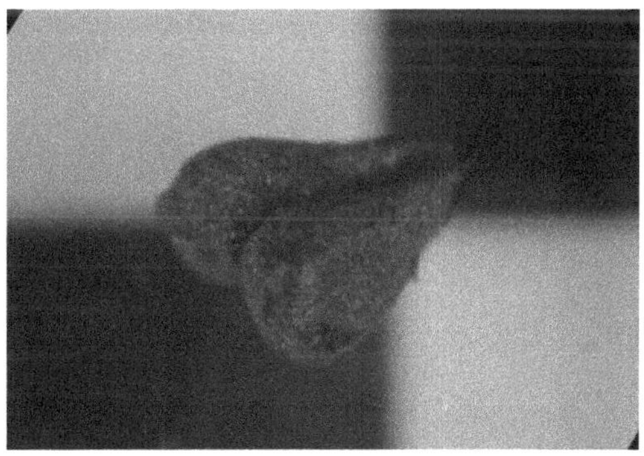

cf. Malvaceae- UKN#44-126

cf. Nyctaginaceae-UKN#41-121.1

UKN#13-176

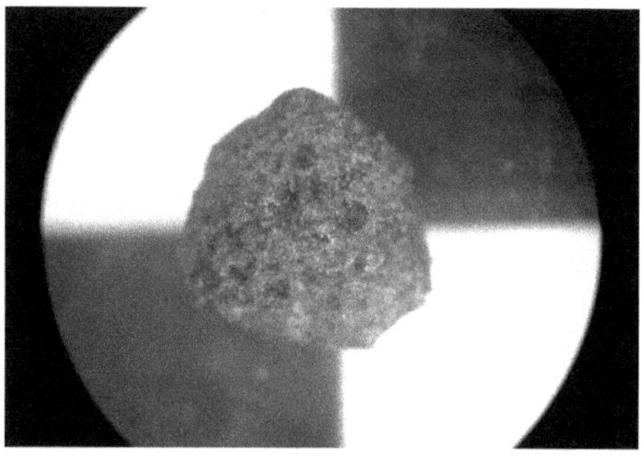

UKN#28-45

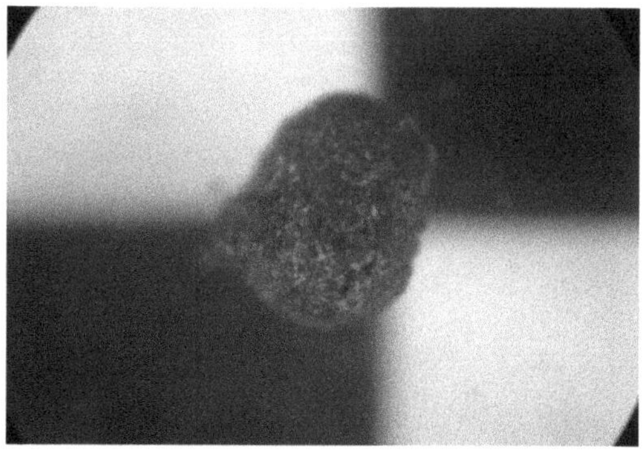

UKN#53-147

APPENDIX E

Recording Forms for Sample Contents
Modified from Dr. D. Pearsall (University of Missouri, Columbia, Anthropology Department)

Proyecto_______________ Procedencia _______
Pagina ___/___

No. de Muestra_______________ Procedencia_______________

Tamaño de Muestra_______________ml

Peso de Muestra _______________g

Porción de Muestra Escaneado _______% 4.0mm _______%2.0mm _______1.0mm%
_______%0.5mm
(por Volumen Total de la Muestra)

Escaneado por _______________ Fecha de Emp._______ Fecha de Term._______

Identificado por _______________ Fecha _______________

Madera/Carbón	Peso(g)	Cantidad	Observaciones

Semillas/Nueces	Peso(g)	Cantidad	Observaciones

Otros	Peso(g)	Cantidad	Observaciones

Otro Lado: ¿si o no?

APPENDIX F

Recording Forms for Wood Anatomy
Taken from Dr. L. A. Newsom (The Pennsylvania State University, Anthropology Department)

SITE:

PROVENIENCE:

IDENTIFICATION:

Vessel elements--

Vessel distribution-- ring-porous:
diffuse-porous:
semi-ring:

Vessel arrangement-- solitary:
short radial series:
chains:
dendritic:
tangential groups:
clusters:
other:

Vessel density*-- few (<25/mm^2):
moderate (>25<75/mm^2):
numerous (>75/mm^2):

Approx. average vessel size*-- v. small (<50 um):
small (51-100 um):
medium (101-150 um):
large (151-200 um):
v. large(>200 um):

Perforation Plates-- simple:
scalariform:
other:

Intervessel pitting-- alternate: approx. size:
opposite: approx. size:
scalariform:
transitional:

Spiral thickenings-- vessel walls:
fiber walls:

Fibers -- libriform:
fiber tracheids:
approximate density:
wall/lumen ratio:

CHARCOAL DATA SHEET, page 2   Specimen/provenience:

Vascular rays-- homocellular:
heterocellular:

ray height:
ray width:

>12 per mm:
pore-widths apart:

storied:

Axial parenchyma-- generally sparse/absent:

apotracheal -diffuse:

-diffuse-in-ag:

-reticulate:

-banded:

paratracheal-sparse:

-aliform:

-vasicentric:

-confluent:

-banded:

storied (w/ rays, vessels, etc):

Illustrations/notes:

*see Panshin & de Zeeuw 1980, p. 535)

APPENDIX G

Wood Anatomy of 14 Major Wood Species from the
La Leche River Valley, Lambayeque, Peru (130X)

Anacardiaceae

Loxopterygium huarango Hualtaco Goldstein 1-2002

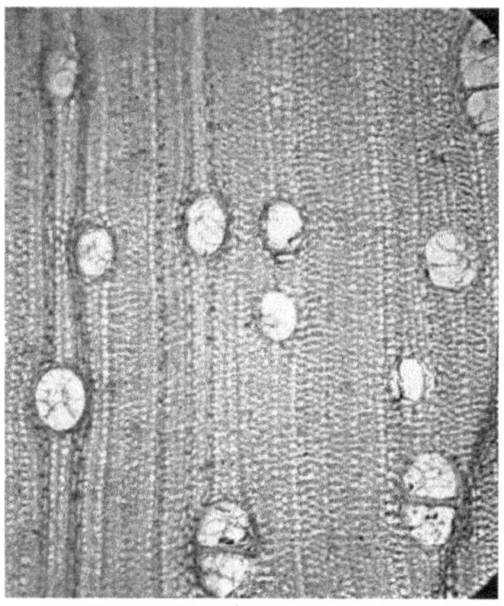

Cross Section

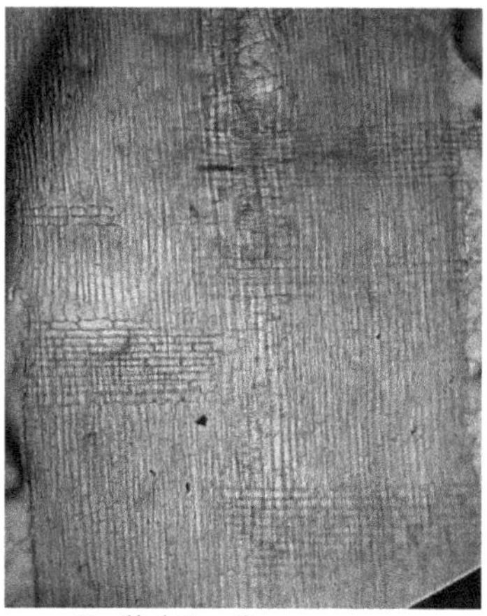

Radial Section

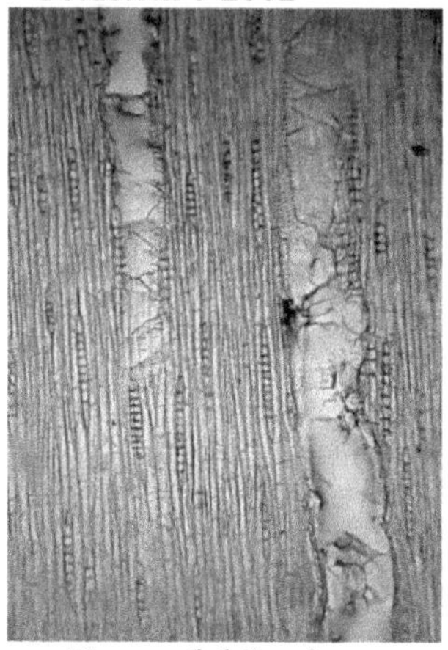

Tangential Section

Cross Section

Diffuse porous wood, with solitary vessels. Pore density less than $25/mm^2$. Perforation plates are simple. Tyloses, gum deposits, present in vessels. Axial parenchyma is vasicentric paratracheal. Vessels are one ray width apart. Rays are occasionally 1-2 cells wide and visible. Tightly packed libriform fibers, wall/lumen ratio 3:1.

Radial Section

Vessels exhibit minute alternate pitting, oblong/rhomboid vessel to ray pitting visible occasionally. Rays are heterocellular with a single row of upright cells at top and bottom of ray. Crystals are present.

Tangential Section

Rays 1-2 cells wide and 9-13 cells high. Rays are not storied. Crystals are visible in section of rays in tangential section. Vessels have gum/resin deposits or tyloses present. Septate fibers are visible on occasion.

Comparative Anatomy

This species exhibits occasional marking of end of growth cycle with thicker fibers in latewood. Marked late wood margins and septate fibers are clearly visible in more mature specimens, juvenile specimens tend not to exhibit these traits. Also in younger specimens, tyloses/gum/resin is less frequent and the pore density can increase to more than $25/mm^2$.

Apocynaceae
Valesia glabra Cuncun Goldstein 19-2002

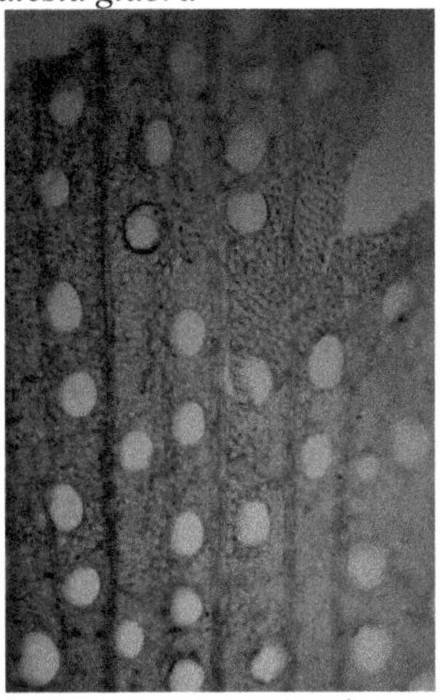

Cross Section

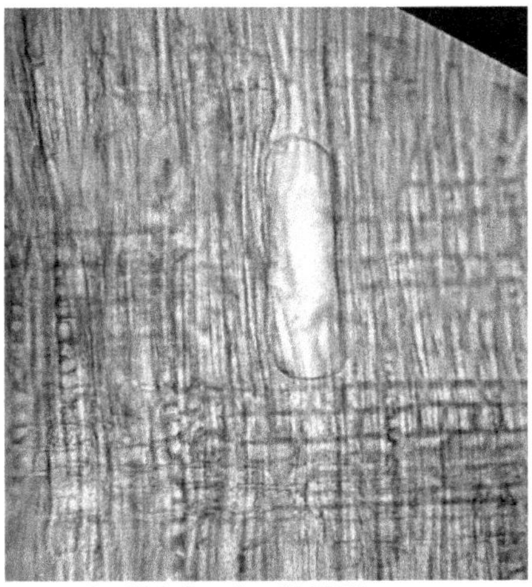

Radial Section(260X)

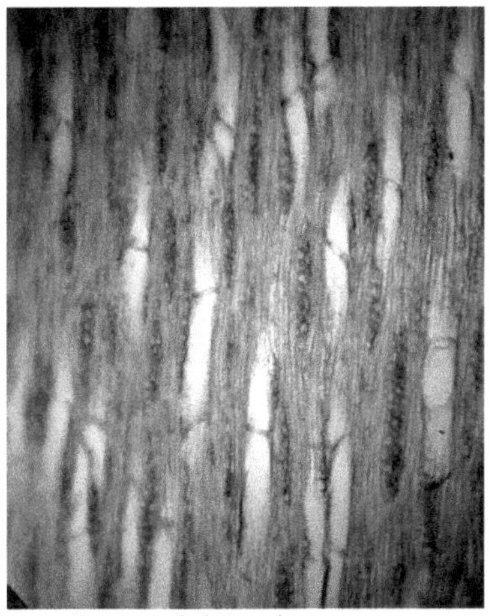

Tangential Section

Cross Section
This is a diffuse porous wood with solitary vessels. Vessels are small, 50 micros, and frequently occur between 25 and 75/mm^2. Vessels fall between rays with 1.5 vessel widths between rays. Perforation plates simple. Rays one to three cells wide. Fibers are libriform and dense with a wall to lumen ratio of 4:1. The axial parenchyma is a combination of scanty paratracheal with unicellular apotracheal dendritic formations between rays.

Radial Section
Rays are heterogeneous with rows of uprights interspersed with rows of procumbent cells. The vessel to ray pitting is distinctly round. Axial parenchyma are crystalliferous.

Tangential Section
Rays are 1-3 cells wide and 8-14 cells high. Vessels pitting is alternate. Vessels tend to be half the height of the rays.

Comparative Anatomy
Of the several diffuse porous woods with 1-2 cell wide heterocellular rays in this assemblage *V. glabra* is very uniform in its distribution of vessels from juvenile to mature growth indicating a steady almost continual grow pattern. While it never appears to be storied, its anatomy is still characterized by a consistently regular distribution of anatomical features.

Asteraceae

Tessaria Integrifolia Pajaro Bobo Goldstein 25-2002

Cross Section

Radial Section

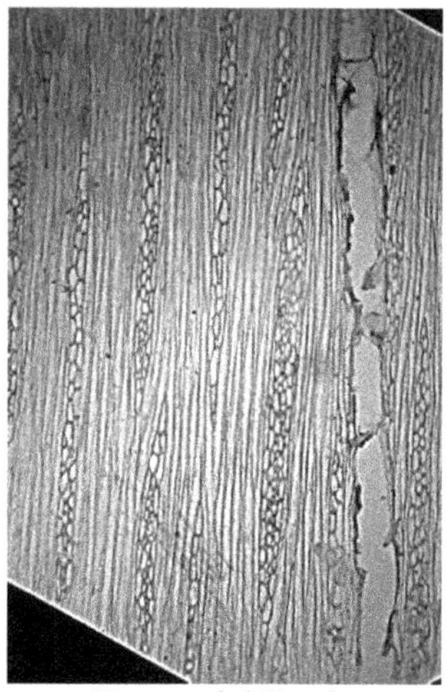

Tangential Section

Cross Section

This is a diffuse porous wood with solitary vessels. The axial parenchyma is paratracheal and markedly not confluent. Rays are 2-3 cells wide and are spaced 2-3 pores apart. Vessel density is less than $25/mm^2$, and have simple perforation plates. Fibers have a wall to lumen ration of 2:1

Radial Section

Rays are heterocellular with rows of upright and procumbent cells interspersed. Occasional rows of uprights border rays, but more commonly are seen throughout the ray. Ray procumbent cells are blocky and markedly broad. Vessel wall pitting is alternate. Ray to vessel pitting is small and round with 4-6 pits/cell.

Tangential Section

Rays are 2-3 cells wide and 35-40 cells high. Gum or resin is occasionally apparent in the vessels. These are some of the longest rays in the assemblage.

Comparative Anatomy

The consistently linear and 2-3 cell wide rays with small and concentrated regions of apotracheal axial parenchyma around vessels are distinct. Additionally, the small vessel to ray pits are consistent throughout the samples, including juvenile samples.

Boraginaceae
Cordia lutea Overo Goldstein 37-2002

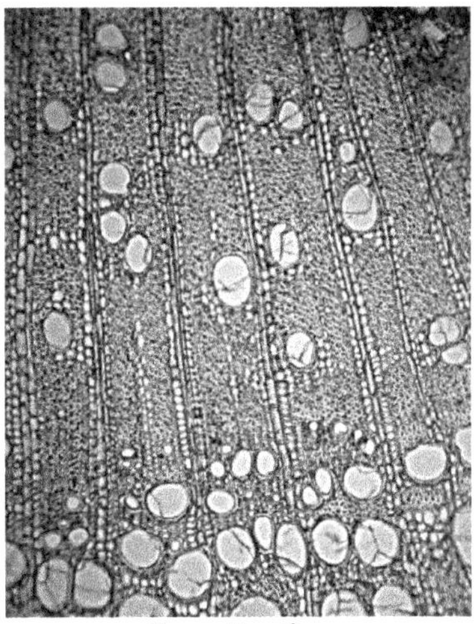

Cross Section

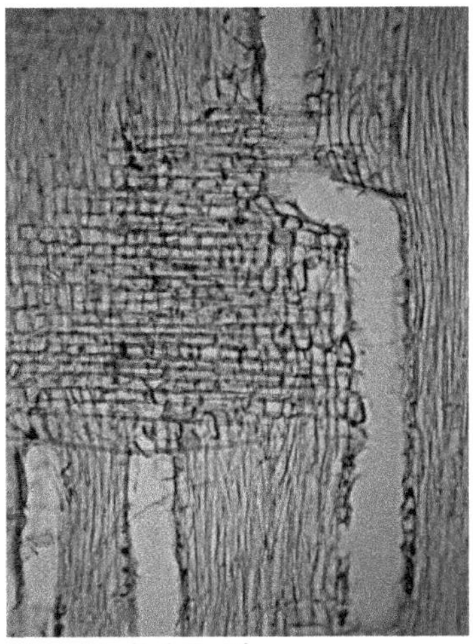

Radial Section

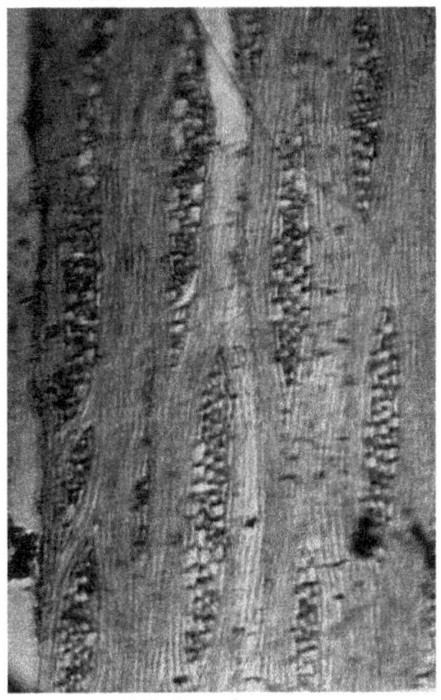

Tangential Section

Cross Section
This wood is diffuse porous, with solitary vessels. However, there is some clustering of vessels in juvenile and branchwood at the beginning of the growth cycle. Also, early wood vessel size tends to be somewhat smaller than late wood vessels. Rays are consistently 3-4 cells wide, and are about 2 pore widths apart. Vessel density is between 25 and 50/mm^2. Vessel perforation plates are simple. Fibers are libriform with wall to lumen ratios of 3:1. Axial parenchyma is predominantly vasicentric paratracheal, with a tendency to confluent axial parenchyma in early wood where vessels are clustered.

Radial Section
Rays are heterocellular with rectangular prismatic crystals appearing predominantly in the procumbent cells. Heterogeneity does not appear to follow in consistent rows, i.e. short series of procumbent and upright cells interspersed; not tile cells. Vessel to ray pitting is simple. Vessel wall pitting is alternate.

Tangential Section
Rays are 2-3 cells wide and 10-20 cells high. Vessels with simple perforation plates and rays tending to be 3-4 vessels high.

Comparative Anatomy
See previous note about early wood growth. Additionally vessel distribution tends to be less regular than in other similar heterogeneous ray, solitary vessel, diffuse porous woods.

Burseraceae

Bursera graveolens Palo Santo Goldstein 3-2002

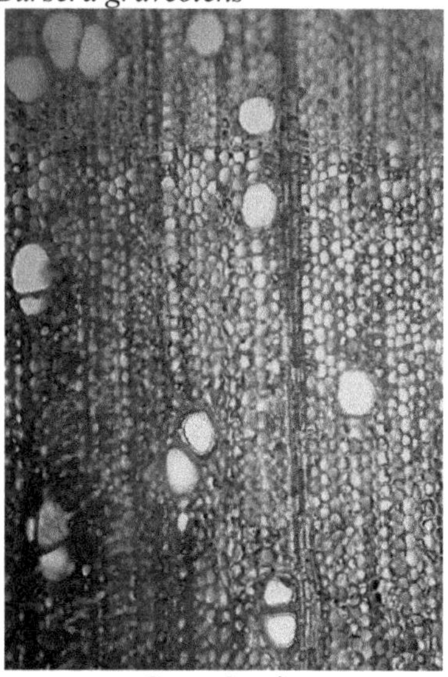

Cross Section

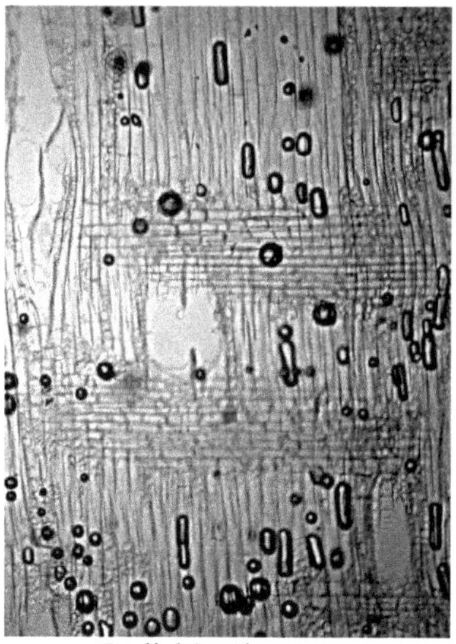

Radial Section

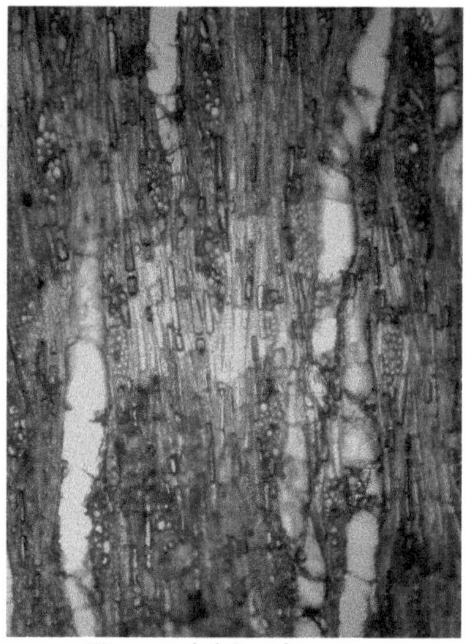

Tangential Section

Cross Section

The wood is diffuse porous with solitary vessels. Rays are 2-4 cells wide and are spaced 2-5 pore widths apart. The vessel perforation plates are simple. Fibers are libriform and are not very thick walled. Axial parenchyma and fibers can be indistinct at times. Fibers appear to be thicker at late wood margins going from wall to lumen ratios of 1:4 to 1:2 in late wood. Axial parenchyma is vasicentric paratracheal and can be sparse at times.

Radial Section

Rays are heterogeneous with rows of uprights interspersed throughout the ray. Oil cells are visible in the parenchyma, and there are occasional resin canals visible in the radial section. Vessel pitting is alternate. Vessel to ray pits are bordered and rectangular. The fibers are septate.

Tangential Section

Rays tend to be 2-4 cells wide consistently, however the height varies between 12 and 25 cells. Rays are generally no more than two vessels high. Resin canals are visible within the rays in tangential section. Also, there are occasional tyloses/gum deposits present in the vessels.

Comparative Anatomy

This is the one non-legume species within the assemblage that has resin canals in its rays, and both the late wood thickening of the fibers and the otherwise indistinct axial parenchyma are distinctive features of this particular wood.

Capparidaceae

Capparis angulata Sapote Goldstein 51-2002

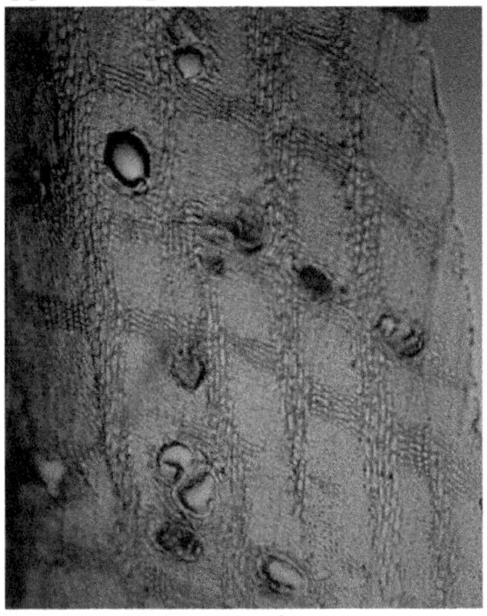

Cross Section

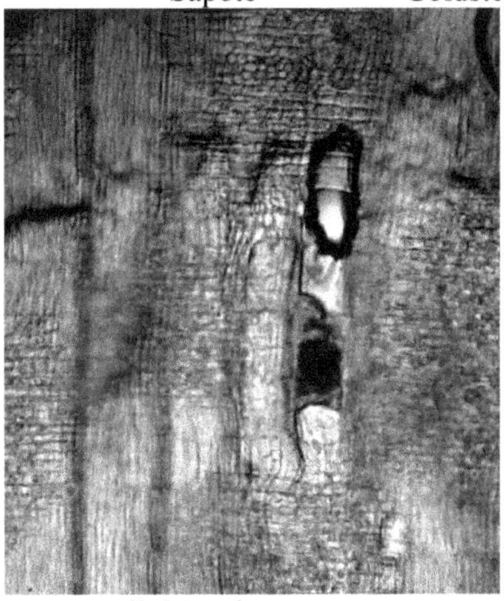

Radial Section

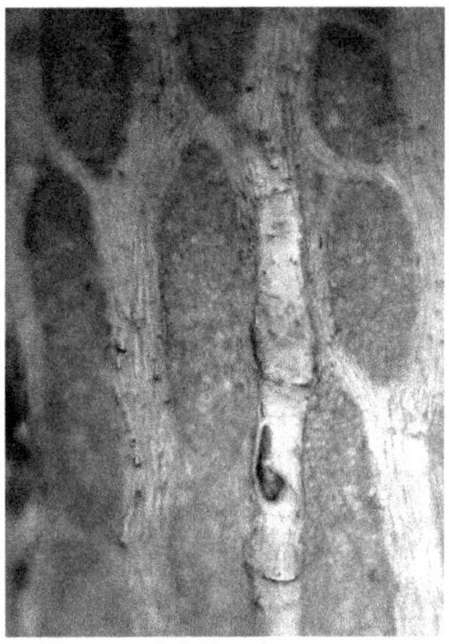

Tangential Section

Cross Section

The wood is diffuse porous, however, it exhibits consistent concentric ring production that is likely tied to some seasonality. Pores are consistently solitary and similarly sized. Ring margins appear to be composed of denser marginal bands of axial parenchyma. Vessels have simple perforation plates with alternate intervessel pitting. Gum inclusions or tyloses are present in some heartwood vessels. Fibers are densely packed with a lumen to wall ratio of 2:1. While some axial parenchyma are paratracheal and vasicentric, the majority of the axial parenchyma appears as apotracheal and banded. Most vessels appear at intersections of the parenchyma bands. Parenchyma bands are 8-10 cells wide.

Radial Section

Rays are heterocellular with rows of 2-3 upright cells at the ray margins. Procumbent cells in the ray bodies contain rhombohedral crystals.

Tangential Section

Rays are unstoried and tend to be 15-45 cells high and 8-10 cells wide. Some rays appear to aggregate. Rays tend to be 1-2 vessel widths apart. Rays are less than 12 per square millimeter.

Comparative Anatomy

The thick and banded parenchyma is the most distinguishing characteristic of the woods in this assemblage. Overall the wood appears, at the anatomical level in all three dimensions, to be 50% parenchyma.

Capparidaceae
Capparis avicennifolia Vichayo Goldstein 54-2002

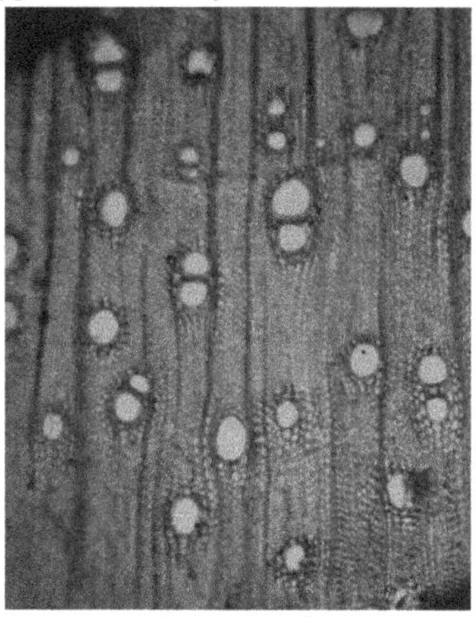
Cross Section

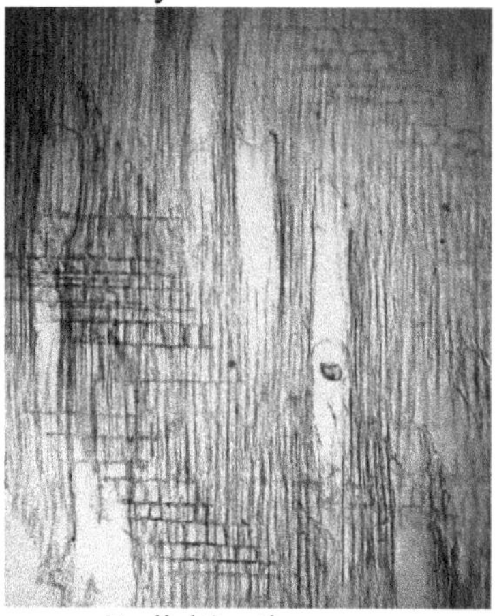
Radial Section

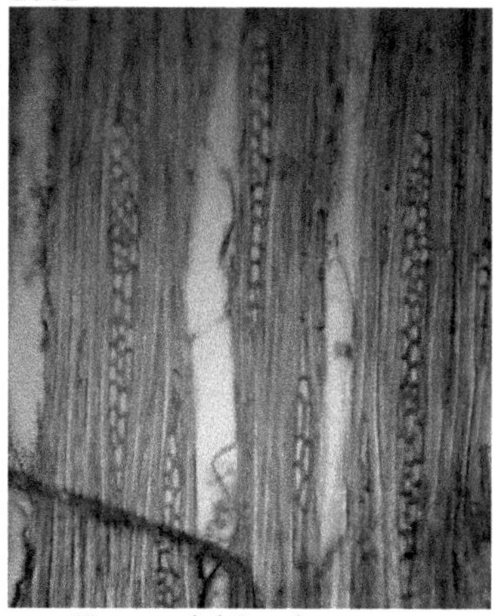
Tangential Section

Cross Section
This wood is also diffuse porous, however, there does appear to be a marked tendency to concentric banding related to seasonal or environmental change. Some latewood (?) pores appear to be smaller than earlier vessels. Vessels are solitary, with simple perforation plates. Vessels tend to be small to very small, less than 50 microns across. Axial parenchyma is predominantly vasicentric paratracheal with a tendency to marginal bands at the end of growth periods and some confluence between close vessel pairs, tending towards aliform. Pores tend to be one rays width apart and rays appear to be 2-3 cells wide. Fibers are dense with a 3:1 wall to lumen ratio

Radial Section
Rays are heterogeneous, with a single row of uprights at the top and bottom of each ray. Occasional crystals are visible in the ray body cells.

Tangential Section
Rays tend to be 1-3 cells wide and 10-30 cells high. They appear to be one pore width apart. Intervessel pitting is alternate. Vessel to ray pitting tends to be simple, occasionally with reduced borders. Upright cells tend to be coupled and embedded in the multiserriate/procumbent section of the rays.

Comparative Anatomy
This wood, in particular, warrants further study in terms of looking at its potential for tracing long term climate change and dendrochronology.

Fabaceae (Mimosoidiae)
Acacia macracantha Faique Goldstein 109-2002

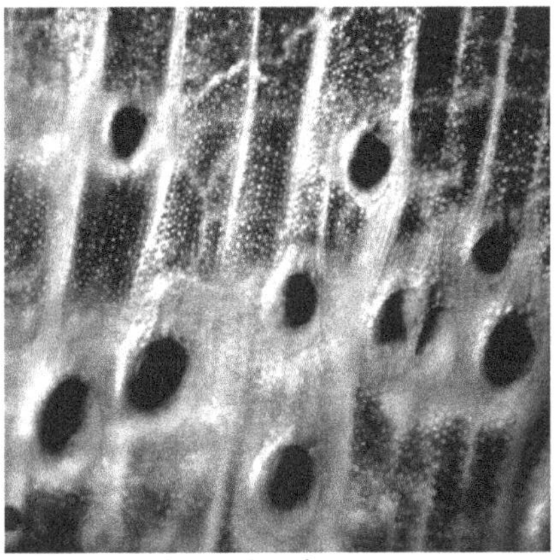
Cross Section

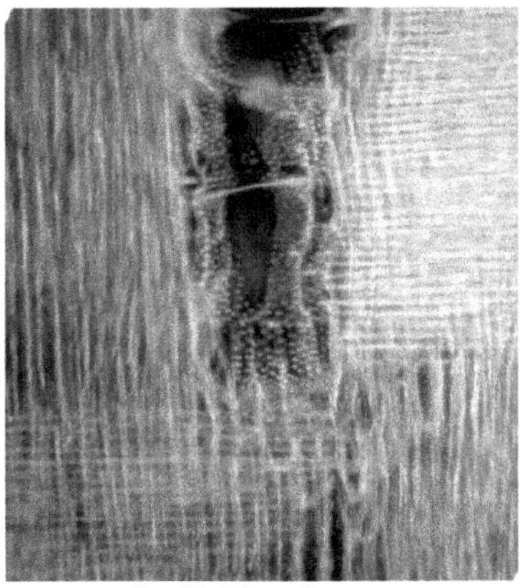
Radial Section

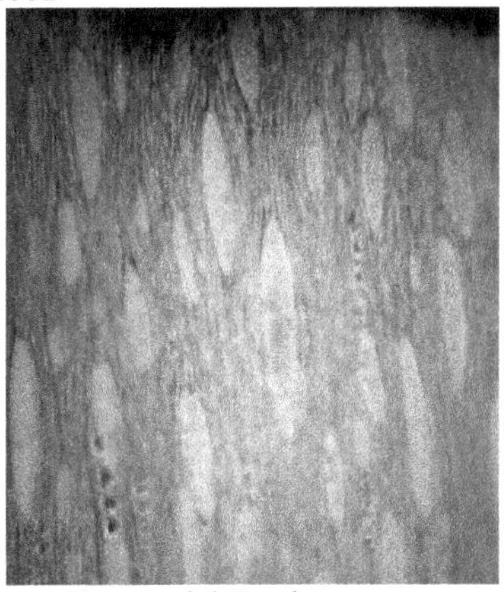
Tangential Section

Cross Section
This wood is diffuse porous with a tendency to marginal concentric bands of vessels interconnected by aliform confluent vasicentric axial parenchyma. Otherwise the axial parenchyma is markedly aliform, but not confluent. Vessels are solitary with simple perforation plates. Rays are 2-3 cells wide. Fibers are libriform and dense with a 4:1 wall to lumen ratio. Vessels tend to be 50-100 microns in diameter. Some tyloses are present with oils/gum/exudates present in heartwood vessels.

Radial Section
Rays are homocellular. Intracellular and vessel to ray pitting is alternate. There are occasional oil/gum/exudate cells present among the ray body cells. Axial parenchyma cells are substantially robust.

Tangential Section
Rays tend to be 1 to 1.5 vessel widths apart. Rays are 2-4 cells wide and 10-25 cells high.

Comparative Anatomy
A. macracantha anatomy is very similar to the anatomy of the *Prosopis* genera in this dataset. The one distinguishing factor used in this analysis is a strong tendency for a pronounced aliform formation of the axial parenchyma in *A. macracantha.* Moutarde (2006) also noted this phenomenon. This trend is visible even in early growth of juvenile wood; visible in branchwood and juvenile boles after the second cycle of growth. Both *A. macracantha* and *Prosopis* sp., however, have very similar juvenile anatomies that is typified in cross section by long chains of young vessels, 4-7 vessels, and marginal bands of axial parenchyma 1-2 cells thick. This pattern tends to be longer lived in *Prosopis*, 3-5 cycles of growth, is formed around the compressed pith of the first conductive tissue formed, and is visible in both branchwood and juvenile boles. Spiral thickenings are not present within juvenile vessels.

Fabaceae (Mimosoidiae)

Pithecellobium multiflorum Angolo Goldstein 107-2002

Cross Section Radial Section Tangential Section

Cross Section

This wood is diffuse porous. Vessels appear in two different arrays, solitary, and short radial series in groups of three or four. Axial parenchyma is paratracheal vasicentric aliform and confluent. Fibers are dense and compact. Fibers have a wall to lumen ratio of 3:1. Rays tend to be 2-3 cells thick, and are two pore widths apart.

Radial Section

Rays are homocellular. Ray to vessel pitting is same as the intervessel pitting that is alternate, and appears vestured. Ray cells appear elongated.

Tangential Section

Rays are unstoried. Rays are 2-3 cells wide and 10-15 cells high. They tend to be 2-3 pore widths apart. Fibers are septate with 5-7 segments.

Comparative Anatomy

The description of the cellular anatomy of this plant is of critical importance as it is likely that this species is a keystone species of the mature stature primary forest of the area. Archaeologically its presence has never been reported.

Fabaceae (Mimosoidiae)

Prosopis pallida Algarrobo (Macho) Goldstein 71-2002

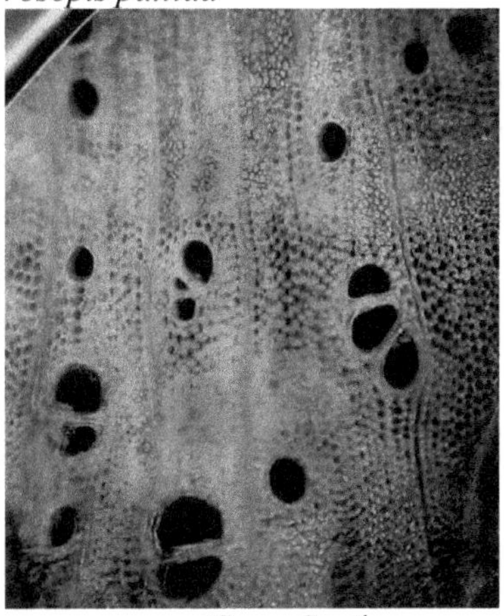

Cross Section

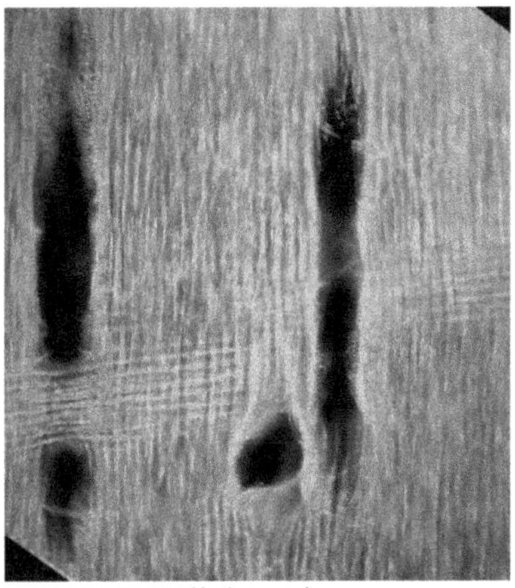

Radial Section

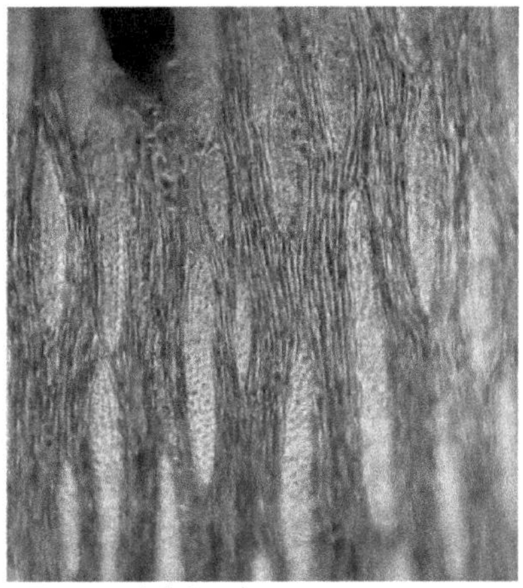

Tangential Section

Cross Section

The wood is diffuse porous. Vessels are solitary and small to very small 20-70 microns. The perforation plates are simple and the vessel pitting is alternate. Record and Hess (1972) state that intervessel pits are vestured, however, the pits are minute and without scanning electron micrographs it is difficult to assess the presence of vestured pits. Fibers are dense and densely compacted with a wall to lumen ratio of 5:1. Rays are 2-5 cells wide. Axial parenchyma is predominantly paratracheal vasicentric and aliform. Some apotracheal axial parenchyma occurs as indistinct marginal or reticulate bands, generally one row of cells.

Radial Section

Rays are homocelluar and unstoried. Fibers are long and dense with almost indistinct slit-like pits. Fibers appear interlocked.

Tangential Section

Rays are 2-5 cells wide and 7-40 cells high. They appear in a density of more than 12 rays per mm squared. Rays tend to be 1.5 to 2 pore widths apart. Fibers appear interlocked.

Comparative Anatomy

Again, the anatomy of this particular wood should be considered relative to that of the *A. macracantha.* As noted above the juvenile and mature anatomy of these two species is difficult to distinguish. One key distinguishing factor of more mature wood in *Prosopis,* not previously mentioned, is that the vessel size in *Prosopis* sp. is considerably smaller, with much more localized vasicentric paratracheal axial parenchyma than as seen in *A. macracantha.* Fibers are finely interlocked and much more densely complied than in *A. macracantha.*

Fabaceae (Caesalpinoideae)
Cassia bicapsularis Alcapparillo Goldstein 1-2000

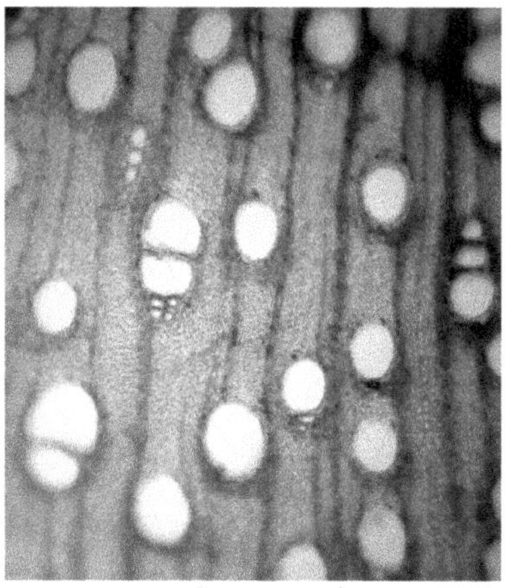
Cross Section

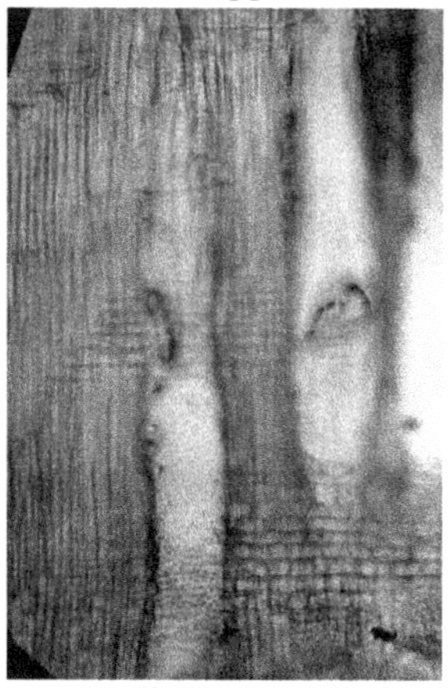
Radial Section

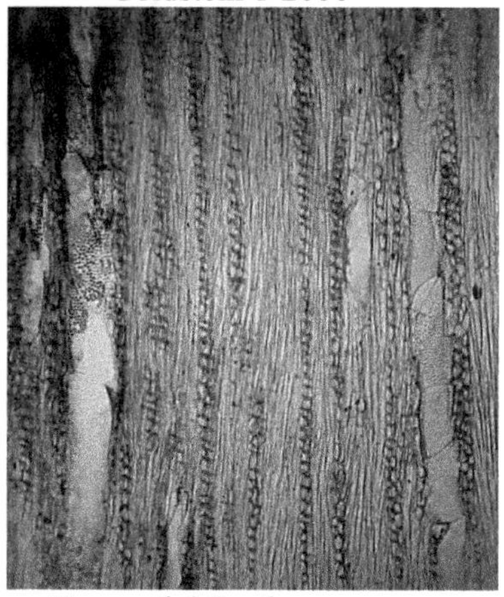
Tangential Section

Cross Section
The wood is diffuse porous. Vessels are predominantly solitary with occasional short radial series of 3-4 smaller vessels. Solitary vessels are medium sized, around 100-150 microns, with simple perforation plates. Fibers are dense, and libriform, with wall to lumen ratios of 4:1. Axial parenchyma is generally sparse to absent. For the most part axial parenchyma tends to be paratracheal. Rays are unicellular.

Radial Section
Rays are homocellular. Vessel pits are minute and alternate. Vessel to ray pitting is same as the intervessel pitting. Some crystals are found in ray body cells.

Tangential Section
Rays are unicellular and long, between 25 and 30 cells high. Rays tend to be one vessel width apart.

Comparative Anatomy
The uniseriate rays are a major distinguishing characteristic among the species in this assemblage.

Fabaceae (Caesalpinoideae)
Cercidium praecox Palo Verde Goldstein 6-2002

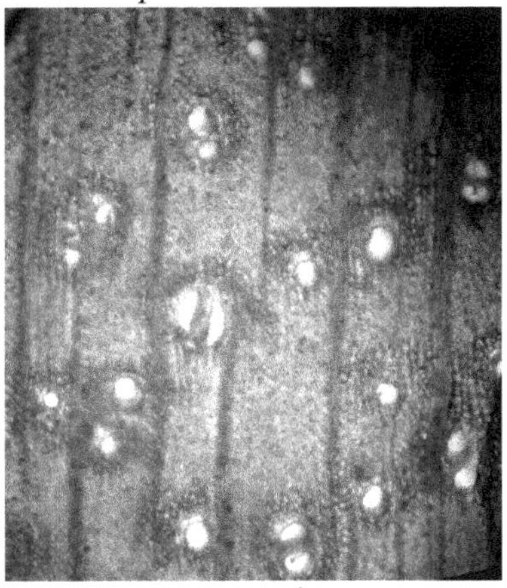
Cross Section

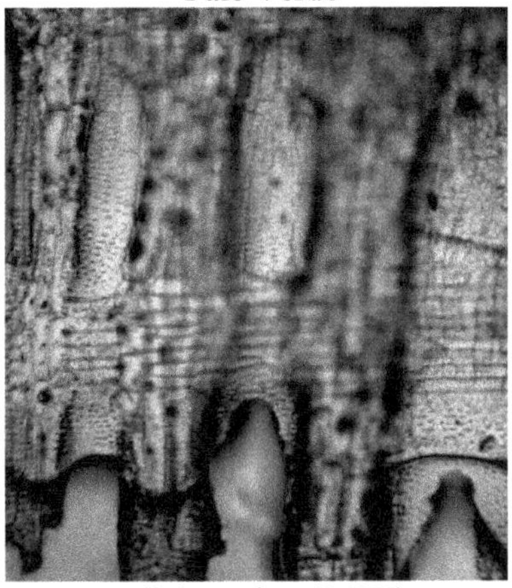
Radial Section (260X)

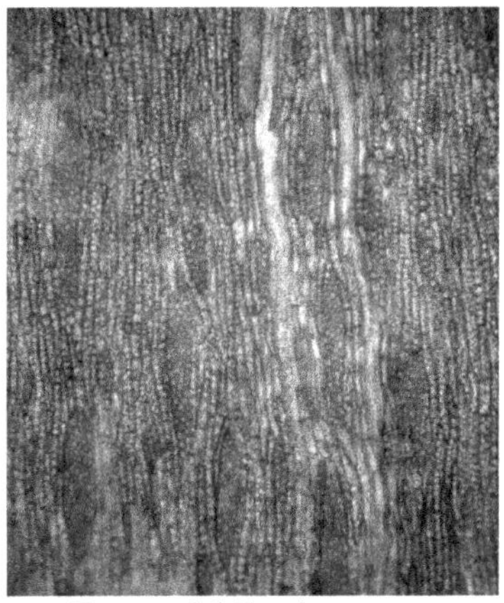
Tangential Section

Cross Section
The wood is diffuse porous. Most vessels are solitary and very small, 40-70 microns. Occasional short series of 2-3 vessels are visible. Vessel density is low, less than 25 per square millimeter. Perforation plates are simple, fibers are dense and libriform with wall to lumen ratios of 5:1. Axial parenchyma is paratracheal vasicentric aliform, and is occasionally confluent; this pattern appears to be more pronounced at growth margins. Rays are 2-3 cells wide.

Radial Section
Rays are homocellular with rhombohedral crystals present in some body cells. Intervessel pitting is alternate. Vessel to ray pitting is same as the intervessel pitting.

Tangential Section
Rays are unstoried and are 2-3 vessel widths apart. Rays are 2-3 cells wide and 6-14 cells high. Rays are sometime aggregate.

Comparative Anatomy
This tree mostly grows along the upper piedmont slopes and in adjacent areas to the research site. Again, as with other species in this assemblage the tendency for marginal patterns in the axial parenchyma may be a good indicator that species could be exploited for paleoenvironmental work.

Fabaceae (Caesalpinoideae)
Inga feuillei Pacay or Guaba Goldstein 157-2002

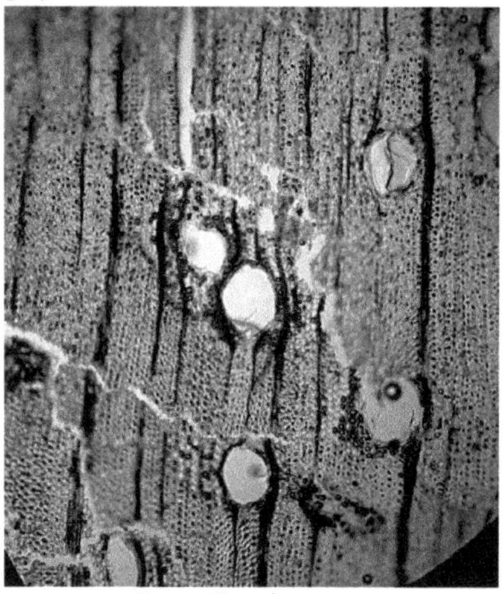
Cross Section

Radial Section

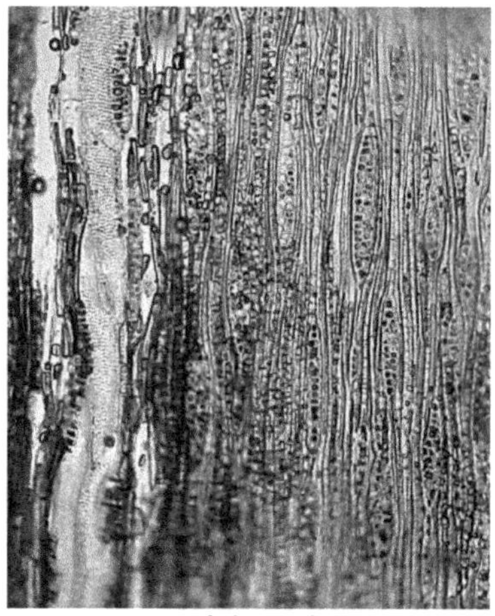
Tangential Section

Cross Section
The wood is diffuse porous. Vessels are solitary and large, 120-160 microns across. Perforation plates are simple. Axial parenchyma is scanty paratracheal and vasicentric. Rays are 2-3 cells wide and border pores. Fibers are libriform and dense with wall to lumen ratios of 5:1.

Radial Section
Intervessel pitting is alternate and tends to be minute. The ray to vessel pitting is the same as the intervessel pitting. Rays are homocellular. Some ray cells have crystals. The fibers are septate with 5-7 divisions.

Tangential Section
Rays are 2-3 cells wide and 15-20 cells high. Rays are mostly one vessel apart and homocellular.

Comparative Anatomy
Of the wood in this assemblage this is the most common tree associated with agricultural systems in the modern landscape. While the seeds and fruit remains have been recovered from several of the sites in the area, none came from Sialupe proper. Most of the plant remains associated with *Inga* sp. come from central coast sites, possibly indicating that its use on the north coast is a later development than the Middle Sícan period.

Fabaceae (Caesalpinoideae)
Parkinsonia acuelata Cabelludo Goldstein 15-2000

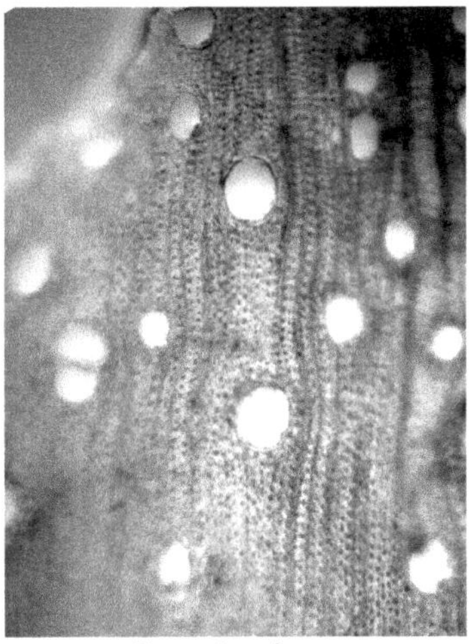

Cross Section

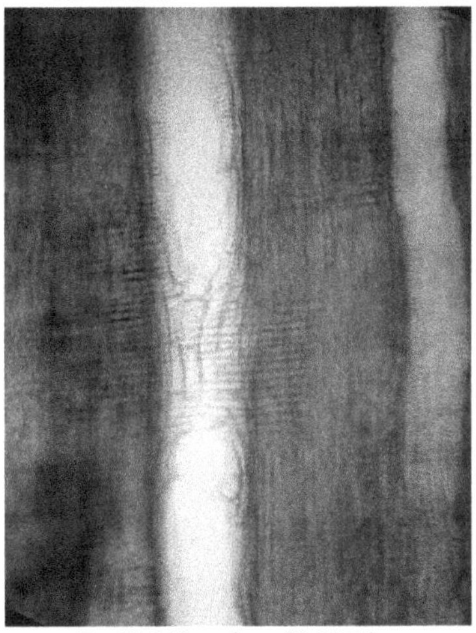

Radial Section (260X)

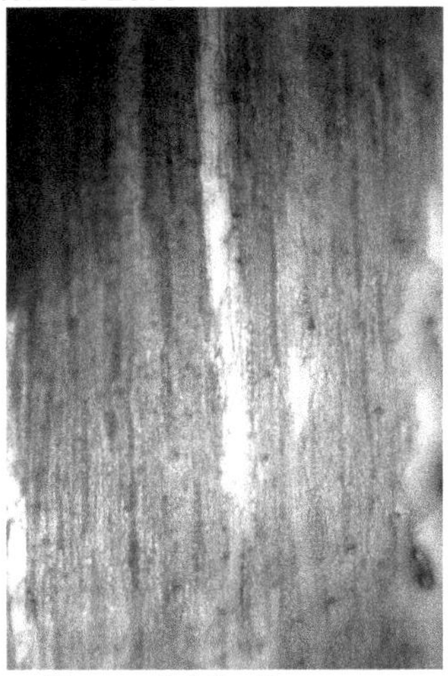

Tangential Section

Cross Section
This wood is diffuse porous. Vessels are solitary, small, and sparsely distributed, less than 25 per square mm. Perforation plates are simple. Fibers are libriform and not densely packed with some intercellular spaces visible in the cross section. Wall to lumen ratio is about 3:1. Axial parenchyma is paratracheal, sparse, and vasicentric. Rays tend to be less than one pore width apart and are 1-2 cells wide.

Radial Section
Rays are homocellular. Intervessel pitting is alternate, and ray to vessel pitting is the same.

Tangential Section
Rays are 1-2 cells wide and 10-15 cells high. Vessels, while not storied are rather evenly distributed with a density that is less than 12 per square millimeter.

Comparative Anatomy
Ethnotaxonomically this species is important as it is also referred to as *Palo Verde* in some areas, and unlike *Cercidium praecox* it grows in the lower valley areas frequently being used as a living fence. Additionally, it is dangerously spiny and as a result is rarely used for much other than fencing.

APPENDIX H

Overall Weight and Volume Data for Huaca Sialupe Excavated Samples

Sample Number	29	30	31	41	44	45	84	85	86	87	94	99	108
Total Weight (Kg)	11.623	12.9625	5.0625	8.5625	10.3625	6.06	4.6125	3.5625	3.2625	3.0625	2.3625	4.4625	1.0625
Total Volume (ml)	6000	10100	4050	6100	5825	4270	3000	3675	3650	2450	3700	3900	3475
Context	SIA-Rm21-F1	SIA-Rm30-F1	SIA-Rm30-Fea63	SIA-Rm30-F3	SIA-Rm25-F1	SIA-Rm24-F3	SIA-Rm23-F2-South	SIA-U3-F7(or Below)-Fea35	SIA-U12/13-3/4-Fea92A	SIA-Rm23-F2-Fea83	SIA-Rm23to35-F2-Fea 84B	SIA-Rm21-F1-Fea49	SIA-MII-U7/8-F4-Fea15
Volume (ml) X>0.425 mm	520	970	500	60	150	250	275	400	475	150	300		325
Weight (g) X>0.425 mm	850	1058	538	67	220	313	337	225	286	152	169		65
Volume (ml) X>1.0 mm	620	520	400	50	150	300	275	225	300	180	250		350
Weight (g) X>1.0 mm	650	492	434	49	143	308	256	120	169	192	130		65
Volume (ml) X>2.0 mm	600	1075	500	475	750	475	375	225	275	250	250		350
Weight (g) X>2.0 mm	623	1053	484	428	749	400	410	122	183	276	166		62
Volume (ml) X>4.0 mm	650	875	600	925	2350	950	375	500	1400	2165	560		2175
Weight (g) X>4.0 mm	643	852	686	905	2675	962	398	294	1110	2091	500		523
SAP Sample No.	T-82	T-88,T-89,T-90	T-91, T-92	T-111, T-115	T-114, T-110	T-109	T-152	T-121	T-151	T-155	T-168	T-126	T-189

Sample Number	109	110	111	113	114	115	116	117	118	118.1	119	120	120.1	121.1	126
Total Weight (Kg)	2.0625	5.3625	2.4625	2.1625	2.7625	4.5625	5.0625	3.3625	4.8	2.9625	3.8469	3.846	2.1625	2.3491	2.9625
Total Volume (ml)	2225	3625	2350	1125	3670	4375	4215	3800	4775	2300	3300	6050	4000	4375	3130
Context	SIA-U51-F2-Wall38(1999)-Fea96	SIA-U54-F1-Fea99B	SIA-U55-Fea98B	SIA-MII-U2-Fea9	SIA-U52/53-F2-Fea97	SIA-U57-Fea100	SIA-U55-Fea101-Below Platform	SIA-MII-F5-Fea8	SIA-MII-U2/7-Fea9-Fogon2	SIA-MII-U5/11-Fea11B	SIA-MII-Fea14	SIA-MII-U2/7-Fea9-Fogon1	SIA-MII-Fea114	SIA-MII-F5-Fea8	SIA-U56-Fea103
Volume (ml) X>0.425 mm	250	400	250	125	425	350	225	225	275	175	275	250	325	550	470
Weight (g) X>0.425 mm	102	249	255	58	241	225	196	227	198	160	193	146	137	165	245
Volume (ml) X>1.0 mm	225	325	225	50	300	225	125	275	350	175	225	350	325	650	525
Weight (g) X>1.0 mm	139	182	200	32	160	134	85	116	178	148	142	120	109	162	265
Volume (ml) X>2.0 mm	200	250	150	60	175	175	90	450	475	210	250	475	450	650	325
Weight (g) X>2.0 mm	138	186	150	30	120	114	67	156	197	179	165	143	135	164	187
Volume (ml) X>4.0 mm	225	1200	250	55	255	350	200	2100	1575	450	1000	4250	3000	1775	500
Weight (g) X>4.0 mm	141	1061	173	49	167	254	150	660	570	403	672	1042	796	412	428
SAP Sample No.	T-183	T-185	T-186	T-172	T-184	T-188	T-187	T-193	C-57	T-173	T-200, C-24	C-56, C-20	C-53	C-52, C-25	T-178

Sample Number	127	128	129	130	131	132	133	134	135	136	137	139	140	121	142
Total Weight (Kg)	4.4625	4.5625	2.4	2.5	2	0.1431	2.9	4.2	5	6.4186	4.2403	3.6	3.0625 (100.4)	2.8	3.3
Total Volume (ml)	3380	3495	3125	3140	3575	350	2355	3156	3570	7350	7025	2725	525	4425	3600
Context	SIA-MII-U11/5-Fea11A	SIA-MII-U11/5-Fea16	SIA-E.Extension-Fea68B	SIA-E.Extension-Fea68C	SIA-E.Extension-Fea68D	SIA-M?-U13-F4	SIA-MII-U11/12/12+-F3	SIA-MII-U11-F3	SIA-MII-U11-Fea21	SIA-MII-U12-Fea19	SIA-MII-U12-Fea20	SIA-MII-U15/18-Rm3-F2	SIA-MII-U2/3-Fea12	SIA-MII-U2/3-Fea12	SIA-MII-U7-Below F4
Volume (ml) X>0.425 mm	300	255	125	140	100	5	300	500	350	375	350	375	1	375	525
Weight (g) X>0.425 mm	223	246	91	111	74	1	336	483	290	196	221	353	1	136	196
Volume (ml) X>1.0 mm	300	250	125	125	100	5	300	475	275	375	425	400	2	475	325
Weight (g) X>1.0 mm	207	230	65	87	56	2	299	450	235	137	174	346	2	124	99
Volume (ml) X>2.0 mm	325	275	225	160	125	5	250	555	275	525	725	350	10	750	500
Weight (g) X>2.0 mm	232	253	87	91	57	2	264	540	247	154	218	330	3	161	131
Volume (ml) X>4.0 mm	675	700	2000	2250	3000	350	250	700	675	4800	5100	325	425	1750	325
Weight (g) X>4.0 mm	501	512	652	903	938	74	242	725	607	1116	1317	302	93.3	363	85
SAP Sample No.	T-174	T-175	C-55	C-59	C-54	C-27	T-278	T-277	T-310	C-49, C-69, T-311	C-50, C-70	T-312	C-19	C-60	T-262

Sample Number	145	146	147	148	149	153	154	155	156	157	159	160	161	162	169
Total Weight (Kg)	1.9	1.4	1.2442	1.5	4.2	4.2	1.7	2.3	2.8	1.2	0.191	3.4	11.8	1.6	0.9
Total Volume (ml)	3975	1975	975	1300	3425	5925	1950	2850	3700	3200	750	4950	10850	3450	775
Context	SIA-Rm23-Fea81	SIA-Rm23to35-F1-Fea84A	SIA-Rm29-F1-Fea67	SIA-Rm30-F4-Fea131	SIA-Rm30-F4	SIA-Rm38-Fea110-Horno 4A	SIA-Rm38-Fea110-Horno 4B	SIA-Rm38-Fea111-Horno5	SIA-Rm39-F4	SIA-Rm39-F4-Fea122A	SIA-E.Extension-Fea68Ca	SIA-Rm39-Fea109-Horno3	SIA-Rm39-Fea117	SIA-Rm39-Fea123	SIA-U2-Fea91C
Volume (ml) X>0.425 mm	150	400	50	125	250	325	260	350	600	6		460	350	60	80
Weight (g) X>0.425 mm	110	170	41	119	304	167	130	202	326	14		260	232	39	65
Volume (ml) X>1.0 mm	150	325	50	70	250	290	200	225	425	10		425	300	60	25
Weight (g) X>1.0 mm	103	137	28	51	273	127	89	141	239	9		200	153	27	20
Volume (ml) X>2.0 mm	350	325	40	55	325	525	190	250	350	10		300	1000	75	10
Weight (g) X>2.0 mm	171	130	28	39	332	201	70	158	205	10		162	503	31	16
Volume (ml) X>4.0 mm	3250	575	65	175	575	750	325	350	675	2900		500	3700	3165	80
Weight (g) X>4.0 mm	908	341	46	78	498	303	88	191	402	884		251	1614	816	65
SAP Sample No.	C-65	T-244	T-280	T-269	T-270	T-283	T-288	T-285	T-289	C-35	C-58	T-287	C-61	C-36	T-238

Sample Number	170	171	172	173	175.1	176	177
Total Weight (Kg)	4.9	4.1	3	1.8	4.5	4.8	4.6
Total Volume (ml)	6000	4475	2625	2175	3450	3325	4000
Context	SIA-U51-F2-Fea96	SIA-U58-F1-Fea106D	SIA-U59-F1-Fea106B.2	SIA-E.Extension-Fea68A	SIA-U58-F1-Fea106B.1	SIA-U58-F1-Fea106A	SIA-U58-F1-Fea106C
Volume (ml) X>0.425 mm	825	375	300	125	325	450	400
Weight (g) X>0.425 mm	377	347	178	86	215	267	292
Volume (ml) X>1.0 mm	675	425	210	140	225	250	250
Weight (g) X>1.0 mm	248	314	109	56	142	201	173
Volume (ml) X>2.0 mm	800	300	225	200	225	325	325
Weight (g) X>2.0 mm	274	222	127	69	144	230	160
Volume (ml) X>4.0 mm	3375	425	350	700	500	700	425
Weight (g) X>4.0 mm	1053	286	247	196	351	442	244
SAP Sample No.	C-67, C-68	T-275	T-274	T-153	T-180	T-181	T-182

APPENDIX I

Complete Spreadsheet of all Huaca Sialupe Raw Archaeobotanical Data

Sample Number	Context	Mound	Room	Floor	Feature	Feature Part	Family	Determination	Organ/Type	Item Description	Raw Count	Wood Weight (g) (estimates in italics)	Density [count or weight/ml (*1E4)]	Feature/floor Volume (L) Estimates in Italics	Volume(ml)	Percent of Actual (sample V/Feature V)
29	Floor	1	21	1			Poaceae	Zea mays	Seed	Flour	3		5	96.25	6000	6
29	Floor	1	21	1			Fabaceae	Prosopis sp.	Seed	Mature	2		3	96.25	6000	6
29	Floor	1	21	1			Fabaceae	cf. Fabaceae	Seed	Mature	1		2	96.25	6000	6
29	Floor	1	21	1				4mm Charcoal	Stem	Branchwood		0.10	0.17	96.25	6000	6
29	Floor	1	21	1			Fabaceae	Prosopis sp.	Stem	Branchwood		2.00	3.33	96.25	6000	6
29	Floor	1	21	1			Asteraceae	Tessaria integrifolia	Stem	Mature		0.13	0.22	96.25	6000	6
29	Floor	1	21	1				2mm Charcoal	Stem	Branchwood		2.95	4.92	96.25	6000	6
30	Floor	1	30	1			Poaceae	Zea mays	Seed	Mature	2		2	45	10100	22
30	Floor	1	30	1			Fabaceae	cf. Fabaceae	Seed	Mature	1		1	45	10100	22
30	Floor	1	30	1				4mm Charcoal	Stem	Branchwood		0.49	0.49	8.075	10100	125
30	Floor	1	30	1				Ceramic Blackware	Ceramic			1.06	1.05	45	10100	22
30	Floor	1	30	1				Mammal	Bone			0.15	0.15	45	10100	22
30	Floor	1	30	1			Fabaceae	Acacia sp.	Stem	Branchwood		0.19	0.19	45	10100	22
30	Floor	1	30	1			Asteraceae	Tessaria integrifolia	Stem	Mature		0.27	0.27	45	10100	22
30	Floor	1	30	1				Llama sp.	Dung			0.10	0.10	45	10100	22
30	Floor	1	30	1			Fabaceae	Prosopis sp.	Stem	Branchwood		5.68	5.62	45	10100	22
30	Floor	1	30	1				2mm Charcoal	Stem	Branchwood		1.84	1.82	45	10100	22
30	Floor	1	30	1				Llama sp.	Dung			*0.38*	0.38	45	10100	22
31	Hearth	1	30	1	63		Fabaceae	Prosopis sp.	Stem	Branchwood		1.93	4.77	4.05	4050	100
31	Hearth	1	30	1	63			2mm Charcoal	Stem	Branchwood		0.42	1.04	4.05	4050	100
41	Floor	1	30	3			Fabaceae	Prosopis sp.	Seed	Mature	8		13	45	6100	14
41	Floor	1	30	3			Euphorbiaceae	Euphorbia sp.	Seed	Mature	1		2	45	6100	14
41	Floor	1	30	3			Verbenaceae	Lippia sp.	Seed	Mature	8		13	45	6100	14
41	Floor	1	30	3			Fabaceae	Acacia sp.	Fruit	Mature	1		2	45	6100	14
41	Floor	1	30	3			Fabaceae	Prosopis sp.	Fruit	Mature	1		2	45	6100	14
41	Floor	1	30	3			Verbenaceae	Lippia sp.	Fruit	Mature	1		2	45	6100	14
41	Floor	1	30	3			Poaceae	Zea mays	Fruit	Cob	1		2	45	6100	14
41	Floor	1	30	3				Wood Fungi	Spore		16		26	45	6100	14
41	Floor	1	30	3				4mm Charcoal	Stem	Branchwood		8.35	13.69	8.075	6100	76
41	Floor	1	30	3			Apocynaceae	Vallesia glabra	Stem	Mature		0.35	0.57	45	6100	14
41	Floor	1	30	3			Capparidaceae	Capparis ovalifolia	Stem	Branchwood		2.03	3.33	45	6100	14
41	Floor	1	30	3			Fabaceae	Prosopis sp.	Stem	Branchwood		3.76	6.16	45	6100	14
41	Floor	1	30	3			Fabaceae	Acacia sp.	Stem	Branchwood		7.64	12.52	45	6100	14
41	Floor	1	30	3				2mm Charcoal	Stem	Branchwood		3.21	5.26	45	6100	14
41	Floor	1	30	3				Llama sp.	Dung			*1.12*	1.84	45	6100	14
44	Floor	1	25	1			Poaceae	Zea mays	Seed	Mature	2		3	11.65	5825	50
44	Floor	1	25	1			Verbenaceae	Lippia sp.	Seed	Mature	14		24	11.65	5825	50
44	Floor	1	25	1			Boraginaceae	Heliotropum sp.	Seed	Mature	1		2	11.65	5825	50
44	Floor	1	25	1			Chenopodiaceae	Chenopodium sp.	Seed	Mature	1		2	11.65	5825	50
44	Floor	1	25	1			Poaceae	Zea mays	Fruit	Cupule	1		2	11.65	5825	50
44	Floor	1	25	1				Slag	Metal	Amorphous	1		2	11.65	5825	50

44	Floor	1	25	1				4mm Charcoal	Stem	Branchwood		2.97	5.10	11.65	5825	50
44	Floor	1	25	1			Fabaceae	Prosopis sp.	Stem	Branchwood		3.42	5.87	11.65	5825	50
44	Floor	1	25	1			Fabaceae	Acacia sp.	Stem	Branchwood		5.98	10.27	11.65	5825	50
44	Floor	1	25	1			Asteraceae	Tessaria integrifolia	Stem	Mature		0.37	0.64	11.65	5825	50
44	Floor	1	25	1			Capparidaceae	Capparis angulata	Stem	Branchwood		1.02	1.75	11.65	5825	50
44	Floor	1	25	1			Poaceae	Zea mays	Stem			0.12	0.21	11.65	5825	50
44	Floor	1	25	1				2mm Charcoal	Stem	Branchwood		2.19	3.76	11.65	5825	50
45	Floor	1	24	3			Verbenaceae	Lippia sp.	Seed	Mature	33		77	6	4270	71
45	Floor	1	24	3			Fabaceae	Prosopis sp.	Seed	Mature	2		5	6	4270	71
45	Floor	1	24	3				UKN#28-45	Seed	Mature	1		2	6	4270	71
45	Floor	1	24	3			Verbenaceae	Lippia sp.	Fruit	Stem	1		2	6	4270	71
45	Floor	1	24	3				2mm Charcoal	Stem	Branchwood		1.02	2.39	6	4270	71
45	Floor	1	24	3			Fabaceae	Prosopis sp.	Stem	Branchwood		4.33	10.14	6	4270	71
45	Floor	1	24	3				Llama sp.	Dung			*0.19*	0.44	6	4270	71
84	Floor	1	23	2			Chenopodiaceae	Chenopodium sp.	Seed	Mature	1		3	6	3000	50
84	Floor	1	23	2			Verbenaceae	Lippia sp.	Seed	Mature	18		60	6	3000	50
84	Floor	1	23	2			Asteraceae	Tessaria integrifolia	Stem	Mature		0.30	1.00	6	3000	50
84	Floor	1	23	2			Fabaceae	Acacia sp.	Stem	Branchwood		0.31	1.03	6	3000	50
84	Floor	1	23	2			Fabaceae	Prosopis sp.	Stem	Branchwood		1.58	5.27	6	3000	50
84	Floor	1	23	2				4mm Charcoal	Stem	Branchwood		2.92	9.73	6	3000	50
84	Floor	1	23	2				2mm Charcoal	Stem	Branchwood		0.72	2.40	6	3000	50
87	Hearth	1	23	2	83		Brassicaceae	Lepidium sp.	seed	Mature	1		4	2.668	2450	92
87	Hearth	1	23	2	83		Fabaceae	Acacia sp.	leaf	spine	1		4	2.668	2450	92
87	Hearth	1	23	2	83			2mm Charcoal	Stem	Branchwood		0.06	0.24	2.668	2450	92
94	Ash Pit	1	35	2	84	B	Amaranthaceae	Amaranthus sp.	Seed	Mature	1		3	4.232	3700	87
94	Ash Pit	1	35	2	84	B	Malvaceae	Gossypium sp.	Seed	Mature	1		3	4.232	3700	87
94	Ash Pit	1	35	2	84	B	Chenopodiaceae	Chenopodium sp.	Seed	Mature	2		5	4.232	3700	87
94	Ash Pit	1	35	2	84	B	Fabaceae	Prosopis sp.	Seed	Mature	15		41	4.232	3700	87
94	Ash Pit	1	35	2	84	B	Fabaceae	Acacia sp.	Flower	Mature	3		8	4.232	3700	87
94	Ash Pit	1	35	2	84	B	Solanaceae	Capsicum sp.	Seed	Mature	1		3	4.232	3700	87
94	Ash Pit	1	35	2	84	B	Fabaceae	Acacia sp.	Seed	Mature	2		5	4.232	3700	87
94	Ash Pit	1	35	2	84	B	Fabaceae	cf. Fabaceae	Seed	Mature	12		32	4.232	3700	87
94	Ash Pit	1	35	2	84	B	Poaceae	Zea mays	Seed	Embryo	1		3	4.232	3700	87
94	Ash Pit	1	35	2	84	B	Poaceae	Zea mays	Seed	Mature	17		46	4.232	3700	87
94	Ash Pit	1	35	2	84	B	Verbenaceae	Lippia sp.	Fruit	Mature	14		38	4.232	3700	87
94	Ash Pit	1	35	2	84	B	Verbenaceae	Lippia sp.	Seed	Mature	1192		3222	4.232	3700	87
94	Ash Pit	1	35	2	84	B		Unid.	Seed	Mature	9		24	4.232	3700	87
94	Ash Pit	1	35	2	84	B	Capparidaceae	Capparis angulata	Flower	Sepal	1		3	4.232	3700	87
94	Ash Pit	1	35	2	84	B		Unid.	Fruit	Stem	3		8	4.232	3700	87
94	Ash Pit	1	35	2	84	B	Fabaceae	cf. Fabaceae	Fruit	Stem	1		3	4.232	3700	87
94	Ash Pit	1	35	2	84	B		Shell 4mm	Shell		1		3	4.232	3700	87
94	Ash Pit	1	35	2	84	B	Poaceae	cf. Poaceae	Root		1		3	4.232	3700	87
94	Ash Pit	1	35	2	84	B	Fabaceae	Acacia sp.	Stem	Branchwood		4.61	12.46	4.232	3700	87
94	Ash Pit	1	35	2	84	B	Fabaceae	Fabaceae	Stem	(Desiccated) Juvenile		0.54	1.46	4.232	3700	87
94	Ash Pit	1	35	2	84	B	Fabaceae	Fabaceae	Stem	Juvenile		0.99	2.68	4.232	3700	87
94	Ash Pit	1	35	2	84	B	Asteraceae	Tessaria integrifolia	Stem	Mature		0.23	0.62	4.232	3700	87
94	Ash Pit	1	35	2	84	B	Poaceae	Gynerium sp.	Stem			0.04	0.11	4.232	3700	87
94	Ash Pit	1	35	2	84	B	Fabaceae	Prosopis sp.	Stem	Branchwood		0.83	2.24	4.232	3700	87
94	Ash Pit	1	35	2	84	B		Llama sp.	Dung			2.10	5.68	4.232	3700	87

94	Ash Pit	1	35	2	84	B		Llama sp.	Dung			*5.53*	14.95	4.232	3700	87
99	Hearth	1	39	1	49		Cucurbitaceae	Cucurbita sp.	Seed	Mature	1		3	16.8	3900	23
99	Hearth	1	39	1	49		Caryophyllaceae	Silene sp.	Seed	Mature	2		5	16.8	3900	23
99	Hearth	1	39	1	49		Fabaceae	Prosopis sp.	Seed	Mature	14		36	16.8	3900	23
99	Hearth	1	39	1	49		Fabaceae	Caesalpinia sp.	Seed	Mature	3		8	16.8	3900	23
99	Hearth	1	39	1	49		Fabaceae	Acacia sp.	Seed	Mature	2		5	16.8	3900	23
99	Hearth	1	39	1	49		Fabaceae	Phaseolus sp.	Seed	Mature	1		3	16.8	3900	23
99	Hearth	1	39	1	49		Fabaceae	Arachis sp.	Seed	Mature	1		3	16.8	3900	23
99	Hearth	1	39	1	49		Chenopodiaceae	Chenopodium sp.	Seed	Mature	2		5	16.8	3900	23
99	Hearth	1	39	1	49		Poaceae	Zea mays	Seed	Embryo	1		3	16.8	3900	23
99	Hearth	1	39	1	49		Verbenaceae	Lippia sp.	Seed	Mature	143		367	16.8	3900	23
99	Hearth	1	39	1	49		Verbenaceae	Lippia sp.	Fruit	Mature	1		3	16.8	3900	23
99	Hearth	1	39	1	49			Unid.	Seed	Mature	9		23	16.8	3900	23
99	Hearth	1	39	1	49			Wood Fungi	Spore		56		144	16.8	3900	23
99	Hearth	1	39	1	49		Fabaceae	cf. Fabaceae	Fruit	Mature	1		3	16.8	3900	23
99	Hearth	1	39	1	49		Fabaceae	Prosopis sp.	Leaf	Spine	2		5	16.8	3900	23
99	Hearth	1	39	1	49			Unid.	Leaf	Spine	1		3	16.8	3900	23
99	Hearth	1	39	1	49		Fabaceae	cf. Fabaceae	Fruit	Stem	2		5	16.8	3900	23
99	Hearth	1	39	1	49			Cavia sp.	Dung		7		18	16.8	3900	23
99	Hearth	1	39	1	49			4mm Charcoal	Stem	Branchwood		13.01	33.36	16.8	3900	23
99	Hearth	1	39	1	49			Ceramic Blackware	Ceramic			1.32	3.38	16.8	3900	23
99	Hearth	1	39	1	49			Donax sp.	Shell			0.56	1.44	16.8	3900	23
99	Hearth	1	39	1	49		Apocynaceae	Vallesia glabra	Stem	Mature		0.31	0.79	16.8	3900	23
99	Hearth	1	39	1	49		Asteraceae	Tessaria integrifolia	Stem	Mature		0.34	0.87	16.8	3900	23
99	Hearth	1	39	1	49			Llama sp.	Dung			*0.75*	1.92	16.8	3900	23
99	Hearth	1	39	1	49		Fabaceae	Fabaceae	Stem	Juvenile		1.70	4.36	16.8	3900	23
99	Hearth	1	39	1	49		Fabaceae	Prosopis sp.	Stem	Branchwood		161.36	413.74	16.8	3900	23
99	Hearth	1	39	1	49			2mm Charcoal	Stem	Branchwood		2.20	5.64	16.8	3900	23
99	Hearth	1	39	1	49			Llama sp.	Dung			*0.47*	1.21	16.8	3900	23
108	Old Porron	2	Open	5	15		Fabaceae	UKN#64-120.1	Fruit	Mature	1		3	3.475	3475	100
108	Old Porron	2	Open	5	15		Fabaceae	Prosopis sp.	Seed	Mature	5		14	3.475	3475	100
108	Old Porron	2	Open	5	15		Amaranthaceae	Amaranthus sp.	Seed	Mature	38		109	3.475	3475	100
108	Old Porron	2	Open	5	15		Portulacaceae	Portulaca sp.	Seed	Mature	8		23	3.475	3475	100
108	Old Porron	2	Open	5	15		Asteraceae	Ambrosia sp	Seed	Mature	2		6	3.475	3475	100
108	Old Porron	2	Open	5	15		Verbenaceae	Lippia sp.	Seed	Mature	6		17	3.475	3475	100
108	Old Porron	2	Open	5	15		Boraginaceae	Heliotropum sp.	Seed	Mature	2		6	3.475	3475	100
108	Old Porron	2	Open	5	15		Asteraceae	cf. Asteraceae	Seed	Mature	1		3	3.475	3475	100
108	Old Porron	2	Open	5	15		Ulmaceae	Celtis sp.	Seed	Mature	1		3	3.475	3475	100
108	Old Porron	2	Open	5	15		Fabaceae	Prosopis sp.	Seed	Mature	3		9	3.475	3475	100
108	Old Porron	2	Open	5	15		Solanaceae	cf. Solanaceae	seed	Mature	1		3	3.475	3475	100
108	Old Porron	2	Open	5	15		Solanaceae	Nicotiana sp.	Seed	Mature	2		6	3.475	3475	100
108	Old Porron	2	Open	5	15		Asteraceae	Taraxacum sp.	Seed	Mature	1		3	3.475	3475	100
108	Old Porron	2	Open	5	15		Brassicaceae	Lepidium sp.	Seed	Mature	1		3	3.475	3475	100
108	Old Porron	2	Open	5	15		Asteraceae	Bidens sp.	Seed	Mature	70		201	3.475	3475	100
108	Old Porron	2	Open	5	15		Poaceae	Cenchrus sp.	Seed	Mature	26		75	3.475	3475	100
108	Old Porron	2	Open	5	15		Fabaceae	UKN#64-120.1	Fruit	Mature	1		3	3.475	3475	100
108	Old Porron	2	Open	5	15		Boraginaceae	Heliotropum sp.	Fruit	Immature	1		3	3.475	3475	100
108	Old Porron	2	Open	5	15		Verbenaceae	Lippia sp.	Fruit	Mature	1		3	3.475	3475	100
108	Old Porron	2	Open	5	15		Fabaceae	Prosopis sp.	Leaf	Spine	1		3	3.475	3475	100

108	Old Porron	2	Open	5	15		Fabaceae	Prosopis sp.	Fruit	Stem	1		3	3.475	3475	100
108	Old Porron	2	Open	5	15		Fabaceae	Prosopis sp.	Leaf		16		46	3.475	3475	100
108	Old Porron	2	Open	5	15		Fabaceae	Fabaceae	Stem	(Desiccated) Juvenile		3.57	10.27	3.475	3475	100
108	Old Porron	2	Open	5	15		Fabaceae	Acacia sp.	Stem	Branchwood		0.19	0.55	3.475	3475	100
108	Old Porron	2	Open	5	15		Asteraceae	Tessaria integrifolia	Stem	Mature		0.06	0.17	3.475	3475	100
108	Old Porron	2	Open	5	15		Poaceae	Gynerium sp.	Stem			34.00	97.84	3.475	3475	100
108	Old Porron	2	Open	5	15		Poaceae	Gynerium sp.	Stem			19.73	56.78	3.475	3475	100
108	Old Porron	2	Open	5	15			Llama sp.	Dung			137.94	396.95	3.475	3475	100
108	Old Porron	2	Open	5	15			Llama sp.	Dung			13.03	37.50	3.475	3475	100
109	Preheating Fire	1	39	2	96		Fabaceae	Prosopis sp.	Seed	Mature	1		4	90	2225	2
109	Preheating Fire	1	39	2	96		Fabaceae	Acacia sp.	Seed	Mature	1		4	90	2225	2
109	Preheating Fire	1	39	2	96		Fabaceae	cf. Fabaceae	Seed	Mature	1		4	90	2225	2
109	Preheating Fire	1	39	2	96		Fabaceae	Prosopis sp.	Stem	Branchwood		11.39	51.19	90	2225	2
109	Preheating Fire	1	39	2	96			Llama sp.	Dung			0.27	1.21	90	2225	2
109	Preheating Fire	1	39	2	96		Capparidaceae	Capparis angulata	Stem	Branchwood		0.11	0.49	90	2225	2
109	Preheating Fire	1	39	2	96		Fabaceae	Fabaceae	Stem	Juvenile		0.12	0.54	90	2225	2
109	Preheating Fire	1	39	2	96			2mm Charcoal	Stem	Branchwood		2.43	10.92	90	2225	2
109	Preheating Fire	1	39	2	96			Llama sp.	Dung			*0.28*	1.26	90	2225	2
110	Ash Pit	1	39	1	99	B	Fabaceae	Prosopis sp.	Seed	Mature	7		19	3.625	3625	100
110	Ash Pit	1	39	1	99	B	Malvaceae	Gossypium sp.	Seed	Mature	92		254	3.625	3625	100
110	Ash Pit	1	39	1	99	B	Erythoxylaceae	Erythoxylon coca	Seed	Mature	1		3	3.625	3625	100
110	Ash Pit	1	39	1	99	B	Chenopodiaceae	Suaeda sp.	Seed	Mature	1		3	3.625	3625	100
110	Ash Pit	1	39	1	99	B	Malvaceae	Malva sp.	Seed	Mature	1		3	3.625	3625	100
110	Ash Pit	1	39	1	99	B		Cavia sp.	Dung		19		52	3.625	3625	100
110	Ash Pit	1	39	1	99	B	Fabaceae	Prosopis sp.	Leaf	Spine	1		3	3.625	3625	100
110	Ash Pit	1	39	1	99	B	Fabaceae	Prosopis sp.	Stem	Branchwood		10.71	29.54	3.625	3625	100
110	Ash Pit	1	39	1	99	B	Fabaceae	Acacia sp.	Stem	Branchwood		3.82	10.54	3.625	3625	100
110	Ash Pit	1	39	1	99	B	Fabaceae	Fabaceae	Stem	(Desiccated) Juvenile		0.48	1.32	3.625	3625	100
110	Ash Pit	1	39	1	99	B		2mm Charcoal	Stem	Branchwood		5.08	14.01	3.625	3625	100
111	Ash Pit	1	38	2	98	B	Fabaceae	cf. Fabaceae	Seed	Mature	8		34	2.35	2350	100
111	Ash Pit	1	38	2	98	B	Fabaceae	Acacia sp.	Seed	Mature	4		17	2.35	2350	100
111	Ash Pit	1	38	2	98	B	Fabaceae	Prosopis sp.	Seed	Mature	6		26	2.35	2350	100
111	Ash Pit	1	38	2	98	B	Poaceae	Zea mays	Seed	Flour	2		9	2.35	2350	100
111	Ash Pit	1	38	2	98	B	Boraginaceae	Heliotropum sp.	Seed	Mature	1		4	2.35	2350	100
111	Ash Pit	1	38	2	98	B	Euphorbiaceae	Euphorbia sp.	Seed	Mature	1		4	2.35	2350	100
111	Ash Pit	1	38	2	98	B	Verbenaceae	Lippia sp.	Seed	Mature	17		72	2.35	2350	100
111	Ash Pit	1	38	2	98	B		Wood Fungi	Spore		1		4	2.35	2350	100
111	Ash Pit	1	38	2	98	B		Fish	Bone	Mandible	1		4	2.35	2350	100
111	Ash Pit	1	38	2	98	B		Mammal	Bone		1		4	2.35	2350	100
111	Ash Pit	1	38	2	98	B		Shell 4mm	Shell		8		34	2.35	2350	100
111	Ash Pit	1	38	2	98	B	Fabaceae	cf. Fabaceae	Fruit	Stem	1		4	2.35	2350	100
111	Ash Pit	1	38	2	98	B	Fabaceae	Acacia sp.	Flower	Mature	1		4	2.35	2350	100
111	Ash Pit	1	38	2	98	B		4mm Charcoal	Stem	Branchwood		12.00	51.06	2.35	2350	100
111	Ash Pit	1	38	2	98	B	Fabaceae	Prosopis sp.	Stem	Branchwood		32.23	137.15	2.35	2350	100
111	Ash Pit	1	38	2	98	B	Asteraceae	Tessaria integrifolia	Stem	Mature		0.17	0.72	2.35	2350	100
111	Ash Pit	1	38	2	98	B		Burnt Clay	Ceramic			2.84	12.09	2.35	2350	100
111	Ash Pit	1	38	2	98	B		Llama sp.	Dung			0.16	0.68	2.35	2350	100
111	Ash Pit	1	38	2	98	B		2mm Charcoal	Stem	Branchwood		7.79	33.15	2.35	2350	100

113	Charcoal Hearth	2	Open	5	9	A	Verbenaceae	Lippia sp.	Seed	Mature	2		18	37.66	1125	20
113	Charcoal Hearth	2	Open	5	9	A	Malvaceae	Gossypium sp.	Seed	Mature	1		9	37.66	1125	20
113	Charcoal Hearth	2	Open	5	9	A	Fabaceae	Prosopis sp.	Stem	Branchwood		0.83	7.38	37.66	1125	20
113	Charcoal Hearth	2	Open	5	9	A	Asteraceae	Tessaria integrifolia	Stem	Mature		0.11	0.98	37.66	1125	20
113	Charcoal Hearth	2	Open	5	9	A		Llama sp.	Dung			*0.09*	0.80	37.66	1125	20
114	Communal Hearth	1	39	2	97		Boraginaceae	Heliotropum sp.	Seed	Mature	2		5	9	3670	41
114	Communal Hearth	1	39	2	97		Poaceae	Zea mays	Seed	Mature	4		11	9	3670	41
114	Communal Hearth	1	39	2	97		Fabaceae	Acacia sp.	Seed	Mature	58		158	9	3670	41
114	Communal Hearth	1	39	2	97		Oxalidaceae	Oxalis sp.	Seed	Mature	1		3	9	3670	41
114	Communal Hearth	1	39	2	97		Chenopodiaceae	Chenopodium sp.	Seed	Mature	10		27	9	3670	41
114	Communal Hearth	1	39	2	97		Poaceae	Zea mays	Seed	Mature	2		5	9	3670	41
114	Communal Hearth	1	39	2	97		Verbenaceae	Lippia sp.	Seed	Mature	92		251	9	3670	41
114	Communal Hearth	1	38	2	97			Wood Fungi	Spore		3		8	9	3670	41
114	Communal Hearth	1	39	2	97		Plantaginaceae	Plantago sp.	Seed	Mature	1		3	9	3670	41
114	Communal Hearth	1	39	2	97		Malvaceae	Sida sp.	Seed	Mature	1		3	9	3670	41
114	Communal Hearth	1	39	2	97		Fabaceae	cf. Fabaceae	Seed	Mature	35		95	9	3670	41
114	Communal Hearth	1	39	2	97			Unid.	Seed	Mature	4		11	9	3670	41
114	Communal Hearth	1	38	2	97			Donax sp.	Shell		1		3	9	3670	41
114	Communal Hearth	1	38	2	97			Cavia sp.	Dung		4		11	9	3670	41
114	Communal Hearth	1	38	2	97			Unid.	Fruit	Stem	7		19	9	3670	41
114	Communal Hearth	1	38	2	97		Fabaceae	Acacia sp.	Flower	Mature	1		3	9	3670	41
114	Communal Hearth	1	39	2	97		Fabaceae	Prosopis sp.	Stem	Branchwood		2.10	5.72	9	3670	41
114	Communal Hearth	1	39	2	97		Fabaceae	Fabaceae	Stem	Juvenile		0.23	0.63	9	3670	41
114	Communal Hearth	1	39	2	97		Asteraceae	Tessaria integrifolia	Stem	Mature		0.33	0.90	9	3670	41
114	Communal Hearth	1	39	2	97			Llama sp.	Dung			2.13	5.80	9	3670	41
114	Communal Hearth	1	39	2	97			Llama sp.	Dung			*3.28*	8.94	9	3670	41
115	Ash Pit	1	38	2	100		Fabaceae	Acacia sp.	Seed	Mature	28		64	5.625	4375	78
115	Ash Pit	1	38	2	100		Fabaceae	cf. Fabaceae	Seed	Mature	89		203	5.625	4375	78
115	Ash Pit	1	38	2	100		Fabaceae	Prosopis sp.	Seed	Mature	41		94	5.625	4375	78
115	Ash Pit	1	38	2	100		Amaranthaceae	Amaranthus sp.	Seed	Mature	1		2	5.625	4375	78
115	Ash Pit	1	38	2	100		Boraginaceae	Heliotropum sp.	Seed	Mature	1		2	5.625	4375	78
115	Ash Pit	1	38	2	100		Portulacaceae	cf. Portulacaceae	Seed	Mature	1		2	5.625	4375	78
115	Ash Pit	1	38	2	100		Verbenaceae	Lippia sp.	Seed	Mature	45		103	5.625	4375	78
115	Ash Pit	1	38	2	100		Verbenaceae	Lippia sp.	Fruit	Mature	1		2	5.625	4375	78
115	Ash Pit	1	38	2	100		Chenopodiaceae	Chenopodium sp.	Seed	Mature	1		2	5.625	4375	78
115	Ash Pit	1	38	2	100		Erythoxylaceae	Erythoxylon coca	Seed	Mature	1		2	5.625	4375	78
115	Ash Pit	1	38	2	100		Chenopodiaceae	Suaeda sp.	Seed	Mature	1		2	5.625	4375	78
115	Ash Pit	1	38	2	100		Poaceae	Zea mays	Seed	Flour	4		9	5.625	4375	78
115	Ash Pit	1	38	2	100			Cavia sp.	Dung		15		34	5.625	4375	78
115	Ash Pit	1	38	2	100		Fabaceae	Acacia sp.	Flower	Mature	2		5	5.625	4375	78
115	Ash Pit	1	38	2	100		Fabaceae	cf. Fabaceae	Fruit	Stem	5		11	5.625	4375	78
115	Ash Pit	1	38	2	100			Mammal	Bone		3		7	5.625	4375	78
115	Ash Pit	1	38	2	100			Donax sp.	Shell		20		46	5.625	4375	78
115	Ash Pit	1	38	2	100			Llama sp.	Dung			9.99	22.83	5.625	4375	78
115	Ash Pit	1	38	2	100		Fabaceae	Prosopis sp.	Stem	Branchwood		7.22	16.50	5.625	4375	78
115	Ash Pit	1	38	2	100			Ceramic Blackware	Ceramic			1.01	2.31	5.625	4375	78
115	Ash Pit	1	38	2	100		Poaceae	Zea mays	Stem			0.17	0.39	5.625	4375	78
115	Ash Pit	1	38	2	100		Fabaceae	Fabaceae	Stem	Branchwood		0.81	1.85	5.625	4375	78

115	Ash Pit	1	38	2	100		Fabaceae	Acacia sp.	Stem	Branchwood		0.80	1.83	5.625	4375	78
115	Ash Pit	1	38	2	100		Asteraceae	Tessaria integrifolia	Stem	Mature		2.40	5.49	5.625	4375	78
115	Ash Pit	1	38	2	100			2mm Charcoal	Stem	Branchwood		12.61	28.82	5.625	4375	78
115	Ash Pit	1	38	2	100			Llama sp.	Dung			*7.98*	18.24	5.625	4375	78
116	Offering	1	38	2	101		Erythoxylaceae	Erythoxylon coca	Seed	Mature	2		5	4.215	4215	100
116	Offering	1	38	2	101		Fabaceae	cf. Fabaceae	Seed	Mature	14		33	4.215	4215	100
116	Offering	1	38	2	101		Fabaceae	Prosopis sp.	Seed	Mature	37		88	4.215	4215	100
116	Offering	1	38	2	101		Euphorbiaceae	Euphorbia sp.	Seed	Mature	1		2	4.215	4215	100
116	Offering	1	38	2	101		Chenopodiaceae	Chenopodium sp.	Seed	Mature	16		38	4.215	4215	100
116	Offering	1	38	2	101		Poaceae	Zea mays	Seed	Mature	2		5	4.215	4215	100
116	Offering	1	38	2	101		Boraginaceae	Heliotropum sp.	Seed	Mature	6		14	4.215	4215	100
116	Offering	1	38	2	101		Solanaceae	Capsicum sp.	Seed	Mature	1		2	4.215	4215	100
116	Offering	1	38	2	101		Verbenaceae	Lippia sp.	Seed	Mature	71		168	4.215	4215	100
116	Offering	1	38	2	101			Unid.	Seed	Mature	6		14	4.215	4215	100
116	Offering	1	38	2	101		Fabaceae	Prosopis sp.	Stem	Branchwood		24.62	58.41	4.215	4215	100
116	Offering	1	38	2	101			2mm Charcoal	Stem	Branchwood		0.71	1.68	4.215	4215	100
117	Porron Contents	2	Open	5	8		Amaranthaceae	Amaranthus sp.	Seed	Mature	2		5	73.593	3800	5
117	Porron Contents	2	Open	5	8		Erythoxylaceae	Erythoxylon coca	Seed	Mature	2		5	73.593	3800	5
117	Porron Contents	2	Open	5	8		Azioiaceae	Trianthema sp.	Seed	Mature	1		3	73.593	3800	5
117	Porron Contents	2	Open	5	8		Rubiaceae	Gallium sp.	Seed	Mature	1		3	73.593	3800	5
117	Porron Contents	2	Open	5	8		Euphorbiaceae	Euphorbia sp.	Seed	Mature	1		3	73.593	3800	5
117	Porron Contents	2	Open	5	8		Asteraceae	Ambrosia sp	Seed	Mature	1		3	73.593	3800	5
117	Porron Contents	2	Open	5	8		Fabaceae	UKN#43-121.1	Fruit	Mature	1		3	73.593	3800	5
117	Porron Contents	2	Open	5	8		Asteraceae	cf. Asteraceae	Seed	Mature	3		8	73.593	3800	5
117	Porron Contents	2	Open	5	8		Verbenaceae	Lippia sp.	Seed	Mature	10		26	73.593	3800	5
117	Porron Contents	2	Open	5	8		Fabaceae	Neptunia sp.	Seed	Mature	1		3	73.593	3800	5
117	Porron Contents	2	Open	5	8		Poaceae	Zea mays	Seed	Flour	1		3	73.593	3800	5
117	Porron Contents	2	Open	5	8		Solanaceae	Physalis sp.	Seed	Mature	4		11	73.593	3800	5
117	Porron Contents	2	Open	5	8		Solanaceae	Nicotiana sp.	Seed	Mature	4		11	73.593	3800	5
117	Porron Contents	2	Open	5	8		Poaceae	Zea mays	Seed	Mature	1		3	73.593	3800	5
117	Porron Contents	2	Open	5	8		Fabaceae	Acacia sp.	Seed	Mature	2		5	73.593	3800	5
117	Porron Contents	2	Open	5	8		Fabaceae	Prosopis sp.	Seed	Mature	44		116	73.593	3800	5
117	Porron Contents	2	Open	5	8		Cucurbitaceae	Cucurbita maxima	Seed	Mature	21		55	73.593	3800	5
117	Porron Contents	2	Open	5	8		Cucurbitaceae	Lagenaria sp.	Seed	Mature	18		47	73.593	3800	5
117	Porron Contents	2	Open	5	8		Cucurbitaceae	Cucurbita sp.	Seed	Mature	13		34	73.593	3800	5
117	Porron Contents	2	Open	5	8		Fabaceae	cf. Fabaceae	Seed	Mature	3		8	73.593	3800	5
117	Porron Contents	2	Open	5	8		Poaceae	cf. Poaceae	Root	Corm	1		3	73.593	3800	5
117	Porron Contents	2	Open	5	8		Capparidaceae	UKN#66-166	Fruit	Mature	1		3	73.593	3800	5
117	Porron Contents	2	Open	5	8		Lauraceae	Persea sp.	Seed	Mature	1		3	73.593	3800	5
117	Porron Contents	2	Open	5	8		Fabaceae	UKN#43-121.1	Fruit	Stem	2		5	73.593	3800	5
117	Porron Contents	2	Open	5	8		Cucurbitaceae	Cucurbitaceae	Seed	Mature	3		8	73.593	3800	5
117	Porron Contents	2	Open	5	8			Unid.	Seed	Mature	6		16	73.593	3800	5
117	Porron Contents	2	Open	5	8		Verbenaceae	Lippia sp.	Fruit	Mature	24		63	73.593	3800	5
117	Porron Contents	2	Open	5	8		Fabaceae	Acacia sp.	Fruit	stem	8		21	73.593	3800	5
117	Porron Contents	2	Open	5	8			Cavia sp.	Dung		30		79	73.593	3800	5
117	Porron Contents	2	Open	5	8			Shell 4mm	Shell		7		18	73.593	3800	5
117	Porron Contents	2	Open	5	8			Donax sp.	Shell		17		45	73.593	3800	5
117	Porron Contents	2	Open	5	8			Fish	Bone		7		18	73.593	3800	5

117	Porron Contents	2	Open	5	8		Poaceae	Zea mays	Fruit	12-row	1		3	73.593	3800	5
117	Porron Contents	2	Open	5	8		Poaceae	Zea mays	Fruit	Cupule	17		45	73.593	3800	5
117	Porron Contents	2	Open	5	8		Fabaceae	Acacia sp.	Leaf	Spine	1		3	73.593	3800	5
117	Porron Contents	2	Open	5	8		Fabaceae	Prosopis sp.	Leaf	Spine	2		5	73.593	3800	5
117	Porron Contents	2	Open	5	8		Cucurbitaceae	Lagenaria sp.	Fruit	Mate	3		8	73.593	3800	5
117	Porron Contents	2	Open	5	8			Unid.	Fruit	Stem	24		63	73.593	3800	5
117	Porron Contents	2	Open	5	8			4mm Charcoal	Stem	Branchwood		127.62	335.84	73.593	3800	5
117	Porron Contents	2	Open	5	8		Capparidaceae	Capparis ovalifolia	Stem	Branchwood		23.12	60.84	73.593	3800	5
117	Porron Contents	2	Open	5	8		Fabaceae	Fabaceae	Stem	Juvenile		12.28	32.32	73.593	3800	5
117	Porron Contents	2	Open	5	8		Fabaceae	Fabaceae	Stem	(Desiccated) Juvenile		3.67	9.66	73.593	3800	5
117	Porron Contents	2	Open	5	8		Apocynaceae	Vallesia glabra	Stem	Mature		0.80	2.11	73.593	3800	5
117	Porron Contents	2	Open	5	8		Capparidaceae	Capparis angulata	Stem	Branchwood		1.52	4.00	73.593	3800	5
117	Porron Contents	2	Open	5	8		Fabaceae	Prosopis sp.	Stem	Branchwood		29.92	78.74	73.593	3800	5
117	Porron Contents	2	Open	5	8		Fabaceae	Acacia sp.	Stem	Branchwood		28.65	75.39	73.593	3800	5
117	Porron Contents	2	Open	5	8			Llama sp.	Dung			0.65	1.71	73.593	3800	5
118	Charcoal Hearth	2	Open	5	9	B	Cucurbitaceae	Cucurbita sp.	Seed	Mature	2		4	49.016	4775	10
118	Charcoal Hearth	2	Open	5	9	B	Poaceae	Zea mays	Seed	Mature	6		13	49.016	4775	10
118	Charcoal Hearth	2	Open	5	9	B	Solanaceae	cf. Solanaceae	Seed	Mature	3		6	49.016	4775	10
118	Charcoal Hearth	2	Open	5	9	B	Fabaceae	cf. Fabaceae	Seed	Mature	7		15	49.016	4775	10
118	Charcoal Hearth	2	Open	5	9	B	Fabaceae	Prosopis sp.	Seed	Mature	6		13	49.016	4775	10
118	Charcoal Hearth	2	Open	5	9	B		Unid.	Seed	Mature	7		15	49.016	4775	10
118	Charcoal Hearth	2	Open	5	9	B	Poaceae	Zea mays	Fruit	Cob	16		34	49.016	4775	10
118	Charcoal Hearth	2	Open	5	9	B		Cavia sp.	Dung		6		13	49.016	4775	10
118	Charcoal Hearth	2	Open	5	9	B		Fish	Bone	Vertebra	6		13	49.016	4775	10
118	Charcoal Hearth	2	Open	5	9	B		Mammal	Bone		2		4	49.016	4775	10
118	Charcoal Hearth	2	Open	5	9	B		Cavia sp.	Bone	Tooth	1		2	49.016	4775	10
118	Charcoal Hearth	2	Open	5	9	B		Shell 4mm	Shell		73		153	49.016	4775	10
118	Charcoal Hearth	2	Open	5	9	B		Ceramic Blackware	Ceramic		1		2	49.016	4775	10
118	Charcoal Hearth	2	Open	5	9	B		Slag	Metal	Amorphous	2		4	49.016	4775	10
118	Charcoal Hearth	2	Open	5	9	B		Cu Alloy	Metal		1		2	49.016	4775	10
118	Charcoal Hearth	2	Open	5	9	B	Fabaceae	Prosopis sp.	Leaf	Spine	4		8	49.016	4775	10
118	Charcoal Hearth	2	Open	5	9	B		Wood Type GG	Stem	Branchwood		21.82	45.70	49.096	4775	10
118	Charcoal Hearth	2	Open	5	9	B	Fabaceae	Fabaceae	Stem	Juvenile		20.08	42.05	49.096	4775	10
118	Charcoal Hearth	2	Open	5	9	B	Apocynaceae	Vallesia glabra	Stem	Mature		4.31	9.03	49.096	4775	10
118	Charcoal Hearth	2	Open	5	9	B	Fabaceae	Prosopis sp.	Stem	Branchwood		16.77	35.12	49.096	4775	10
118	Charcoal Hearth	2	Open	5	9	B	Capparidaceae	Capparis ovalifolia	Stem	Branchwood		8.50	17.80	49.096	4775	10
118	Charcoal Hearth	2	Open	5	9	B	Fabaceae	Acacia sp.	Stem	Branchwood		41.71	87.35	49.096	4775	10
118	Charcoal Hearth	2	Open	5	9	B		Llama sp.	Dung			*0.09*	0.19	49.096	4775	10
118	Charcoal Hearth	2	Open	5	9	B		4mm Charcoal	Stem	Juvenile		37.53	78.60	49.096	4775	10
118	Charcoal Hearth	2	Open	5	9	B		2mm Charcoal	Stem	Branchwood		110.00	230.37	49.096	4775	10
118	Charcoal Hearth	2	Open	5	9	B		Llama sp.	Dung			*0.84*	1.76	49.096	4775	10
118	Hearth	2	3	2	11	B	Fabaceae	Prosopis sp.	Seed	Mature	14		61	36	2300	6
118	Hearth	2	3	2	11	B	Fabaceae	Phaseolus sp.	Seed	Mature	1		4	36	2300	6
118	Hearth	2	3	2	11	B	Poaceae	cf. Poaceae	Seed	Mature	2		9	36	2300	6
118	Hearth	2	3	2	11	B	Fabaceae	Acacia sp.	Seed	Mature	1		4	36	2300	6
118	Hearth	2	3	2	11	B	Poaceae	Zea mays	Seed	Mature	11		48	36	2300	6
118	Hearth	2	3	2	11	B	Poaceae	Zea mays	Seed	Embryo	5		22	36	2300	6
118	Hearth	2	3	2	11	B	Poaceae	Zea mays	Seed	Mature	17		74	36	2300	6

118	Hearth	2	3	2	11	B	Verbenaceae	Lippia sp.	Seed	Mature	1		4	36	2300	6
118	Hearth	2	3	2	11	B	Solanaceae	cf. Solanaceae	Seed	Mature	1		4	36	2300	6
118	Hearth	2	3	2	11	B		Unid.	Seed	Mature	3		13	36	2300	6
118	Hearth	2	3	2	11	B		Cavia sp.	Dung		2		9	36	2300	6
118	Hearth	2	3	2	11	B	Poaceae	cf. Poaceae	Leaf		1		4	36	2300	6
118	Hearth	2	3	2	11	B	Poaceae	Zea mays	Fruit	12-row wet	1		4	36	2300	6
118	Hearth	2	3	2	11	B	Poaceae	Zea mays	Fruit	Cob	10		43	36	2300	6
118	Hearth	2	3	2	11	B	Poaceae	Cenchrus sp.	Fruit	Mature	1		4	36	2300	6
118	Hearth	2	3	2	11	B		4mm Charcoal	Stem	Branchwood		1.99	8.65	36	2300	6
118	Hearth	2	3	2	11	B	Fabaceae	Prosopis sp.	Stem	(Desiccated) Chip		0.29	1.26	36	2300	6
118	Hearth	2	3	2	11	B	Fabaceae	Fabaceae	Stem	(Desiccated) Juvenile		0.08	0.35	36	2300	6
118	Hearth	2	3	2	11	B		Llama sp.	Dung			*0.08*	0.35	36	2300	6
118	Hearth	2	3	2	11	B	Capparidaceae	Capparis ovalifolia	Stem	Branchwood		0.12	0.52	36	2300	6
118	Hearth	2	3	2	11	B	Asteraceae	Tessaria integrifolia	Stem	Mature		0.40	1.74	36	2300	6
118	Hearth	2	3	2	11	B		2mm Charcoal	Stem	Branchwood		0.19	0.83	36	2300	6
118	Hearth	2	3	2	11	B		Llama sp.	Dung			*0.28*	1.22	36	2300	6
119	Old Porron	2	3	2	14		Fabaceae	Phaseolus sp.	Seed	Mature	2		6	20.698	3300	16
119	Old Porron	2	3	2	14		Cucurbitaceae	Lagenaria sp.	Seed	Mature	1		3	20.698	3300	16
119	Old Porron	2	3	2	14		Fabaceae	Prosopis sp.	Fruit	Mature	32		97	20.698	3300	16
119	Old Porron	2	3	2	14		Fabaceae	Prosopis sp.	Seed	Mature	13		39	20.698	3300	16
119	Old Porron	2	3	2	14		Fabaceae	Acacia sp.	Seed	Mature	7		21	20.698	3300	16
119	Old Porron	2	3	2	14		Poaceae	Zea mays	Seed	Flour	4		12	20.698	3300	16
119	Old Porron	2	3	2	14		Malvaceae	Gossypium sp.	Seed	Mature	43		130	20.698	3300	16
119	Old Porron	2	3	2	14		Fabaceae	Arachis sp.	Seed	Mature	2		6	20.698	3300	16
119	Old Porron	2	3	2	14		Asteraceae	cf. Asteraceae	Seed	Mature	1		3	20.698	3300	16
119	Old Porron	2	3	2	14		Verbenaceae	Lippia sp.	Seed	Mature	3		9	20.698	3300	16
119	Old Porron	2	3	2	14		Fabaceae	cf. Fabaceae	Seed	Mature	1		3	20.698	3300	16
119	Old Porron	2	3	2	14		Chenopodiaceae	Chenopodium sp.	Seed	Mature	1		3	20.698	3300	16
119	Old Porron	2	3	2	14		Euphorbiaceae	Euphorbia sp.	Seed	Mature	1		3	20.698	3300	16
119	Old Porron	2	3	2	14		Solanaceae	Capsicum sp.	Seed	Mature	6		18	20.698	3300	16
119	Old Porron	2	3	2	14		Solanaceae	Physalis sp.	Seed	Mature	4		12	20.698	3300	16
119	Old Porron	2	3	2	14		Solanaceae	Nicotiana sp.	Seed	Mature	1		3	20.698	3300	16
119	Old Porron	2	3	2	14		Solanaceae	cf. Solanaceae	Seed	Mature	12		36	20.698	3300	16
119	Old Porron	2	3	2	14		Myrtaceae	cf. Myrtaceae	Fruit	Mature	1		3	20.698	3300	16
119	Old Porron	2	3	2	14			Unid.	Seed	Mature	12		36	20.698	3300	16
119	Old Porron	2	3	2	14			Cavia sp.	Dung		47		142	20.698	3300	16
119	Old Porron	2	3	2	14		Fabaceae	Prosopis sp.	Fruit	Mature	9		27	20.698	3300	16
119	Old Porron	2	3	2	14		Fabaceae	Acacia sp.	Fruit	Mature	3		9	20.698	3300	16
119	Old Porron	2	3	2	14		Fabaceae	cf. Fabaceae	Fruit	Mature	3		9	20.698	3300	16
119	Old Porron	2	3	2	14		Cucurbitaceae	Lagenaria sp.	Fruit	Mate	1		3	20.698	3300	16
119	Old Porron	2	3	2	14		Poaceae	Zea mays	Fruit	Cob	1		3	20.698	3300	16
119	Old Porron	2	3	2	14		Fabaceae	Acacia sp.	Leaf	Spine	1		3	20.698	3300	16
119	Old Porron	2	3	2	14		Fabaceae	Prosopis sp.	Leaf	Spine	1		3	20.698	3300	16
119	Old Porron	2	3	2	14		Fabaceae	cf. Fabaceae	Fruit	Stem	10		30	20.698	3300	16
119	Old Porron	2	3	2	14			Wood Fungi	Spore		65		197	20.698	3300	16
119	Old Porron	2	3	2	14			Wood Fungi	Spore		9		27	20.698	3300	16
119	Old Porron	2	3	2	14			4mm Charcoal	Stem	Branchwood		20.15	61.06	20.698	3300	16
119	Old Porron	2	3	2	14		Fabaceae	Acacia sp.	Stem	Branchwood		20.05	60.76	20.698	3300	16
119	Old Porron	2	3	2	14		Fabaceae	Prosopis sp.	Stem	Branchwood		52.42	158.85	20.698	3300	16

119	Old Porron	2	3	2	14		Capparidaceae	Capparis ovalifolia	Stem	Branchwood		15.66	47.45	20.698	3300	16
119	Old Porron	2	3	2	14		Fabaceae	Fabaceae	Stem	(Desiccated) Juvenile		0.58	1.76	20.698	3300	16
119	Old Porron	2	3	2	14		Poaceae	Gynerium sp.	Stem			0.14	0.42	20.698	3300	16
119	Old Porron	2	3	2	14		Capparidaceae	Capparis angulata	Stem	Branchwood		5.22	15.82	20.698	3300	16
119	Old Porron	2	3	2	14			2mm Charcoal	Stem	Branchwood		3.40	10.30	20.698	3300	16
119	Old Porron	2	3	2	14			Llama sp.	Dung			*1.03*	3.12	20.698	3300	16
120	Charcoal Hearth	2	Open	5	9	B	Malvaceae	Gossypium sp.	Seed	Mature	10		17	49.016	6050	12
120	Charcoal Hearth	2	Open	5	9	B	Verbenaceae	Lippia sp.	Seed	Mature	3		5	49.016	6050	12
120	Charcoal Hearth	2	Open	5	9	B	Euphorbiaceae	Euphorbia sp.	Seed	Mature	1		2	49.016	6050	12
120	Charcoal Hearth	2	Open	5	9	B	Chenopodiaceae	Chenopodium sp.	Seed	Mature	1		2	49.016	6050	12
120	Charcoal Hearth	2	Open	5	9	B	Solanaceae	Nicotiana sp.	Seed	Mature	3		5	49.016	6050	12
120	Charcoal Hearth	2	Open	5	9	B	Malvaceae	Malva sp.	Seed	Mature	1		2	49.016	6050	12
120	Charcoal Hearth	2	Open	5	9	B	Poaceae	Zea mays	Seed	Embryo	1		2	49.016	6050	12
120	Charcoal Hearth	2	Open	5	9	B	Solanaceae	Capsicum sp.	Seed	Mature	1		2	49.016	6050	12
120	Charcoal Hearth	2	Open	5	9	B	Poaceae	Zea mays	Seed	Mature	1		2	49.016	6050	12
120	Charcoal Hearth	2	Open	5	9	B	Bixaceae	Bixa sp.	Seed	Mature	1		2	49.016	6050	12
120	Charcoal Hearth	2	Open	5	9	B		Unid.	Seed	Mature	12		20	49.016	6050	12
120	Charcoal Hearth	2	Open	5	9	B		Donax sp.	Shell		17		28	49.016	6050	12
120	Charcoal Hearth	2	Open	5	9	B		Shell 4mm	Shell		7		12	49.016	6050	12
120	Charcoal Hearth	2	Open	5	9	B		Cavia sp.	Dung		6		10	49.016	6050	12
120	Charcoal Hearth	2	Open	5	9	B		Unid.	Dung		35		58	49.016	6050	12
120	Charcoal Hearth	2	Open	5	9	B	Fabaceae	Prosopis sp.	Leaf	Spine	14		23	49.016	6050	12
120	Charcoal Hearth	2	Open	5	9	B		Fish	Bone		1		2	49.016	6050	12
120	Charcoal Hearth	2	Open	5	9	B	Fabaceae	Prosopis sp.	Fruit	Mature	2		3	49.016	6050	12
120	Charcoal Hearth	2	Open	5	9	B		4mm Charcoal	Stem	Branchwood		205.34	339.40	49.016	6050	12
120	Charcoal Hearth	2	Open	5	9	B	Fabaceae	Acacia sp.	Stem	Branchwood		98.1	162.15	49.016	6050	12
120	Charcoal Hearth	2	Open	5	9	B	Fabaceae	Fabaceae	Stem	Juvenile		12.18	20.13	49.016	6050	12
120	Charcoal Hearth	2	Open	5	9	B	Capparidaceae	Capparis angulata	Stem	Branchwood		13.68	22.61	49.016	6050	12
120	Charcoal Hearth	2	Open	5	9	B	Fabaceae	Prosopis sp.	Stem	Branchwood		57.47	94.99	49.016	6050	12
120	Charcoal Hearth	2	Open	5	9	B	Poaceae	Zea mays	Stem			0.17	0.28	49.016	6050	12
120	Charcoal Hearth	2	Open	5	9	B		2mm Charcoal	Stem	Branchwood		60.93	100.71	49.016	6050	12
120	Old Porron	2	3	2	14		Fabaceae	Prosopis sp.	Seed	Mature	29		73	20.698	4000	19
120	Old Porron	2	3	2	14		Fabaceae	Parkinsonia sp.	Seed	Mature	1		3	20.698	4000	19
120	Old Porron	2	3	2	14		Fabaceae	Phaseolus sp.	Seed	Mature	1		3	20.698	4000	19
120	Old Porron	2	3	2	14		Fabaceae	cf. Fabaceae	Seed	Mature	23		58	20.698	4000	19
120	Old Porron	2	3	2	14		Boraginaceae	Cordia sp.	Fruit	Mature	1		3	20.698	4000	19
120	Old Porron	2	3	2	14		Chenopodiaceae	Chenopodium sp.	Seed	Mature	3		8	20.698	4000	19
120	Old Porron	2	3	2	14		Amaranthaceae	Amaranthus sp.	Seed	Mature	2		5	20.698	4000	19
120	Old Porron	2	3	2	14		Fabaceae	Acacia sp.	Flower	Mature	4		10	20.698	4000	19
120	Old Porron	2	3	2	14		Poaceae	Panicum sp.	Seed	Mature	2		5	20.698	4000	19
120	Old Porron	2	3	2	14		Poaceae	cf. Poaceae	Seed	Mature	4		10	20.698	4000	19
120	Old Porron	2	3	2	14		Boraginaceae	Heliotropum sp.	Seed	Mature	1		3	20.698	4000	19
120	Old Porron	2	3	2	14		Fabaceae	UKN#64-120.1	Fruit	Mature	14		35	20.698	4000	19
120	Old Porron	2	3	2	14		Amaranthaceae	Amaranthus sp.	Seed	Mature	2		5	20.698	4000	19
120	Old Porron	2	3	2	14		Malvaceae	Gossypium sp.	Seed	Mature	2		5	20.698	4000	19
120	Old Porron	2	3	2	14		Fabaceae	UKN#64-120.1	Fruit	Mature	3		8	20.698	4000	19
120	Old Porron	2	3	2	14		Verbenaceae	Lippia sp.	Seed	Mature	4		10	20.698	4000	19
120	Old Porron	2	3	2	14		Verbenaceae	Lippia sp.	Seed	Mature	5		13	20.698	4000	19

120	Old Porron	2	3	2	14			Unid.	Seed	Mature	19		48	20.698	4000	19
120	Old Porron	2	3	2	14		Poaceae	Zea mays	Seed	Mature	1		3	20.698	4000	19
120	Old Porron	2	3	2	14			Mica Bead	Bead		1		3	20.698	4000	19
120	Old Porron	2	3	2	14			Slag	Metal	Amorphous	8		20	20.698	4000	19
120	Old Porron	2	3	2	14			Cu Alloy	Metal		1		3	20.698	4000	19
120	Old Porron	2	3	2	14			Unid.	Leaf	Spine	3		8	20.698	4000	19
120	Old Porron	2	3	2	14		Poaceae	Zea mays	Fruit	12-row	1		3	20.698	4000	19
120	Old Porron	2	3	2	14			Unid.	Fruit	Stem	13		33	20.698	4000	19
120	Old Porron	2	3	2	14			Cavia sp.	Dung		20		50	20.698	4000	19
120	Old Porron	2	3	2	14			Fish	Bone	Vertebra	5		13	20.698	4000	19
120	Old Porron	2	3	2	14			Llama sp.	Dung			*0.28*	0.70	20.698	4000	19
120	Old Porron	2	3	2	14			2mm Charcoal	Stem	Branchwood		106.50	266.25	20.698	4000	19
120	Old Porron	2	3	2	14			2mm Charcoal	Stem	Branchwood		146.00	365.00	20.698	4000	19
120	Old Porron	2	3	2	14			2mm Charcoal	Stem	Branchwood		186.00	465.00	20.698	4000	19
120	Old Porron	2	3	2	14		Fabaceae	Prosopis sp.	Stem	Branchwood		391.23	978.08	20.698	4000	19
120	Old Porron	2	3	2	14			Scoria	Metal	Amorphous		2.02	5.05	20.698	4000	19
120	Old Porron	2	3	2	14			Llama sp.	Dung			*0.28*	0.70	20.698	4000	19
120	Old Porron	2	3	2	14			Wood Type CC	Stem	Branchwood		0.05	0.13	20.698	4000	19
120	Old Porron	2	3	2	14			2mm Charcoal	Stem	Branchwood		90.00	225.00	20.698	4000	19
120	Old Porron	2	3	2	14		Fabaceae	Fabaceae	Stem	Juvenile		2.69	6.73	20.698	4000	19
121	Old Porron	2	Open	5	12		Fabaceae	Prosopis sp.	Seed	Mature	7		16	16.031	4425	28
121	Old Porron	2	Open	5	12		Fabaceae	Prosopis sp.	Fruit	Mature	2		5	16.031	4425	28
121	Old Porron	2	Open	5	12		Chenopodiaceae	Chenopodium sp.	Seed	Mature	3		7	16.031	4425	28
121	Old Porron	2	Open	5	12		Euphorbiaceae	Euphorbia sp.	Seed	Mature	1		2	16.031	4425	28
121	Old Porron	2	Open	5	12		Caryophyllaceae	Silene sp.	Seed	Mature	1		2	16.031	4425	28
121	Old Porron	2	Open	5	12		Fabaceae	cf. Fabaceae	Seed	Mature	5		11	16.031	4425	28
121	Old Porron	2	Open	5	12		Fabaceae	Acacia sp.	Seed	Mature	3		7	16.031	4425	28
121	Old Porron	2	Open	5	12		Solanaceae	cf. Solanaceae	Seed	Mature	2		5	16.031	4425	28
121	Old Porron	2	Open	5	12		Cactaceae	cf. Cactaceae	Seed	Mature	1		2	16.031	4425	28
121	Old Porron	2	Open	5	12		Verbenaceae	Lippia sp.	Seed	Mature	1		2	16.031	4425	28
121	Old Porron	2	Open	5	12		Fabaceae	Prosopis sp.	Leaf	Spine	15		34	16.031	4425	28
121	Old Porron	2	Open	5	12			Fish	Bone	Vertebra	3		7	16.031	4425	28
121	Old Porron	2	Open	5	12			Ceramic Redware	Ceramic		1		2	16.031	4425	28
121	Old Porron	2	Open	5	12			Shell 4mm	Shell		52		118	16.031	4425	28
121	Old Porron	2	Open	5	12			Cavia sp.	Dung		6		14	16.031	4425	28
121	Old Porron	2	Open	5	12		Fabaceae	Acacia sp.	Flower	Mature	1		2	16.031	4425	28
121	Old Porron	2	Open	5	12		Poaceae	Zea mays	Fruit	Cob	32		72	16.031	4425	28
121	Old Porron	2	Open	5	12		Poaceae	Zea mays	Fruit	Cupule	5		11	16.031	4425	28
121	Old Porron	2	Open	5	12		Fabaceae	Fabaceae	Stem	Juvenile		107.37	242.64	16.031	4425	28
121	Old Porron	2	Open	5	12		Capparidaceae	Capparis ovalifolia	Stem	Branchwood		12.86	29.06	16.031	4425	28
121	Old Porron	2	Open	5	12		Fabaceae	Acacia sp.	Stem	Branchwood		13.44	30.37	16.031	4425	28
121	Old Porron	2	Open	5	12		Fabaceae	Prosopis sp.	Stem	Branchwood		5.00	11.30	16.031	4425	28
121	Old Porron	2	Open	5	12		Fabaceae	Fabaceae	Stem	Juvenile		0.15	0.34	16.031	4425	28
121	Old Porron	2	Open	5	12		Poaceae	Zea mays	Stem			0.20	0.45	16.031	4425	28
121	Old Porron	2	Open	5	12			2mm Charcoal	Stem	Branchwood		84.00	189.83	16.031	4425	28
121	Old Porron	2	Open	5	12			Llama sp.	Dung			*0.28*	0.64	16.031	4425	28
121	Porron Contents	2	Open	5	8		Fabaceae	Acacia sp.	Seed	Mature	12		27	73.593	4375	6
121	Porron Contents	2	Open	5	8		Poaceae	Zea mays	Fruit	Cob	1		2	73.593	4375	6
121	Porron Contents	2	Open	5	8		Fabaceae	Acacia sp.	Seed	Mature	8		18	73.593	4375	6

121	Porron Contents	2	Open	5	8		Fabaceae	Prosopis sp.	Seed	Mature	78		178	73.593	4375	6
121	Porron Contents	2	Open	5	8		Azioiaceae	Trianthema sp.	Seed	Mature	2		5	73.593	4375	6
121	Porron Contents	2	Open	5	8		Poaceae	Bromus sp.	Seed	Mature	2		5	73.593	4375	6
121	Porron Contents	2	Open	5	8		Chenopodiaceae	Chenopodium sp.	Seed	Mature	6		14	73.593	4375	6
121	Porron Contents	2	Open	5	8		Amaranthaceae	Amaranthus sp.	Seed	Mature	3		7	73.593	4375	6
121	Porron Contents	2	Open	5	8		Nyctaginaceae	UKN#41-121.1	Fruit	Mature	1		2	73.593	4375	6
121	Porron Contents	2	Open	5	8		Fabaceae	Acacia sp.	Flower	UKN#42-121.1	1		2	73.593	4375	6
121	Porron Contents	2	Open	5	8		Poaceae	Zea mays	Seed	Embryo	1		2	73.593	4375	6
121	Porron Contents	2	Open	5	8		Verbenaceae	Lippia sp.	Seed	Mature	3		7	73.593	4375	6
121	Porron Contents	2	Open	5	8		Fabaceae	UKN#43-121.1	Seed	Mature	1		2	73.593	4375	6
121	Porron Contents	2	Open	5	8		Asteraceae	cf. Asteraceae	Seed	Mature	1		2	73.593	4375	6
121	Porron Contents	2	Open	5	8		Boraginaceae	Heliotropum sp.	Seed	Mature	1		2	73.593	4375	6
121	Porron Contents	2	Open	5	8		Brassicaceae	Lepidium sp.	Seed	Mature	1		2	73.593	4375	6
121	Porron Contents	2	Open	5	8		Malvaceae	Gossypium sp.	Seed	Mature	6		14	73.593	4375	6
121	Porron Contents	2	Open	5	8		Poaceae	Zea mays	Seed	Mature	4		9	73.593	4375	6
121	Porron Contents	2	Open	5	8		Fabaceae	Neptunia sp.	Seed	Mature	28		64	73.593	4375	6
121	Porron Contents	2	Open	5	8		Fabaceae	cf. Fabaceae	Seed	Mature	2		5	73.593	4375	6
121	Porron Contents	2	Open	5	8		Cucurbitaceae	Cucurbitaceae	Seed	Mature	1		2	73.593	4375	6
121	Porron Contents	2	Open	5	8			Unid.	Seed	Mature	20		46	73.593	4375	6
121	Porron Contents	2	Open	5	8			Ceramic Redware	ceramic		4		9	73.593	4375	6
121	Porron Contents	2	Open	5	8			Donax sp.	Shell		20		46	73.593	4375	6
121	Porron Contents	2	Open	5	8			Shell 4mm	Shell		10		23	73.593	4375	6
121	Porron Contents	2	Open	5	8			Fish	Bone		3		7	73.593	4375	6
121	Porron Contents	2	Open	5	8			Cavia sp.	Dung		21		48	73.593	4375	6
121	Porron Contents	2	Open	5	8		Fabaceae	Prosopis sp.	Fruit	stem	11		25	73.593	4375	6
121	Porron Contents	2	Open	5	8		Fabaceae	Acacia sp.	Fruit	Stem	1		2	73.593	4375	6
121	Porron Contents	2	Open	5	8		Verbenaceae	Lippia sp.	Fruit	Mature	8		18	73.593	4375	6
121	Porron Contents	2	Open	5	8		Boraginaceae	Heliotropum sp.	Fruit	Mature	23		53	73.593	4375	6
121	Porron Contents	2	Open	5	8		Poaceae	Zea mays	Fruit	Cupule	24		55	73.593	4375	6
121	Porron Contents	2	Open	5	8		Piperaceae	Piper sp.	Flower	Mature	5		11	73.593	4375	6
121	Porron Contents	2	Open	5	8			4mm Charcoal	Stem	Branchwood		76.04	173.81	73.593	4375	6
121	Porron Contents	2	Open	5	8		Fabaceae	Fabaceae	Stem	Juvenile		8.77	20.05	73.593	4375	6
121	Porron Contents	2	Open	5	8		Fabaceae	Acacia sp.	Stem	Branchwood		19.37	44.27	73.593	4375	6
121	Porron Contents	2	Open	5	8		Capparidaceae	Capparis ovalifolia	Stem	Branchwood		11.13	25.44	73.593	4375	6
121	Porron Contents	2	Open	5	8		Fabaceae	Prosopis sp.	Stem	Branchwood		27.57	63.02	73.593	4375	6
121	Porron Contents	2	Open	5	8		Poaceae	Zea mays	Stem			0.09	0.21	73.593	4375	6
121	Porron Contents	2	Open	5	8		Capparidaceae	Capparis angulata	Stem	Branchwood		3.95	9.03	73.593	4375	6
121	Porron Contents	2	Open	5	8			Scoria	Metal	Amorphous		9.33	21.33	73.593	4375	6
121	Porron Contents	2	Open	5	8			2mm Charcoal	Stem	Branchwood		88.50	202.29	73.593	4375	6
121	Porron Contents	2	Open	5	8			Llama sp.	Dung			0.94	2.15	73.593	4375	6
126	Old Porron	1	38	1	103		Solanaceae	Nicotiana sp.	Seed	Mature	1		3	3.532	3130	89
126	Old Porron	1	38	1	103		Poaceae	Andropogon sp.	Seed	Mature	1		3	3.532	3130	89
126	Old Porron	1	38	1	103		Fabaceae	cf. Fabaceae	Seed	Mature	17		54	3.532	3130	89
126	Old Porron	1	38	1	103		Malvaceae	UKN#44-126	Seed	Mature	2		6	3.532	3130	89
126	Old Porron	1	38	1	103		Cyperaceae	Cyperus sp.	Seed	Mature	1		3	3.532	3130	89
126	Old Porron	1	38	1	103		Poaceae	Zea mays	Seed	Flour	12		38	3.532	3130	89
126	Old Porron	1	38	1	103		Fabaceae	Prosopis sp.	Seed	Mature	10		32	3.532	3130	89
126	Old Porron	1	38	1	103		Fabaceae	Acacia sp.	Seed	Mature	14		45	3.532	3130	89

126	Old Porron	1	38	1	103			Unid.	Seed	Mature	2		6	3.532	3130	89
126	Old Porron	1	38	1	103		Fabaceae	Prosopis sp.	Leaf		15		48	3.532	3130	89
126	Old Porron	1	38	1	103			Donax sp.	Shell		1		3	3.532	3130	89
126	Old Porron	1	38	1	103			Mammal	Bone		1		3	3.532	3130	89
126	Old Porron	1	38	1	103		Poaceae	cf. Poaceae	Leaf		3		10	3.532	3130	89
126	Old Porron	1	38	1	103		Asteraceae	Tessaria integrifolia	Stem	Mature		0.13	0.42	3.532	3130	89
126	Old Porron	1	38	1	103		Poaceae	Wood Type KK	Stem			0.03	0.10	3.532	3130	89
126	Old Porron	1	38	1	103		Poaceae	Gynerium sp.	Stem			0.15	0.48	3.532	3130	89
126	Old Porron	1	38	1	103			Llama sp.	Dung			*0.05*	0.16	3.532	3130	89
126	Old Porron	1	38	1	103			4mm Charcoal	Stem	Branchwood		0.16	0.51	3.532	3130	89
126	Old Porron	1	38	1	103			2mm Charcoal	Stem	Branchwood		0.09	0.29	3.532	3130	89
126	Old Porron	1	38	1	103			llama sp.	Dung			*0.38*	1.21	3.532	3130	89
127	Hearth	2	3	2	11	A	Fabaceae	Phaseolus sp.	Seed	Mature	5		15	36	3380	9
127	Hearth	2	3	2	11	A	Cucurbitaceae	Lagenaria sp.	Seed	Mature	5		15	36	3380	9
127	Hearth	2	3	2	11	A	Erythoxylaceae	Erythoxylon coca	Seed	Mature	1		3	36	3380	9
127	Hearth	2	3	2	11	A	Boraginaceae	Cordia sp.	Seed	UKN#46-127	1		3	36	3380	9
127	Hearth	2	3	2	11	A	Solanaceae	Capsicum sp.	Seed	Mature	3		9	36	3380	9
127	Hearth	2	3	2	11	A	Solanaceae	Physalis sp.	Seed	Mature	1		3	36	3380	9
127	Hearth	2	3	2	11	A	Verbenaceae	Lippia sp.	Seed	Mature	4		12	36	3380	9
127	Hearth	2	3	2	11	A	Malvaceae	Gossypium sp.	Seed	Mature	21		62	36	3380	9
127	Hearth	2	3	2	11	A	Poaceae	Zea mays	Seed	Mature	28		83	36	3380	9
127	Hearth	2	3	2	11	A	Fabaceae	Acacia sp.	Seed	Mature	7		21	36	3380	9
127	Hearth	2	3	2	11	A	Fabaceae	Prosopis sp.	Seed	Mature	9		27	36	3380	9
127	Hearth	2	3	2	11	A		Unid.	Seed	Mature	6		18	36	3380	9
127	Hearth	2	3	2	11	A		Cavia sp.	Dung		7		21	36	3380	9
127	Hearth	2	3	2	11	A	Poaceae	Zea mays	Fruit	12-row dry	6		18	36	3380	9
127	Hearth	2	3	2	11	A	Cucurbitaceae	Lagenaria sp.	Fruit	Mate	2		6	36	3380	9
127	Hearth	2	3	2	11	A		Cu Alloy	Metal	Prill	1		3	36	3380	9
127	Hearth	2	3	2	11	A	Fabaceae	Acacia sp.	Fruit	Mature	6		18	36	3380	9
127	Hearth	2	3	2	11	A	Fabaceae	Prosopis sp.	Fruit	Mature	14		41	36	3380	9
127	Hearth	2	3	2	11	A		4mm Charcoal	Stem	Branchwood		8.62	25.50	36	3380	9
127	Hearth	2	3	2	11	A	Fabaceae	Prosopis sp.	Stem	(Desiccated) Chip		0.64	1.89	36	3380	9
127	Hearth	2	3	2	11	A	Capparidaceae	Capparis ovalifolia	Stem	Branchwood		0.22	0.65	36	3380	9
127	Hearth	2	3	2	11	A	Fabaceae	Prosopis sp.	Stem	Branchwood		2.78	8.22	36	3380	9
127	Hearth	2	3	2	11	A	Fabaceae	Fabaceae	Stem	(Desiccated) Juvenile		2.07	6.12	36	3380	9
127	Hearth	2	3	2	11	A		2mm Charcoal	Stem	Branchwood		0.29	0.86	36	3380	9
127	Hearth	2	3	2	11	A		Llama sp.	Dung			*0.75*	2.22	36	3380	9
128	Hearth	2	3	2	16			Unid.	Nut	Mature	1		3	12.6	3495	28
128	Hearth	2	3	2	16		Convulvulaceae	Ipomoea sp.	Seed	Mature	1		3	12.6	3495	28
128	Hearth	2	3	2	16		Verbenaceae	Lippia sp.	Seed	Mature	1		3	12.6	3495	28
128	Hearth	2	3	2	16		Malvaceae	Gossypium sp.	Seed	Mature	2		6	12.6	3495	28
128	Hearth	2	3	2	16		Fabaceae	Acacia sp.	Seed	Mature	3		9	12.6	3495	28
128	Hearth	2	3	2	16		Fabaceae	Prosopis sp.	Seed	Mature	1		3	12.6	3495	28
128	Hearth	2	3	2	16		Fabaceae	cf. Fabaceae	Seed	Mature	3		9	12.6	3495	28
128	Hearth	2	3	2	16		Poaceae	Zea mays	Seed	Mature	3		9	12.6	3495	28
128	Hearth	2	3	2	16		Solanaceae	cf. Solanaceae	Seed	Mature	1		3	12.6	3495	28
128	Hearth	2	3	2	16			Unid.	Seed	Mature	2		6	12.6	3495	28
128	Hearth	2	3	2	16		Fabaceae	Acacia sp.	Fruit	Mature	3		9	12.6	3495	28
128	Hearth	2	3	2	16		Fabaceae	cf. Fabaceae	Fruit	Stem	1		3	12.6	3495	28

128	Hearth	2	3	2	16			Unid.	Fruit	Stem	1		3	12.6	3495	28
128	Hearth	2	3	2	16			Lump	Amorphous		1		3	12.6	3495	28
128	Hearth	2	3	2	16		Fabaceae	Acacia sp.	Leaf	Spine	1		3	12.6	3495	28
128	Hearth	2	3	2	16		Fabaceae	Prosopis sp.	Leaf	Spine	2		6	12.6	3495	28
128	Hearth	2	3	2	16		Fabaceae	Fabaceae	Stem	(Desiccated) Juvenile		0.96	2.75	12.6	3495	28
128	Hearth	2	3	2	16			4mm Charcoal	Stem	Branchwood		10.45	29.90	12.6	3495	28
128	Hearth	2	3	2	16		Fabaceae	Prosopis sp.	Stem	Branchwood		5.44	15.57	12.6	3495	28
128	Hearth	2	3	2	16		Fabaceae	Acacia sp.	Stem	Branchwood		13.60	38.91	12.6	3495	28
128	Hearth	2	3	2	16		Poaceae	Zea mays	Stem			0.16	0.46	12.6	3495	28
128	Hearth	2	3	2	16			2mm Charcoal	Stem	Branchwood		1.30	3.72	12.6	3495	28
128	Hearth	2	3	2	16		Fabaceae	Fabaceae	Stem	Juvenile		0.80	2.29	12.6	3495	28
128	Hearth	2	3	2	16			Llama sp.	Dung			*0.19*	0.54	12.6	3495	28
129	Kiln	1	39	2	68	B	Poaceae	Zea mays	Fruit	16-row	1		3	3.125	3125	100
129	Kiln	1	39	2	68	B	Verbenaceae	Lippia sp.	Seed	Mature	3		10	3.125	3125	100
129	Kiln	1	39	2	68	B	Fabaceae	Prosopis sp.	Fruit	Mature	1		3	3.125	3125	100
129	Kiln	1	39	2	68	B	Poaceae	Zea mays	Seed	Mature	2		6	3.125	3125	100
129	Kiln	1	39	2	68	B	Fabaceae	cf. Fabaceae	Seed	Mature	2		6	3.125	3125	100
129	Kiln	1	39	2	68	B	Poaceae	Zea mays	Fruit	12-row	2		6	3.125	3125	100
129	Kiln	1	39	2	68	B	Poaceae	Zea mays	Fruit	Cupule	8		26	3.125	3125	100
129	Kiln	1	39	2	68	B	Fabaceae	Prosopis sp.	Leaf	Spine	3		10	3.125	3125	100
129	Kiln	1	39	2	68	B	Fabaceae	Acacia sp.	Leaf	Spine	2		6	3.125	3125	100
129	Kiln	1	39	2	68	B		Slag	Metal	Amorphous	3		10	3.125	3125	100
129	Kiln	1	39	2	68	B		Unid.	Dung		5		16	3.125	3125	100
129	Kiln	1	39	2	68	B		Wood Fungi	Spore		35		112	3.125	3125	100
129	Kiln	1	39	2	68	B		Shell 4mm	Shell		27		86	3.125	3125	100
129	Kiln	1	39	2	68	B		Donax sp.	Shell		16		51	3.125	3125	100
129	Kiln	1	39	2	68	B		Fish	Bone	Vertebra	9		29	3.125	3125	100
129	Kiln	1	39	2	68	B		Mammal	Bone		7		22	3.125	3125	100
129	Kiln	1	39	2	68	B		4mm Charcoal	Stem	Branchwood		87.50	280.00	3.125	3125	100
129	Kiln	1	39	2	68	B	Fabaceae	Prosopis sp.	Stem	Branchwood		89.43	286.18	3.125	3125	100
129	Kiln	1	39	2	68	B	Fabaceae	Acacia sp.	Stem	Branchwood		27.13	86.82	3.125	3125	100
129	Kiln	1	39	2	68	B	Capparidaceae	Capparis angulata	Stem	Juvenile		0.45	1.44	3.125	3125	100
129	Kiln	1	39	2	68	B	Capparidaceae	Capparis angulata	Stem	Branchwood		2.77	8.86	3.125	3125	100
129	Kiln	1	39	2	68	B		2mm Charcoal	Stem	Branchwood		50.90	162.88	3.125	3125	100
130	Kiln	1	39	2	68	C	Solanaceae	cf. Solanaceae	Seed	Mature	3		10	3.14	3140	100
130	Kiln	1	39	2	68	C	Chenopodiaceae	Chenopodium sp.	Seed	Mature	4		13	3.14	3140	100
130	Kiln	1	39	2	68	C	Boraginaceae	Heliotropum sp.	Seed	Mature	2		6	3.14	3140	100
130	Kiln	1	39	2	68	C	Verbenaceae	Lippia sp.	Seed	Mature	11		35	3.14	3140	100
130	Kiln	1	39	2	68	C	Fabaceae	Acacia sp.	Seed	Mature	2		6	3.14	3140	100
130	Kiln	1	39	2	68	C	Fabaceae	cf. Fabaceae	Seed	Mature	3		10	3.14	3140	100
130	Kiln	1	39	2	68	C	Poaceae	Zea mays	Seed	Mature	4		13	3.14	3140	100
130	Kiln	1	39	2	68	C		Unid.	Seed	Mature	2		6	3.14	3140	100
130	Kiln	1	39	2	68	C	Poaceae	Zea mays	Fruit	16-row	2		6	3.14	3140	100
130	Kiln	1	39	2	68	C		Cu Alloy	Metal	Prill	1		3	3.14	3140	100
130	Kiln	1	39	2	68	C		Fish	Bone	Vertebra	6		19	3.14	3140	100
130	Kiln	1	39	2	68	C		Shell 4mm	Shell		4		13	3.14	3140	100
130	Kiln	1	39	2	68	C		Mammal	Bone		3		10	3.14	3140	100
130	Kiln	1	39	2	68	C		Cavia sp.	Dung		3		10	3.14	3140	100
130	Kiln	1	39	2	68	C	Fabaceae	Prosopis sp.	Leaf	Spine	2		6	3.14	3140	100

130	Kiln	1	39	2	68	C		Slag	Metal	Amorphous	1		3	3.14	3140	100
130	Kiln	1	39	2	68	C	Fabaceae	cf. Fabaceae	Fruit	Stem	2		6	3.14	3140	100
130	Kiln	1	39	2	68	C	Poaceae	Zea mays	Fruit	Glume	17		54	3.14	3140	100
130	Kiln	1	39	2	68	C		4mm Charcoal	Stem	Branchwood		161.79	515.25	3.14	3140	100
130	Kiln	1	39	2	68	C	Fabaceae	Acacia sp.	Stem	Mature		21.77	69.33	3.14	3140	100
130	Kiln	1	39	2	68	C	Capparidaceae	Capparis angulata	Stem	Mature		5.42	17.26	3.14	3140	100
130	Kiln	1	39	2	68	C		Wood Type BB	Stem	Mature		1.85	5.89	3.14	3140	100
130	Kiln	1	39	2	68	C		Wood Type AA	Stem	Mature		3.61	11.50	3.14	3140	100
130	Kiln	1	39	2	68	C	Fabaceae	Prosopis sp.	Stem	Mature		360.20	1147.13	3.14	3140	100
130	Kiln	1	39	2	68	C		2mm Charcoal	Stem	Branchwood		69.50	221.34	3.14	3140	100
131	Kiln	1	39	2	68	D	Fabaceae	Prosopis sp.	Seed	Mature	1		3	3.575	3575	100
131	Kiln	1	39	2	68	D	Poaceae	Zea mays	Seed	Mature	4		11	3.575	3575	100
131	Kiln	1	39	2	68	D	Asteraceae	cf. Asteraceae	Seed	Mature	2		6	3.575	3575	100
131	Kiln	1	39	2	68	D	Verbenaceae	Lippia sp.	Seed	Mature	4		11	3.575	3575	100
131	Kiln	1	39	2	68	D	Poaceae	cf. Poaceae	Seed	Mature	1		3	3.575	3575	100
131	Kiln	1	39	2	68	D		Ceramic Redware	Ceramic		2		6	3.575	3575	100
131	Kiln	1	39	2	68	D		Shell 4mm	Shell		23		64	3.575	3575	100
131	Kiln	1	39	2	68	D		Fish	Bone	Vertebra	4		11	3.575	3575	100
131	Kiln	1	39	2	68	D		Mammal	Bone		9		25	3.575	3575	100
131	Kiln	1	39	2	68	D		Lump	Amorphous		1		3	3.575	3575	100
131	Kiln	1	39	2	68	D		Cavia sp.	Dung		12		34	3.575	3575	100
131	Kiln	1	39	2	68	D	Fabaceae	Prosopis sp.	Leaf	Spine	1		3	3.575	3575	100
131	Kiln	1	39	2	68	D	Fabaceae	Prosopis sp.	Stem	Branchwood		352.80	986.85	3.575	3575	100
131	Kiln	1	39	2	68	D		4mm Charcoal	Stem	Branchwood		55.20	154.41	3.575	3575	100
131	Kiln	1	39	2	68	D		2mm Charcoal	Stem	Branchwood		21.60	60.42	3.575	3575	100
133	Floor	2	Open	5				4mm Charcoal	Stem	Branchwood		0.66	2.80	12.5	2355	19
134	Floor	2	Open	5				2mm Charcoal	Stem	Branchwood		0.11	0.35	6.25	3156	50
135	Old Porron	2	3	2	21		Cucurbitaceae	Lagenaria sp.	Seed	Mature	3		8	9.408	3570	38
135	Old Porron	2	3	2	21		Fabaceae	Acacia sp.	Seed	Mature	3		8	9.408	3570	38
135	Old Porron	2	3	2	21		Cucurbitaceae	Cucurbita maxima	Seed	Mature	5		14	9.408	3570	38
135	Old Porron	2	3	2	21		Solanaceae	Nicotiana sp.	Seed	Mature	1		3	9.408	3570	38
135	Old Porron	2	3	2	21		Euphorbiaceae	Euphorbia sp.	Seed	Mature	2		6	9.408	3570	38
135	Old Porron	2	3	2	21		Solanaceae	Capsicum sp.	Seed	Mature	2		6	9.408	3570	38
135	Old Porron	2	3	2	21		Malvaceae	Gossypium sp.	Seed	Mature	4		11	9.408	3570	38
135	Old Porron	2	3	2	21		Fabaceae	Prosopis sp.	Seed	Mature	17		48	9.408	3570	38
135	Old Porron	2	3	2	21		Verbenaceae	Lippia sp.	Seed	Mature	13		36	9.408	3570	38
135	Old Porron	2	3	2	21			Unid.	Seed	Mature	7		20	9.408	3570	38
135	Old Porron	2	3	2	21		Cucurbitaceae	Lagenaria sp.	Fruit	Mate	2		6	9.408	3570	38
135	Old Porron	2	3	2	21		Fabaceae	Acacia sp.	Fruit	Mature	13		36	9.408	3570	38
135	Old Porron	2	3	2	21		Poaceae	Zea mays	Fruit	Cupule	1		3	9.408	3570	38
135	Old Porron	2	3	2	21			Donax sp.	Shell		2		6	9.408	3570	38
135	Old Porron	2	3	2	21		Verbenaceae	Lippia sp.	Fruit	Mature	2		6	9.408	3570	38
135	Old Porron	2	3	2	21			4mm Charcoal	Stem	Branchwood		4.66	13.05	9.408	3570	38
135	Old Porron	2	3	2	21		Fabaceae	Acacia sp.	Stem	Branchwood		12.29	34.43	9.408	3570	38
135	Old Porron	2	3	2	21		Capparidaceae	Capparis ovalifolia	Stem	Branchwood		6.44	18.04	9.408	3570	38
135	Old Porron	2	3	2	21		Fabaceae	Fabaceae	Stem	Juvenile		0.84	2.35	9.408	3570	38
135	Old Porron	2	3	2	21		Fabaceae	Fabaceae	Stem	(Desiccated) Juvenile		1.54	4.31	9.408	3570	38
135	Old Porron	2	3	2	21		Asteraceae	Tessaria integrifolia	Stem	Desiccated		1.78	4.99	9.408	3570	38
135	Old Porron	2	3	2	21		Asteraceae	Tessaria integrifolia	Stem	Mature		0.15	0.42	9.408	3570	38

135	Old Porron	2	3	2	21		Fabaceae	Acacia sp.	Fruit			0.15	0.42	9.408	3570	38
135	Old Porron	2	3	2	21		Capparidaceae	Capparis angulata	Stem	Branchwood		1.28	3.59	9.408	3570	38
135	Old Porron	2	3	2	21		Fabaceae	Prosopis sp.	Stem	Branchwood		1.17	3.28	9.408	3570	38
135	Old Porron	2	3	2	21			2mm Charcoal	Stem	Branchwood		1.42	3.98	9.408	3570	38
135	Old Porron	2	3	2	21			Llama sp.	Dung			*0.38*	1.06	9.408	3570	38
136	Old Porron	2	Open	5	19		Boraginaceae	Heliotropum sp.	Seed	Mature	3		4	10.289	7350	71
136	Old Porron	2	Open	5	19		Fabaceae	cf. Fabaceae	Seed	Mature	1		1	10.289	7350	71
136	Old Porron	2	Open	5	19		Malvaceae	Gossypium sp.	Seed	Mature	5		7	10.289	7350	71
136	Old Porron	2	Open	5	19		Verbenaceae	Lippia sp.	Seed	Mature	7		10	10.289	7350	71
136	Old Porron	2	Open	5	19		Solanaceae	Capsicum sp.	Seed	Mature	9		12	10.289	7350	71
136	Old Porron	2	Open	5	19		Asteraceae	cf. Asteraceae	Seed	Mature	1		1	10.289	7350	71
136	Old Porron	2	Open	5	19		Chenopodiaceae	Suaeda sp.	Seed	Mature	1		1	10.289	7350	71
136	Old Porron	2	Open	5	19		Fabaceae	Acacia sp.	Flower	Mature	1		1	10.289	7350	71
136	Old Porron	2	Open	5	19		Solanaceae	Nicotiana sp.	Seed	Mature	1		1	10.289	7350	71
136	Old Porron	2	Open	5	19		Convulvulaceae	Ipomoea sp.	Seed	Mature	1		1	10.289	7350	71
136	Old Porron	2	Open	5	19		Cucurbitaceae	Lagenaria sp.	Seed	Mature	5		7	10.289	7350	71
136	Old Porron	2	Open	5	19		Cucurbitaceae	Cucurbitaceae	Seed	Mature	8		11	10.289	7350	71
136	Old Porron	2	Open	5	19		Cucurbitaceae	Cucurbita maxima	Seed	Mature	7		10	10.289	7350	71
136	Old Porron	2	Open	5	19		Cucurbitaceae	Cucurbita sp.	Seed	Mature	3		4	10.289	7350	71
136	Old Porron	2	Open	5	19		Fabaceae	Acacia sp.	Seed	Mature	2		3	10.289	7350	71
136	Old Porron	2	Open	5	19		Fabaceae	Prosopis sp.	Seed	Mature	40		54	10.289	7350	71
136	Old Porron	2	Open	5	19			Unid.	Seed	Mature	5		7	10.289	7350	71
136	Old Porron	2	Open	5	19			Ceramic Redware	ceramic		6		8	10.289	7350	71
136	Old Porron	2	Open	5	19			Donax sp.	Shell		25		34	10.289	7350	71
136	Old Porron	2	Open	5	19			Shell 4mm	Shell		18		24	10.289	7350	71
136	Old Porron	2	Open	5	19			Fish	Bone	Vertebra	1		1	10.289	7350	71
136	Old Porron	2	Open	5	19			Fish	Bone		10		14	10.289	7350	71
136	Old Porron	2	Open	5	19			Cavia sp.	Dung		136		185	10.289	7350	71
136	Old Porron	2	Open	5	19		Capparidaceae	Capparis sp.	Fruit	Juvenile	1		1	10.289	7350	71
136	Old Porron	2	Open	5	19			Unid.	Flower	Mature	1		1	10.289	7350	71
136	Old Porron	2	Open	5	19		Poaceae	Zea mays	Fruit	16-row dry	1		1	10.289	7350	71
136	Old Porron	2	Open	5	19		Fabaceae	Prosopis sp.	Leaf	Spine (juvenile)	4		5	10.289	7350	71
136	Old Porron	2	Open	5	19		Poaceae	Zea mays	Fruit	Cob	2		3	10.289	7350	71
136	Old Porron	2	Open	5	19		Fabaceae	cf. Fabaceae	Fruit	Mature	15		20	10.289	7350	71
136	Old Porron	2	Open	5	19		Fabaceae	Acacia sp.	Fruit	Mature	3		4	10.289	7350	71
136	Old Porron	2	Open	5	19			Scoria	Metal	Amorphous		14.54	19.78	10.289	7350	71
136	Old Porron	2	Open	5	19		Capparidaceae	Capparis angulata	Stem	Branchwood		16.67	22.68	10.289	7350	71
136	Old Porron	2	Open	5	19		Fabaceae	Acacia sp.	Stem	Branchwood		27.53	37.46	10.289	7350	71
136	Old Porron	2	Open	5	19		Fabaceae	Prosopis sp.	Stem	Branchwood		126.94	172.71	10.289	7350	71
136	Old Porron	2	Open	5	19			4mm Charcoal	Stem	Branchwood		280.50	381.63	10.289	7350	71
136	Old Porron	2	Open	5	19			4mm Charcoal	Stem	Branchwood		5.98	8.14	10.289	7350	71
136	Old Porron	2	Open	5	19			2mm Charcoal	Stem	Branchwood		77.85	105.92	10.289	7350	71
136	Old Porron	2	Open	5	19			Llama sp	Dung			*0.19*	0.26	10.289	7350	71
137	Old Porron	2	3	2	20		Poaceae	cf. Poaceae	Seed	Mature	1		1	12.902	7025	54
137	Old Porron	2	3	2	20		Poaceae	Andropogon sp.	Seed	Mature	1		1	12.902	7025	54
137	Old Porron	2	3	2	20		Solanaceae	Capsicum sp.	Seed	Mature	1		1	12.902	7025	54
137	Old Porron	2	3	2	20		Verbenaceae	Lippia sp.	Seed	Mature	5		7	12.902	7025	54
137	Old Porron	2	3	2	20		Solanaceae	Physalis sp.	Seed	Mature	1		1	12.902	7025	54
137	Old Porron	2	3	2	20		Asteraceae	UKN#25-37	Seed	Mature	1		1	12.902	7025	54

137	Old Porron	2	3	2	20			Unid.	Seed	Mature	6		9	12.902	7025	54
137	Old Porron	2	3	2	20			Cosu Blanco	Bone		1		1	12.902	7025	54
137	Old Porron	2	3	2	20			Shell 4mm	Shell		5		7	12.902	7025	54
137	Old Porron	2	3	2	20			Ceramic Redware	ceramic		4		6	12.902	7025	54
137	Old Porron	2	3	2	20			Slag	Metal	Amorphous	2		3	12.902	7025	54
137	Old Porron	2	3	2	20		Poaceae	Zea mays	Fruit	Cupule	1		1	12.902	7025	54
137	Old Porron	2	3	2	20			2mm Charcoal	Stem	Branchwood		2.22	3.16	12.902	7025	54
139	Floor	2	3	2			Verbenaceae	Lippia sp.	Seed	Mature	1		4	47.2	2725	6
139	Floor	2	3	2			Poaceae	Zea mays	Seed	Mature	1		4	47.2	2725	6
139	Floor	2	3	2			Fabaceae	Prosopis sp.	Seed	Mature	5		18	47.2	2725	6
139	Floor	2	3	2			Fabaceae	cf. Fabaceae	Seed	Mature	2		7	47.2	2725	6
139	Floor	2	3	2				4mm Charcoal	Stem	Branchwood		0.17	0.62	47.2	2725	6
139	Floor	2	3	2				2mm Charcoal	Stem	Branchwood		0.19	0.70	47.2	2725	6
140	Old Porron	2	Open	5	12			Wood Fungi	Spore		36		686	2.884	525	18
140	Old Porron	2	Open	5	12			Cavia sp.	Dung		4		76	2.884	525	18
140	Old Porron	2	Open	5	12		Fabaceae	cf. Fabaceae	Fruit	Mature	2		38	2.884	525	18
140	Old Porron	2	Open	5	12			4mm Charcoal	Stem	Branchwood		31.60	601.90	2.884	525	18
140	Old Porron	2	Open	5	12		Fabaceae	Fabaceae	Stem	Juvenile		17.99	342.67	2.884	525	18
140	Old Porron	2	Open	5	12		Fabaceae	Prosopis sp.	Stem	Branchwood		15.65	298.10	2.884	525	18
140	Old Porron	2	Open	5	12		Fabaceae	Acacia sp.	Stem	Branchwood		8.82	168.00	2.884	525	18
140	Old Porron	2	Open	5	12		Capparidaceae	Capparis angulata	Stem	Branchwood		16.09	306.48	2.884	525	18
140	Old Porron	2	Open	5	12		Poaceae	Zea mays	Stem			0.65	12.38	2.884	525	18
140	Old Porron	2	Open	5	12		Capparidaceae	Capparis ovalifolia	Stem	Branchwood		2.39	45.52	2.884	525	18
140	Old Porron	2	Open	5	12			2mm Charcoal	Stem	Branchwood		2.39	45.52	2.884	525	18
140	Old Porron	2	Open	5	12			Llama sp.	Dung			*0.09*	1.71	2.884	525	18
142	Floor	2	Open	5			Fabaceae	UKN#43-121.1	Seed	Mature	11		31	50	3600	7
142	Floor	2	Open	5			Poaceae	Zea mays	Seed	Embryo	6		17	50	3600	7
142	Floor	2	Open	5			Poaceae	Zea mays	Seed	Mature	4		11	50	3600	7
142	Floor	2	Open	5			Fabaceae	Prosopis sp.	Seed	Mature	123		342	50	3600	7
142	Floor	2	Open	5			Boraginaceae	Heliotropum sp.	Seed	Mature	2		6	50	3600	7
142	Floor	2	Open	5			Verbenaceae	Lippia sp.	Seed	Mature	21		58	50	3600	7
142	Floor	2	Open	5			Malvaceae	Gossypium sp.	Seed	Mature	4		11	50	3600	7
142	Floor	2	Open	5			Fabaceae	Acacia sp.	Flower	Mature	9		25	50	3600	7
142	Floor	2	Open	5			Solanaceae	Capsicum sp.	Seed	Mature	4		11	50	3600	7
142	Floor	2	Open	5			Chenopodiaceae	Chenopodium sp.	Seed	Mature	10		28	50	3600	7
142	Floor	2	Open	5			Convulvulaceae	Ipomoea sp.	Seed	Mature	5		14	50	3600	7
142	Floor	2	Open	5			Asteraceae	Ambrosia sp	Seed	Mature	7		19	50	3600	7
142	Floor	2	Open	5			Solanaceae	cf. Solanaceae	Seed	Mature	8		22	50	3600	7
142	Floor	2	Open	5			Euphorbiaceae	Euphorbia sp.	Seed	Mature	38		106	50	3600	7
142	Floor	2	Open	5			Poaceae	Zea mays	Seed	Embryo	1		3	50	3600	7
142	Floor	2	Open	5			Solanaceae	Nicotiana sp.	Seed	Mature	2		6	50	3600	7
142	Floor	2	Open	5			Portulacaceae	Portulaca sp.	Seed	Mature	2		6	50	3600	7
142	Floor	2	Open	5			Poaceae	Zea mays	Seed	Mature	5		14	50	3600	7
142	Floor	2	Open	5			Amaranthaceae	Amaranthus sp.	Seed	Mature	2		6	50	3600	7
142	Floor	2	Open	5			Azioiaceae	Trianthema sp.	Seed	Mature	1		3	50	3600	7
142	Floor	2	Open	5			Brassicaceae	Lepidium sp.	Seed	Mature	1		3	50	3600	7
142	Floor	2	Open	5			Fabaceae	cf. Fabaceae	Seed	Mature	3		8	50	3600	7
142	Floor	2	Open	5			Asteraceae	UKN#50-142	Seed	Mature	1		3	50	3600	7
142	Floor	2	Open	5			Solanaceae	Physalis sp.	Seed	UKN#51-142	1		3	50	3600	7

142	Floor	2	Open	5			Cactaceae	cf. Cactaceae	Seed	Mature	5		14	50	3600	7
142	Floor	2	Open	5			Myrtaceae	Psidium sp.	Seed	Mature	1		3	50	3600	7
142	Floor	2	Open	5				Unid.	Seed	Mature	14		39	50	3600	7
142	Floor	2	Open	5				Donax sp.	Shell		18		50	50	3600	7
142	Floor	2	Open	5				Shell 4mm	Shell		2		6	50	3600	7
142	Floor	2	Open	5			Poaceae	Zea mays	Fruit	Cupule	18		50	50	3600	7
142	Floor	2	Open	5				Unid.	Dung		6		17	50	3600	7
142	Floor	2	Open	5			Boraginaceae	Heliotropum sp.	Fruit	Mature	18		50	50	3600	7
142	Floor	2	Open	5			Verbenaceae	Lippia sp.	Fruit	Mature	1		3	50	3600	7
142	Floor	2	Open	5			Poaceae	Zea mays	Fruit	Cob	17		47	50	3600	7
142	Floor	2	Open	5			Fabaceae	Acacia sp.	Fruit	Mature	2		6	50	3600	7
142	Floor	2	Open	5			Fabaceae	cf. Fabaceae	Fruit	Mature	8		22	50	3600	7
142	Floor	2	Open	5				4mm Charcoal	Stem	Branchwood		1.98	5.50	50	3600	7
142	Floor	2	Open	5			Fabaceae	Fabaceae	Stem	Juvenile		3.92	10.89	50	3600	7
142	Floor	2	Open	5			Fabaceae	Prosopis sp.	Stem	Branchwood		6.47	17.97	50	3600	7
142	Floor	2	Open	5				2mm Charcoal	Stem	Branchwood		31.41	87.25	50	3600	7
142	Floor	2	Open	5				Llama sp.	Dung			48.68	135.22	50	3600	7
145	Kiln	1	23	2	81		Verbenaceae	Lippia sp.	Seed	Mature	3		8	27.611	3975	14
145	Kiln	1	23	2	81		Poaceae	Zea mays	Seed	Embryo	1		3	27.611	3975	14
145	Kiln	1	23	2	81		Fabaceae	Prosopis sp.	Seed	Mature	7		18	27.611	3975	14
145	Kiln	1	23	2	81		Fabaceae	cf. Fabaceae	Seed	Mature	1		3	27.611	3975	14
145	Kiln	1	23	2	81			Ceramic Redware	ceramic		2		5	27.611	3975	14
145	Kiln	1	23	2	81		Fabaceae	Prosopis sp.	Leaf	spine	2		5	27.611	3975	14
145	Kiln	1	23	2	81			Donax sp.	Shell		5		13	27.611	3975	14
145	Kiln	1	23	2	81		Cucurbitaceae	Lagenaria sp.	Fruit	Mate	1		3	27.611	3975	14
145	Kiln	1	23	2	81		Fabaceae	Prosopis sp.	Stem	Branchwood		73.83	185.74	27.611	3975	14
145	Kiln	1	23	2	81		Capparidaceae	Capparis angulata	Stem	Branchwood		33.53	84.35	27.611	3975	14
145	Kiln	1	23	2	81		Fabaceae	Acacia sp.	Stem	Branchwood		25.55	64.28	27.611	3975	14
145	Kiln	1	23	2	81		Fabaceae	Fabaceae	Stem	Juvenile		187.12	470.74	27.611	3975	14
145	Kiln	1	23	2	81			2mm Charcoal	stem	Branchwood		68.10	171.32	27.611	3975	14
146	Ash Pit	1	35	1	84	A	Poaceae	Zea mays	Seed	Embryo	1		5	3.328	1975	59
146	Ash Pit	1	35	1	84	A	Poaceae	Zea mays	Seed	Mature	5		25	3.328	1975	59
146	Ash Pit	1	35	1	84	A	Fabaceae	Acacia sp.	Seed	Mature	1		5	3.328	1975	59
146	Ash Pit	1	35	1	84	A	Fabaceae	cf. Fabaceae	Seed	Mature	1		5	3.328	1975	59
146	Ash Pit	1	35	1	84	A	Euphorbiaceae	Euphorbia sp.	Seed	Mature	1		5	3.328	1975	59
146	Ash Pit	1	35	1	84	A		Unid.	Seed	Mature	2		10	3.328	1975	59
146	Ash Pit	1	35	1	84	A		Shell 4mm	Shell		1		5	3.328	1975	59
146	Ash Pit	1	35	1	84	A	Poaceae	Zea mays	Fruit	Cupule	2		10	3.328	1975	59
146	Ash Pit	1	35	1	84	A	Fabaceae	Prosopis sp.	Stem	Branchwood		0.38	1.92	3.328	1975	59
146	Ash Pit	1	35	1	84	A	Fabaceae	Fabaceae	Stem	Juvenile		0.40	2.03	3.328	1975	59
146	Ash Pit	1	35	1	84	A		Llama sp.	Dung			0.19	0.96	3.328	1975	59
147	Vessel Contents	1	29	1	67		Cucurbitaceae	Cucurbita sp.	Seed	Mature	1		10	0.975	975	100
147	Vessel Contents	1	29	1	67		Fabaceae	Prosopis sp.	Seed	Mature	4		41	0.975	975	100
147	Vessel Contents	1	29	1	67		Poaceae	Zea mays	Seed	Mature	2		21	0.975	975	100
147	Vessel Contents	1	29	1	67		Boraginaceae	Heliotropum sp.	Seed	Mature	4		41	0.975	975	100
147	Vessel Contents	1	29	1	67		Verbenaceae	Lippia sp.	Seed	Mature	41		421	0.975	975	100
147	Vessel Contents	1	29	1	67			UKN#53-147	Seed	Mature	2		21	0.975	975	100
147	Vessel Contents	1	29	1	67		Chenopodiaceae	Suaeda sp.	Seed	Mature	1		10	0.975	975	100

147	Vessel Contents	1	29	1	67			Unid.	Seed	Mature	5		51	0.975	975	100
147	Vessel Contents	1	29	1	67		Oxalidaceae	Oxalis sp.	Root		1		10	0.975	975	100
147	Vessel Contents	1	29	1	67		Cucurbitaceae	Lagenaria sp.	Fruit	Mate	1		10	0.975	975	100
147	Vessel Contents	1	29	1	67		Fabaceae	Prosopis sp.	Leaf		7		72	0.975	975	100
147	Vessel Contents	1	29	1	67		Fabaceae	Prosopis sp.	Stem	Branchwood		0.86	8.82	0.975	975	100
147	Vessel Contents	1	29	1	67		Chenopodiaceae	Wood Type LL	Stem			0.05	0.51	0.975	975	100
147	Vessel Contents	1	29	1	67			Llama sp.	Dung			0.07	0.72	0.975	975	100
147	Vessel Contents	1	29	1	67			2mm Charcoal	Stem	Branchwood		0.19	1.95	0.975	975	100
147	Vessel Contents	1	29	1	67			Llama sp.	Dung			*0.94*	9.64	0.975	975	100
148	Kiln	1	30	4	131		Fabaceae	Prosopis sp.	Seed	Mature	1		8	6.75	1300	19
148	Kiln	1	30	4	131		Chenopodiaceae	Chenopodium sp.	Seed	Mature	1		8	6.75	1300	19
148	Kiln	1	30	4	131			Wood Fungi	Spore		6		46	6.75	1300	19
148	Kiln	1	30	4	131		Verbenaceae	Lippia sp.	Seed	Mature	501		3854	6.75	1300	19
148	Kiln	1	30	4	131		Verbenaceae	Lippia sp.	Fruit	Mature	12		92	6.75	1300	19
148	Kiln	1	30	4	131		Boraginaceae	Heliotropum sp.	Seed	Mature	6		46	6.75	1300	19
148	Kiln	1	30	4	131			Unid.	Seed	Mature	1		8	6.75	1300	19
148	Kiln	1	30	4	131			Shell 4mm	Shell		3		23	6.75	1300	19
148	Kiln	1	30	4	131		Fabaceae	Prosopis sp.	Fruit	Mature	3		23	6.75	1300	19
148	Kiln	1	30	4	131		Verbenaceae	Lippia sp.	Fruit	Stem	2		15	6.75	1300	19
148	Kiln	1	30	4	131			Unid.	Fruit	Stem	2		15	6.75	1300	19
148	Kiln	1	30	4	131		Asteraceae	Tessaria integrifolia	Stem	Mature		1.35	10.38	6.75	1300	19
148	Kiln	1	30	4	131		Fabaceae	Fabaceae	Stem	Juvenile		0.04	0.31	6.75	1300	19
148	Kiln	1	30	4	131			Llama sp.	Dung			0.14	1.08	6.75	1300	19
148	Kiln	1	30	4	131			Burnt Clay	Ceramic			1.79	13.77	6.75	1300	19
148	Kiln	1	30	4	131		Fabaceae	Acacia sp.	Stem	Branchwood		7.29	56.08	6.75	1300	19
148	Kiln	1	30	4	131		Fabaceae	Prosopis sp.	Stem	Adze marks		14.19	109.15	6.75	1300	19
148	Kiln	1	30	4	131			4mm Charcoal	Stem	Branchwood		13.49	103.77	6.75	1300	19
148	Kiln	1	30	4	131			2mm Charcoal	Stem	Branchwood		2.46	18.92	6.75	1300	19
148	Kiln	1	30	4	131			Llama sp.	Dung			*0.09*	0.69	6.75	1300	19
148	Kiln	1	30	4	131			Llama sp.	Dung			*0.09*	0.69	6.75	1300	19
149	Floor	1	30	4			Verbenaceae	Lippia sp.	Seed	Mature	21		61	8.075	3425	42
149	Floor	1	30	4			Portulacaceae	Portulaca sp.	Seed	Mature	1		3	8.075	3425	42
149	Floor	1	30	4			Chenopodiaceae	Chenopodium sp.	Seed	Mature	2		6	8.075	3425	42
149	Floor	1	30	4				Donax sp.	Shell		1		3	8.075	3425	42
149	Floor	1	30	4				4mm Charcoal	Stem	Branchwood		1.54	4.50	8.075	3425	42
149	Floor	1	30	4			Fabaceae	Acacia sp.	Stem	Branchwood		0.89	2.60	8.075	3425	42
149	Floor	1	30	4			Fabaceae	Prosopis sp.	Stem	Branchwood		5.64	16.47	8.075	3425	42
149	Floor	1	30	4				2mm Charcoal	Stem	Branchwood		2.69	7.85	8.075	3425	42
153	Kiln	1	39	4	110	A	Verbenaceae	Lippia sp.	Seed	Mature	366		618	44.763	5925	13
153	Kiln	1	39	4	110	A	Capparidaceae	Capparis sp.	Seed	Mature	7		12	44.763	5925	13
153	Kiln	1	39	4	110	A	Poaceae	Zea mays	Seed	Flour	1		2	44.763	5925	13
153	Kiln	1	39	4	110	A	Fabaceae	Acacia sp.	Seed	Mature	5		8	44.763	5925	13
153	Kiln	1	39	4	110	A	Euphorbiaceae	Euphorbia sp.	Seed	Mature	2		3	44.763	5925	13
153	Kiln	1	39	4	110	A	Chenopodiaceae	Chenopodium sp.	Seed	Mature	19		32	44.763	5925	13
153	Kiln	1	39	4	110	A	Solanaceae	Nicotiana sp.	Seed	Mature	1		2	44.763	5925	13
153	Kiln	1	39	4	110	A	Solanaceae	Physalis sp.	Seed	Mature	3		5	44.763	5925	13
153	Kiln	1	39	4	110	A	Boraginaceae	Heliotropum sp.	Seed	Mature	2		3	44.763	5925	13
153	Kiln	1	39	4	110	A	Fabaceae	cf. Fabaceae	Seed	Mature	1		2	44.763	5925	13

153	Kiln	1	39	4	110	A	Fabaceae	Prosopis sp.	Seed	Mature	89		150	44.763	5925	13
153	Kiln	1	39	4	110	A		Unid.	Seed	Mature	3		5	44.763	5925	13
153	Kiln	1	39	4	110	A		Shell 4mm	Shell		5		8	44.763	5925	13
153	Kiln	1	39	4	110	A		Donax sp.	Shell		17		29	44.763	5925	13
153	Kiln	1	39	4	110	A	Poaceae	Zea mays	Fruit	12-row	1		2	44.763	5925	13
153	Kiln	1	39	4	110	A	Fabaceae	Prosopis sp.	Leaf	Spine	8		14	44.763	5925	13
153	Kiln	1	39	4	110	A	Fabaceae	Acacia sp.	Leaf	Spine	6		10	44.763	5925	13
153	Kiln	1	39	4	110	A	Verbenaceae	Lippia sp.	Fruit	Mature	3		5	44.763	5925	13
153	Kiln	1	39	4	110	A		Cosu Blanco	Bone		2		3	44.763	5925	13
153	Kiln	1	39	4	110	A		Slag	Metal	Amorphous	16		27	44.763	5925	13
153	Kiln	1	39	4	110	A	Capparidaceae	Capparis angulata	Stem	Branchwood		0.72	1.22	44.763	5925	13
153	Kiln	1	39	4	110	A	Poaceae	Zea mays	Stem			1.48	2.50	44.763	5925	13
153	Kiln	1	39	4	110	A		Slag	Metal	Amorphous		2.85	4.81	44.763	5925	13
153	Kiln	1	39	4	110	A	Fabaceae	Prosopis sp.	Stem	Mature		16.57	27.97	44.763	5925	13
153	Kiln	1	39	4	110	A	Fabaceae	Prosopis sp.	Stem	Juvenile		74.76	126.18	44.763	5925	13
153	Kiln	1	39	4	110	A	Fabaceae	Prosopis sp.	Stem	Juvenile		14.74	24.88	44.763	5925	13
153	Kiln	1	39	4	110	A		Llama sp.	Dung			*4.22*	7.12	44.763	5925	13
153	Kiln	1	39	4	110	A		Llama sp.	Dung			*10.32*	17.42	44.763	5925	13
154	Kiln	1	39	4	110	B	Verbenaceae	Lippia sp.	Seed	Mature	213		1092	44.452	1950	4
154	Kiln	1	39	4	110	B	Capparidaceae	Capparis sp.	Seed	Mature	1		5	44.452	1950	4
154	Kiln	1	39	4	110	B	Cyperaceae	Scirpus sp.	Seed	Mature	2		10	44.452	1950	4
154	Kiln	1	39	4	110	B	Chenopodiaceae	Chenopodium sp.	Seed	Mature	9		46	44.452	1950	4
154	Kiln	1	39	4	110	B	Boraginaceae	Heliotropum sp.	Seed	Mature	4		21	44.452	1950	4
154	Kiln	1	39	4	110	B	Fabaceae	Acacia sp.	Seed	Mature	2		10	44.452	1950	4
154	Kiln	1	39	4	110	B	Fabaceae	Prosopis sp.	Seed	Mature	31		159	44.452	1950	4
154	Kiln	1	39	4	110	B	Fabaceae	cf. Fabaceae	Seed	Mature	10		51	44.452	1950	4
154	Kiln	1	39	4	110	B	Azioiaceae	Trianthema sp.	Seed	Mature	1		5	44.452	1950	4
154	Kiln	1	39	4	110	B		Unid.	Seed	Mature	1		5	44.452	1950	4
154	Kiln	1	39	4	110	B	Verbenaceae	Lippia sp.	Fruit	Mature	3		15	44.452	1950	4
154	Kiln	1	39	4	110	B	Fabaceae	Prosopis sp.	Leaf	Spine	7		36	44.452	1950	4
154	Kiln	1	39	4	110	B	Fabaceae	Acacia sp.	Leaf	Spine	1		5	44.452	1950	4
154	Kiln	1	39	4	110	B	Fabaceae	cf. Fabaceae	Fruit	Stem	2		10	44.452	1950	4
154	Kiln	1	39	4	110	B		Donax sp.	Shell		2		10	44.452	1950	4
154	Kiln	1	39	4	110	B		Shell 4mm	Shell		1		5	44.452	1950	4
154	Kiln	1	39	4	110	B	Fabaceae	Prosopis sp.	Stem	Branchwood		1.34	6.87	44.452	1950	4
154	Kiln	1	39	4	110	B		Wood Type AA	Stem	Branchwood		0.35	1.79	44.452	1950	4
154	Kiln	1	39	4	110	B	Capparidaceae	Capparis angulata	Stem	Branchwood		19.25	98.72	44.452	1950	4
154	Kiln	1	39	4	110	B		4mm Charcoal	Stem	Branchwood		14.02	71.90	44.452	1950	4
154	Kiln	1	39	4	110	B		Ceramic Redware	ceramic			3.03	15.54	44.452	1950	4
154	Kiln	1	39	4	110	B		2mm Charcoal	Stem	Branchwood		3.74	19.18	44.452	1950	4
154	Kiln	1	39	4	110	B		Llama sp.	Dung			*1.59*	8.15	44.452	1950	4
154	Kiln	1	39	4	110	B		Llama sp.	Dung			*1.22*	6.26	44.452	1950	4
155	Kiln	1	39	4	111		Verbenaceae	Lippia sp.	Seed	Mature	85		298	30.395	2850	9
155	Kiln	1	39	4	111		Boraginaceae	Heliotropum sp.	Seed	Mature	2		7	30.395	2850	9
155	Kiln	1	39	4	111		Fabaceae	cf. Fabaceae	Seed	Mature	16		56	30.395	2850	9
155	Kiln	1	39	4	111		Chenopodiaceae	Chenopodium sp.	Seed	Mature	3		11	30.395	2850	9
155	Kiln	1	39	4	111		Fabaceae	Acacia sp.	Seed	Mature	2		7	30.395	2850	9
155	Kiln	1	39	4	111		Fabaceae	Prosopis sp.	Seed	Mature	8		28	30.395	2850	9
155	Kiln	1	39	4	111		Poaceae	Zea mays	Seed	Mature	2		7	30.395	2850	9

155	Kiln	1	39	4	111		Fabaceae	Arachis sp.	Seed	Mature	1		4	30.395	2850	9	
155	Kiln	1	39	4	111			Unid.	Seed	Mature	5		18	30.395	2850	9	
155	Kiln	1	39	4	111			Cavia sp.	Dung		5		18	30.395	2850	9	
155	Kiln	1	39	4	111			Shell 4mm	Shell		4		14	30.395	2850	9	
155	Kiln	1	39	4	111			Unid.	Leaf	Spine	1		4	30.395	2850	9	
155	Kiln	1	39	4	111		Fabaceae	cf. Fabaceae	Fruit	Stem	1		4	30.395	2850	9	
155	Kiln	1	39	4	111			4mm Charcoal	Stem	Branchwood		32.49	114.00	30.395	2850	9	
155	Kiln	1	39	4	111		Fabaceae	Prosopis sp.	Stem	Branchwood		20.97	73.58	30.395	2850	9	
155	Kiln	1	39	4	111		Capparidaceae	Capparis ovalifolia	Stem	Branchwood		4.72	16.56	30.395	2850	9	
155	Kiln	1	39	4	111		Fabaceae	Fabaceae	Stem	Juvenile		0.80	2.81	30.395	2850	9	
155	Kiln	1	39	4	111		Capparidaceae	Capparis angulata	Stem	Branchwood		1.01	3.54	30.395	2850	9	
155	Kiln	1	39	4	111		Fabaceae	Acacia sp.	Stem	Branchwood		0.98	3.44	30.395	2850	9	
155	Kiln	1	39	4	111			Burnt Clay	Ceramic			4.41	15.47	30.395	2850	9	
155	Kiln	1	39	4	111			Llama sp.	Dung			0.16	0.56	30.395	2850	9	
155	Kiln	1	39	4	111			2mm Charcoal	Stem	Branchwood		6.90	24.21	30.395	2850	9	
156	Floor	1	39	4			Fabaceae	Prosopis sp.	Seed	Mature	1		3	190	3700	2	
156	Floor	1	39	4			Verbenaceae	Lippia sp.	Seed	Mature	77		208	190	3700	2	
156	Floor	1	39	4			Cyperaceae	Cyperus sp.	Seed	Mature	2		5	190	3700	2	
156	Floor	1	39	4			Azioiaceae	Trianthema sp.	Seed	Mature	1		3	190	3700	2	
156	Floor	1	39	4			Fabaceae	Prosopis sp.	Seed	Mature	19		51	190	3700	2	
156	Floor	1	39	4			Fabaceae	Acacia sp.	Seed	Mature	5		14	190	3700	2	
156	Floor	1	39	4			Fabaceae	cf. Fabaceae	Seed	Mature	8		22	190	3700	2	
156	Floor	1	39	4			Poaceae	Panicum sp.	Seed	Mature	1		3	190	3700	2	
156	Floor	1	39	4			Solanaceae	Nicotiana sp.	Seed	Mature	4		11	190	3700	2	
156	Floor	1	39	4			Chenopodiaceae	Chenopodium sp.	Seed	Mature	7		19	190	3700	2	
156	Floor	1	39	4			Euphorbiaceae	Euphorbia sp.	Seed	Mature	2		5	190	3700	2	
156	Floor	1	39	4			Poaceae	Zea mays	Seed	Mature	1		3	190	3700	2	
156	Floor	1	39	4			Poaceae	Zea mays	Seed	Embryo	1		3	190	3700	2	
156	Floor	1	39	4			Myrtaceae	Psidium sp.	Seed	UKN#55-156	1		3	190	3700	2	
156	Floor	1	39	4			Boraginaceae	Heliotropum sp.	Seed	Mature	3		8	190	3700	2	
156	Floor	1	39	4				Unid.	Seed	Mature	3		8	190	3700	2	
156	Floor	1	39	4				Cavia sp.	Dung		1		3	190	3700	2	
156	Floor	1	39	4			Fabaceae	Prosopis sp.	Leaf	spine	6		16	190	3700	2	
156	Floor	1	39	4				Donax sp.	Shell		1		3	190	3700	2	
156	Floor	1	39	4				Slag	Metal	Amorphous	4		11	190	3700	2	
156	Floor	1	39	4				4mm Charcoal	Stem	Branchwood		1.94	5.24	190	3700	2	
156	Floor	1	39	4				Llama sp.	Dung			0.20	0.54	190	3700	2	
156	Floor	1	39	4			Asteraceae	Tessaria integrifolia	Stem	Mature		0.25	0.68	190	3700	2	
156	Floor	1	39	4			Fabaceae	Prosopis sp.	Stem	Branchwood		2.87	7.76	190	3700	2	
156	Floor	1	39	4			Fabaceae	Acacia sp.	Stem	Branchwood		1.31	3.54	190	3700	2	
156	Floor	1	39	4			Fabaceae	Fabaceae	Stem	Juvenile		4.12	11.14	190	3700	2	
156	Floor	1	39	4				2mm Charcoal	Stem	Branchwood		0.43	1.16	190	3700	2	
156	Floor	1	39	4				Llama sp.	Dung			1.97	5.32	190	3700	2	
157	Hearth	1	39	4	122	A		Donax sp.	Shell		1		3	12.214	3200	26	
157	Hearth	1	39	4	122	A	Fabaceae	Prosopis sp.	Seed	Mature	4		13	12.214	3200	26	
157	Hearth	1	39	4	122	A	Portulacaceae	Portulaca sp.	Seed	Mature	1		3	12.214	3200	26	
157	Hearth	1	39	4	122	A	Malvaceae	cf. Malvaceae	Seed	Mature	1		3	12.214	3200	26	
157	Hearth	1	39	4	122	A	Cactaceae	Opuntia sp.	Seed	Mature	1		3	12.214	3200	26	
157	Hearth	1	39	4	122	A	Verbenaceae	Lippia sp.	Seed	Mature	42		131	12.214	3200	26	

157	Hearth	1	39	4	122	A	Boraginaceae	Heliotropum sp.	Seed	Mature	2		6	12.214	3200	26
157	Hearth	1	39	4	122	A		Unid.	Seed	Mature	4		13	12.214	3200	26
157	Hearth	1	39	4	122	A	Poaceae	Zea mays	Fruit	Glume	5		16	12.214	3200	26
157	Hearth	1	39	4	122	A	Poaceae	Zea mays	Fruit	Cupule	2		6	12.214	3200	26
157	Hearth	1	39	4	122	A		Slag	Metal	Amorphous	1		3	12.214	3200	26
157	Hearth	1	39	4	122	A		Shell 4mm	Shell		3		9	12.214	3200	26
157	Hearth	1	39	4	122	A		Fish	Bone	Vertebra	9		28	12.214	3200	26
157	Hearth	1	39	4	122	A	Poaceae	Zea mays	Fruit	12-row	1		3	12.214	3200	26
157	Hearth	1	39	4	122	A		4mm Charcoal	Stem	Branchwood		158.45	495.16	12.214	3200	26
157	Hearth	1	39	4	122	A	Fabaceae	Acacia sp.	Stem	Branchwood		129.75	405.47	12.214	3200	26
157	Hearth	1	39	4	122	A	Fabaceae	Prosopis sp.	Stem	Branchwood		113.08	353.38	12.214	3200	26
157	Hearth	1	39	4	122	A		Ceramic Redware	Ceramic			23.69	74.03	12.214	3200	26
157	Hearth	1	39	4	122	A	Fabaceae	Fabaceae	Stem	Juvenile		5.63	17.59	12.214	3200	26
157	Hearth	1	39	4	122	A	Apocynaceae	Vallesia glabra	Stem	Mature		2.28	7.13	12.214	3200	26
157	Hearth	1	39	4	122	A	Poaceae	Wood Type FF	Stem			0.74	2.31	12.214	3200	26
157	Hearth	1	39	4	122	A		2mm Charcoal	Stem	Branchwood		11.40	35.63	12.214	3200	26
159	Kiln	1	39	4	68	C	Fabaceae	Prosopis sp.	Stem	Branchwood		6.58	87.73	0.75	750	100
160	Kiln	1	39	4	109		Fabaceae	Prosopis sp.	Seed	Mature	6		12	78.5	4950	6
160	Kiln	1	39	4	109			Scoria	Metal	Amorphous	2		4	78.5	4950	6
160	Kiln	1	39	4	109		Verbenaceae	Lippia sp.	Seed	Mature	176		356	78.5	4950	6
160	Kiln	1	39	4	109		Boraginaceae	Heliotropum sp.	Seed	Mature	3		6	78.5	4950	6
160	Kiln	1	39	4	109		Capparidaceae	Capparis sp.	Flower	Sepal	1		2	78.5	4950	6
160	Kiln	1	39	4	109		Cucurbitaceae	Cucurbitaceae	Seed	Mature	2		4	78.5	4950	6
160	Kiln	1	39	4	109		Chenopodiaceae	Chenopodium sp.	Seed	Mature	11		22	78.5	4950	6
160	Kiln	1	39	4	109		Solanaceae	Nicotiana sp.	Seed	Mature	1		2	78.5	4950	6
160	Kiln	1	39	4	109		Fabaceae	cf. Fabaceae	Seed	Mature	63		127	78.5	4950	6
160	Kiln	1	39	4	109		Euphorbiaceae	Euphorbia sp.	Seed	Mature	4		8	78.5	4950	6
160	Kiln	1	39	4	109		Poaceae	Zea mays	Seed	Mature	6		12	78.5	4950	6
160	Kiln	1	39	4	109		Capparidaceae	Capparis sp.	Seed	Mature	5		10	78.5	4950	6
160	Kiln	1	39	4	109		Malvaceae	Gossypium sp.	Seed	Mature	4		8	78.5	4950	6
160	Kiln	1	39	4	109		Poaceae	Zea mays	Seed	Mature	2		4	78.5	4950	6
160	Kiln	1	39	4	109		Fabaceae	Acacia sp.	Seed	Mature	16		32	78.5	4950	6
160	Kiln	1	39	4	109		Fabaceae	Prosopis sp.	Seed	Mature	72		145	78.5	4950	6
160	Kiln	1	39	4	109		Fabaceae	Prosopis sp.	Leaf		2		4	78.5	4950	6
160	Kiln	1	39	4	109		Verbenaceae	Lippia sp.	Fruit	Mature	4		8	78.5	4950	6
160	Kiln	1	39	4	109			Unid.	Fruit	Stem	18		36	78.5	4950	6
160	Kiln	1	39	4	109		Fabaceae	Acacia sp.	Fruit	Stem	1		2	78.5	4950	6
160	Kiln	1	39	4	109			Donax sp.	Shell		9		18	78.5	4950	6
160	Kiln	1	39	4	109			Shell 4mm	Shell		1		2	78.5	4950	6
160	Kiln	1	39	4	109			Slag	Metal	Amorphous	4		8	78.5	4950	6
160	Kiln	1	39	4	109		Fabaceae	Fabaceae	Stem	Juvenile		5.72	11.56	78.5	4950	6
160	Kiln	1	39	4	109		Asteraceae	Tessaria integrifolia	Stem	Mature		0.59	1.19	78.5	4950	6
160	Kiln	1	39	4	109			Llama sp.	Dung			4.33	8.75	78.5	4950	6
160	Kiln	1	39	4	109		Fabaceae	Prosopis sp.	Stem	Branchwood		18.95	38.28	78.5	4950	6
160	Kiln	1	39	4	109			Wood Type JJ	Stem	Branchwood		0.07	0.14	78.5	4950	6
160	Kiln	1	39	4	109		Capparidaceae	Capparis ovalifolia	Stem	Branchwood		0.19	0.38	78.5	4950	6
160	Kiln	1	39	4	109		Fabaceae	Acacia sp.	Stem	Branchwood		3.68	7.43	78.5	4950	6
160	Kiln	1	39	4	109		Capparidaceae	Capparis angulata	Stem	Branchwood		3.97	8.02	78.5	4950	6
160	Kiln	1	39	4	109			4mm Charcoal	Stem	Branchwood		16.31	32.95	78.5	4950	6

160	Kiln	1	39	4	109			2mm Charcoal	Stem	Branchwood		4.94	9.98	78.5	4950	6
160	Kiln	1	39	4	109			Llama sp.	Dung			*3.00*	6.06	78.5	4950	6
160	Kiln	1	39	4	109			Llama sp.	Dung			*2.81*	5.68	78.5	4950	6
161	Kiln	1	39	4	117		Poaceae	Zea mays	Fruit	16-row	1		1	*44.452*	10850	24
161	Kiln	1	39	4	117		Boraginaceae	Heliotropum sp.	Seed	Mature	97		89	*44.452*	10850	24
161	Kiln	1	39	4	117		Fabaceae	UKN#12-177	Seed	Mature	1		1	*44.452*	10850	24
161	Kiln	1	39	4	117		Fabaceae	Acacia sp.	Seed	Mature	11		10	*44.452*	10850	24
161	Kiln	1	39	4	117		Fabaceae	Prosopis sp.	Seed	Mature	20		18	*44.452*	10850	24
161	Kiln	1	39	4	117		Fabaceae	cf. Fabaceae	Seed	Mature	22		20	*44.452*	10850	24
161	Kiln	1	39	4	117		Malvaceae	Gossypium sp.	Seed	Mature	8		7	*44.452*	10850	24
161	Kiln	1	39	4	117		Poaceae	Zea mays	Seed	Embryo	9		8	*44.452*	10850	24
161	Kiln	1	39	4	117		Poaceae	Zea mays	Seed	Mature	2		2	*44.452*	10850	24
161	Kiln	1	39	4	117		Erythoxylaceae	Erythoxylon coca	Seed	Mature	1		1	*44.452*	10850	24
161	Kiln	1	39	4	117		Asteraceae	cf. Asteraceae	Seed	Mature	2		2	*44.452*	10850	24
161	Kiln	1	39	4	117		Chenopodiaceae	Chenopodium sp.	Seed	Mature	16		15	*44.452*	10850	24
161	Kiln	1	39	4	117		Fabaceae	Acacia sp.	Flower	Mature	1		1	*44.452*	10850	24
161	Kiln	1	39	4	117		Euphorbiaceae	Euphorbia sp.	Seed	Mature	1		1	*44.452*	10850	24
161	Kiln	1	39	4	117		Portulacaceae	Portulaca sp.	Seed	Mature	1		1	*44.452*	10850	24
161	Kiln	1	39	4	117		Solanaceae	cf. Solanaceae	Seed	Mature	6		6	*44.452*	10850	24
161	Kiln	1	39	4	117		Solanaceae	Capsicum sp.	Seed	Mature	4		4	*44.452*	10850	24
161	Kiln	1	39	4	117		Solanaceae	Nicotiana sp.	Seed	Mature	8		7	*44.452*	10850	24
161	Kiln	1	39	4	117		Portulacaceae	cf. Portulacaceae	Seed	Mature	2		2	*44.452*	10850	24
161	Kiln	1	39	4	117		Amaranthaceae	Amaranthus sp.	Seed	Mature	1		1	*44.452*	10850	24
161	Kiln	1	39	4	117		Verbenaceae	Lippia sp.	Seed	Mature	901		830	*44.452*	10850	24
161	Kiln	1	39	4	117		Verbenaceae	Lippia sp.	Fruit	Mature	19		18	*44.452*	10850	24
161	Kiln	1	39	4	117			Unid.	Seed	Mature	23		21	*44.452*	10850	24
161	Kiln	1	39	4	117			Ceramic Blackware	Ceramic		1		1	*44.452*	10850	24
161	Kiln	1	39	4	117		Poaceae	Zea mays	Fruit	12-row	4		4	*44.452*	10850	24
161	Kiln	1	39	4	117			Ceramic Redware	Ceramic		1		1	*44.452*	10850	24
161	Kiln	1	39	4	117			Shell 4mm	Shell		42		39	*44.452*	10850	24
161	Kiln	1	39	4	117			Donax sp.	Shell		18		17	*44.452*	10850	24
161	Kiln	1	39	4	117			Mammal	Bone		28		26	*44.452*	10850	24
161	Kiln	1	39	4	117			Cavia sp.	Dung		8		7	*44.452*	10850	24
161	Kiln	1	39	4	117			Unid.	Dung		1		1	*44.452*	10850	24
161	Kiln	1	39	4	117		Capparidaceae	Capparis angulata	Flower	Mature	4		4	*44.452*	10850	24
161	Kiln	1	39	4	117		Poaceae	Zea mays	Fruit	Cupule	10		9	*44.452*	10850	24
161	Kiln	1	39	4	117			Unid.	Leaf	Spine	8		7	*44.452*	10850	24
161	Kiln	1	39	4	117			Slag	Metal	Amorphous	9		8	*44.452*	10850	24
161	Kiln	1	39	4	117		Fabaceae	cf. Fabaceae	Fruit	Stem	6		6	*44.452*	10850	24
161	Kiln	1	39	4	117			Unid.	Fruit	Stem	3		3	*44.452*	10850	24
161	Kiln	1	39	4	117		Fabaceae	cf. Fabaceae	Fruit	Mature	1		1	*44.452*	10850	24
161	Kiln	1	39	4	117			Unid.	Fruit	Mature	1		1	*44.452*	10850	24
161	Kiln	1	39	4	117			4mm Charcoal	Stem	Branchwood		190.22	175.32	*44.452*	10850	24
161	Kiln	1	39	4	117			Mammal	Bone			1.81	1.67	*44.452*	10850	24
161	Kiln	1	39	4	117			Wood Type CC	Stem	Branchwood		3.83	3.53	*44.452*	10850	24
161	Kiln	1	39	4	117		Capparidaceae	Capparis angulata	Stem	Branchwood		4.86	4.48	*44.452*	10850	24
161	Kiln	1	39	4	117		Fabaceae	Acacia sp.	Stem	Branchwood		21.59	19.90	*44.452*	10850	24
161	Kiln	1	39	4	117		Fabaceae	Prosopis sp.	Stem	Branchwood		137.64	126.86	*44.452*	10850	24
161	Kiln	1	39	4	117			2mm Charcoal	Stem	Branchwood		224.30	206.73	*44.452*	10850	24

161	Kiln	1	39	4	117			Llama sp.	Dung			0.75	0.69	44.452	10850	24
162	Hearth	1	39	4	123		Amaranthaceae	Amaranthus sp.	Seed	Mature	1		3	19.625	3450	18
162	Hearth	1	39	4	123		Boraginaceae	Heliotropum sp.	Seed	Mature	8		23	19.625	3450	18
162	Hearth	1	39	4	123		Fabaceae	Prosopis sp.	Seed	Mature	1		3	19.625	3450	18
162	Hearth	1	39	4	123		Fabaceae	cf. Fabaceae	Seed	Mature	1		3	19.625	3450	18
162	Hearth	1	39	4	123		Chenopodiaceae	Chenopodium sp.	Seed	Mature	2		6	19.625	3450	18
162	Hearth	1	39	4	123		Azioiaceae	Trianthema sp.	Seed	Mature	1		3	19.625	3450	18
162	Hearth	1	39	4	123		Poaceae	cf. Poaceae	Seed	Mature	1		3	19.625	3450	18
162	Hearth	1	39	4	123		Cyperaceae	Scirpus sp.	Seed	Mature	2		6	19.625	3450	18
162	Hearth	1	39	4	123		Verbenaceae	Lippia sp.	Seed	Mature	310		899	19.625	3450	18
162	Hearth	1	39	4	123		Fabaceae	Prosopis sp.	Leaf	Spine	4		12	19.625	3450	18
162	Hearth	1	39	4	123			Unid.	Leaf	Spine	4		12	19.625	3450	18
162	Hearth	1	39	4	123			Fish	Bone		1		3	19.625	3450	18
162	Hearth	1	39	4	123			Mammal	Bone		1		3	19.625	3450	18
162	Hearth	1	39	4	123			Shell 4mm	Shell		1		3	19.625	3450	18
162	Hearth	1	39	4	123		Poaceae	Zea mays	Fruit	Cupule	3		9	19.625	3450	18
162	Hearth	1	39	4	123			2mm Charcoal	Stem	Branchwood		21.10	61.16	19.625	3450	18
162	Hearth	1	39	4	123			4mm Charcoal	Stem	Branchwood		94.80	274.78	19.625	3450	18
162	Hearth	1	39	4	123		Fabaceae	Prosopis sp.	Stem	Branchwood		95.20	275.94	19.625	3450	18
162	Hearth	1	39	4	123		Fabaceae	Acacia sp.	Stem	Branchwood		107.50	311.59	19.625	3450	18
162	Hearth	1	39	4	123		Capparidaceae	Capparis angulata	Stem	Branchwood		14.60	42.32	19.625	3450	18
170	Preheating Fire	1	39	4	96		Boraginaceae	Heliotropum sp.	Seed	Mature	13		22	90	6000	7
170	Preheating Fire	1	39	4	96		Poaceae	Zea mays	Seed	Embryo	2		3	90	6000	7
170	Preheating Fire	1	39	4	96		Fabaceae	Acacia sp.	Seed	Mature	1		2	90	6000	7
170	Preheating Fire	1	39	4	96		Asteraceae	Taraxacum sp.	Seed	Mature	2		3	90	6000	7
170	Preheating Fire	1	39	4	96		Solanaceae	Capsicum sp.	Seed	Mature	1		2	90	6000	7
170	Preheating Fire	1	39	4	96		Solanaceae	cf. Solanaceae	Seed	Mature	2		3	90	6000	7
170	Preheating Fire	1	39	4	96		Euphorbiaceae	Euphorbia sp.	Seed	Mature	1		2	90	6000	7
170	Preheating Fire	1	39	4	96		Chenopodiaceae	Chenopodium sp.	Seed	Mature	10		17	90	6000	7
170	Preheating Fire	1	39	4	96		Fabaceae	Prosopis sp.	Seed	Mature	3		5	90	6000	7
170	Preheating Fire	1	39	4	96		Fabaceae	Acacia sp.	Flower	Mature	32		53	90	6000	7
170	Preheating Fire	1	39	4	96		Erythoxylaceae	Erythoxylon coca	Seed	Mature	1		2	90	6000	7
170	Preheating Fire	1	39	4	96		Cucurbitaceae	Cucurbita sp.	Seed	Mature	1		2	90	6000	7
170	Preheating Fire	1	39	4	96		Fabaceae	cf. Fabaceae	Seed	Mature	1		2	90	6000	7
170	Preheating Fire	1	39	4	96		Capparidaceae	Capparis crotonides	Seed	Mature	28		47	90	6000	7
170	Preheating Fire	1	39	4	96		Capparidaceae	Capparis ovalifolia	Seed	Mature	9		15	90	6000	7
170	Preheating Fire	1	39	4	96		Verbenaceae	Lippia sp.	Seed	Mature	###		13177	90	6000	7
170	Preheating Fire	1	39	4	96		Verbenaceae	Lippia sp.	Fruit	Mature	131		218	90	6000	7
170	Preheating Fire	1	39	4	96			Unid.	Seed	Mature	5		8	90	6000	7
170	Preheating Fire	1	39	4	96			Cavia sp.	Dung		9		15	90	6000	7
170	Preheating Fire	1	39	4	96			Unid.	Dung		2		3	90	6000	7
170	Preheating Fire	1	39	4	96			Slag	Metal	Amorphous	62		103	90	6000	7
170	Preheating Fire	1	39	4	96		Fabaceae	Prosopis sp.	Leaf		10		17	90	6000	7
170	Preheating Fire	1	39	4	96		Fabaceae	Acacia sp.	Leaf		1		2	90	6000	7
170	Preheating Fire	1	39	4	96			Shell 4mm	Shell		4		7	90	6000	7
170	Preheating Fire	1	39	4	96			Mammal	Bone		1		2	90	6000	7
170	Preheating Fire	1	39	4	96		Fabaceae	Prosopis sp.	Leaf	Spine	12		20	90	6000	7
170	Preheating Fire	1	39	4	96		Fabaceae	Acacia sp.	Leaf	Spine	8		13	90	6000	7

170	Preheating Fire	1	39	4	96		Fabaceae	cf. Fabaceae	Fruit	Stem	7		12	90	6000	7
170	Preheating Fire	1	39	4	96			Unid.	Fruit	Stem	1		2	90	6000	7
170	Preheating Fire	1	39	4	96		Verbenaceae	Lippia sp.	Leaf		5		8	90	6000	7
170	Preheating Fire	1	39	4	96		Capparidaceae	Capparis ovalifolia	Fruit	Stem	16		27	90	6000	7
170	Preheating Fire	1	39	4	96		Fabaceae	cf. Fabaceae	Fruit	Mature	1		2	90	6000	7
170	Preheating Fire	1	39	4	96		Capparidaceae	Capparis angulata	Flower	Mature	2		3	90	6000	7
170	Preheating Fire	1	39	4	96			Wood Type CC	Stem	Branchwood		10.71	17.85	90	6000	7
170	Preheating Fire	1	39	4	96		Fabaceae	Fabaceae	Stem	Juvenile		102.54	170.90	90	6000	7
170	Preheating Fire	1	39	4	96		Fabaceae	Prosopis sp.	Stem	Juvenile		145.21	242.02	90	6000	7
170	Preheating Fire	1	39	4	96		Fabaceae	Acacia sp.	Stem	Juvenile		23.41	39.02	90	6000	7
170	Preheating Fire	1	39	4	96		Poaceae	Zea mays	Stem			0.51	0.85	90	6000	7
170	Preheating Fire	1	39	4	96			Scoria	Metal	Amorphous		3.58	5.97	90	6000	7
170	Preheating Fire	1	39	4	96			Llama sp.	Dung			0.32	0.53	90	6000	7
170	Preheating Fire	1	39	4	96			2mm Charcoal	Stem	Branchwood		42.90	71.50	90	6000	7
170	Preheating Fire	1	39	4	96			Llama sp.	Dung			*0.65*	1.08	90	6000	7
171	Communal Hearth	1	38	1	106		Fabaceae	Acacia sp.	Seed	Mature	6		8	7.1	7100	100
171	Communal Hearth	1	38	1	106		Malvaceae	Gossypium sp.	Seed	Mature	1		1	7.1	7100	100
171	Communal Hearth	1	38	1	106		Amaranthaceae	Amaranthus sp.	Seed	Mature	1		1	7.1	7100	100
171	Communal Hearth	1	38	1	106		Fabaceae	Prosopis sp.	Seed	Mature	2		3	7.1	7100	100
171	Communal Hearth	1	38	1	106		Verbenaceae	Lippia sp.	Seed	Mature	1		1	7.1	7100	100
171	Communal Hearth	1	38	1	106			Wood Fungi	Spore		4		6	7.1	7100	100
171	Communal Hearth	1	38	1	106			Cavia sp.	Dung		4		6	7.1	7100	100
171	Communal Hearth	1	38	1	106			Ceramic Redware	Ceramic		1		1	7.1	7100	100
171	Communal Hearth	1	38	1	106			Shell 4mm	Shell		8		11	7.1	7100	100
171	Communal Hearth	1	38	1	106			4mm Charcoal	Stem	Branchwood		3.87	5.45	7.1	7100	100
171	Communal Hearth	1	38	1	106		Capparidaceae	Capparis angulata	Stem	Branchwood		1.52	2.14	7.1	7100	100
171	Communal Hearth	1	38	1	106		Fabaceae	Prosopis sp.	Stem	Branchwood		2.46	3.46	7.1	7100	100
171	Communal Hearth	1	38	1	106		Fabaceae	Acacia sp.	Stem	Branchwood		4.08	5.75	7.1	7100	100
171	Communal Hearth	1	38	1	106			2mm Charcoal	Stem	Branchwood		39.91	56.21	7.1	7100	100
172	Communal Hearth	1	38	1	106		Fabaceae	Acacia sp.	Seed	Mature	5		7	7.1	7100	100
172	Communal Hearth	1	38	1	106		Malvaceae	Gossypium sp.	Seed	Mature	4		6	7.1	7100	100
172	Communal Hearth	1	38	1	106			Wood Fungi	Spore		5		7	7.1	7100	100
172	Communal Hearth	1	38	1	106		Fabaceae	Prosopis sp.	Seed	Mature	12		17	7.1	7100	100
172	Communal Hearth	1	38	1	106		Fabaceae	cf. Fabaceae	Seed	Mature	1		1	7.1	7100	100
172	Communal Hearth	1	38	1	106		Boraginaceae	Heliotropum sp.	Seed	Mature	1		1	7.1	7100	100
172	Communal Hearth	1	38	1	106		Verbenaceae	Lippia sp.	Seed	Mature	12		17	7.1	7100	100
172	Communal Hearth	1	38	1	106		Verbenaceae	Lippia sp.	Fruit	Mature	1		1	7.1	7100	100
172	Communal Hearth	1	38	1	106		Poaceae	Zea mays	Seed	Embryo	1		1	7.1	7100	100
172	Communal Hearth	1	38	1	106		Euphorbiaceae	Euphorbia sp.	Seed	Mature	3		4	7.1	7100	100
172	Communal Hearth	1	38	1	106		Chenopodiaceae	Chenopodium sp.	Seed	Mature	7		10	7.1	7100	100
172	Communal Hearth	1	38	1	106		Oxalidaceae	Oxalis sp.	Seed	Mature	1		1	7.1	7100	100
172	Communal Hearth	1	38	1	106		Fabaceae	Phaseolus sp.	Seed	Mature	2		3	7.1	7100	100
172	Communal Hearth	1	38	1	106			Unid.	Seed	Mature	4		6	7.1	7100	100
172	Communal Hearth	1	38	1	106			Shell 4mm	Shell		196		276	7.1	7100	100
172	Communal Hearth	1	38	1	106			Fish	Bone	Vertebra	1		1	7.1	7100	100
172	Communal Hearth	1	38	1	106			Ceramic Redware	Ceramic		1		1	7.1	7100	100
172	Communal Hearth	1	38	1	106			Cavia sp.	Dung		3		4	7.1	7100	100
172	Communal Hearth	1	38	1	106			4mm Charcoal	Stem	Branchwood		32.09	45.20	7.1	7100	100

172	Communal Hearth	1	38	1	106			2mm Charcoal	Stem	Branchwood		15.25	21.48	7.1	7100	100
172	Communal Hearth	1	38	1	106		Asteraceae	Tessaria integrifolia	Stem	Mature		20.18	28.42	7.1	7100	100
172	Communal Hearth	1	38	1	106		Fabaceae	Acacia sp.	Stem	Branchwood		3.08	4.34	7.1	7100	100
172	Communal Hearth	1	38	1	106			Llama sp.	Dung			*0.65*	0.92	7.1	7100	100
173	Kiln	1	39	4	68	A	Verbenaceae	Lippia sp.	Seed	Mature	33		152	2.175	2175	100
173	Kiln	1	39	4	68	A	Boraginaceae	Heliotropum sp.	Seed	Mature	1		5	2.175	2175	100
173	Kiln	1	39	4	68	A	Solanaceae	cf. Solanaceae	Seed	Mature	5		23	2.175	2175	100
173	Kiln	1	39	4	68	A	Solanaceae	Capsicum sp.	Seed	Mature	2		9	2.175	2175	100
173	Kiln	1	39	4	68	A	Erythoxylaceae	Erythoxylon coca	Seed	Mature	1		5	2.175	2175	100
173	Kiln	1	39	4	68	A	Fabaceae	cf. Fabaceae	Seed	Mature	4		18	2.175	2175	100
173	Kiln	1	39	4	68	A	Fabaceae	Acacia sp.	Seed	Mature	3		14	2.175	2175	100
173	Kiln	1	39	4	68	A	Fabaceae	Prosopis sp.	Seed	Mature	2		9	2.175	2175	100
173	Kiln	1	39	4	68	A	Chenopodiaceae	Chenopodium sp.	Seed	Mature	1		5	2.175	2175	100
173	Kiln	1	39	4	68	A	Poaceae	Zea mays	Seed	Mature	7		32	2.175	2175	100
173	Kiln	1	39	4	68	A	Poaceae	Zea mays	Seed	Embryo	1		5	2.175	2175	100
173	Kiln	1	39	4	68	A		Unid.	Seed	Mature	7		32	2.175	2175	100
173	Kiln	1	39	4	68	A		Cavia sp.	Dung		7		32	2.175	2175	100
173	Kiln	1	39	4	68	A	Fabaceae	cf. Fabaceae	Fruit	Stem	1		5	2.175	2175	100
173	Kiln	1	39	4	68	A		Shell 4mm	Shell		8		37	2.175	2175	100
173	Kiln	1	39	4	68	A		Wood Fungi	Spore		55		253	2.175	2175	100
173	Kiln	1	39	4	68	A		4mm Charcoal	Stem	Branchwood		75.69	348.00	2.175	2175	100
173	Kiln	1	39	4	68	A	Fabaceae	Acacia sp.	Stem	Branchwood		15.98	73.47	2.175	2175	100
173	Kiln	1	39	4	68	A	Fabaceae	Prosopis sp.	Stem	Branchwood		35.45	162.99	2.175	2175	100
173	Kiln	1	39	4	68	A	Capparidaceae	Capparis angulata	Stem	Branchwood		3.45	15.86	2.175	2175	100
173	Kiln	1	39	4	68	A		2mm Charcoal	Stem	Branchwood		7.70	35.40	2.175	2175	100
173	Kiln	1	39	4	68	A		Llama sp.	Dung			*1.22*	5.61	2.175	2175	100
175	Communal Hearth	1	38	1	106	B	Poaceae	Bromus sp.	Seed	Mature	3		9	3.45	3450	100
175	Communal Hearth	1	38	1	106	B	Fabaceae	UKN#64-120.1	Fruit	Mature	6		17	3.45	3450	100
175	Communal Hearth	1	38	1	106	B	Poaceae	Zea mays	Seed	Embryo	2		6	3.45	3450	100
175	Communal Hearth	1	38	1	106	B	Fabaceae	Acacia sp.	Seed	Mature	12		35	3.45	3450	100
175	Communal Hearth	1	38	1	106	B	Fabaceae	cf. Fabaceae	Seed	Mature	11		32	3.45	3450	100
175	Communal Hearth	1	38	1	106	B	Boraginaceae	Heliotropum sp.	Seed	Mature	1		3	3.45	3450	100
175	Communal Hearth	1	38	1	106	B	Verbenaceae	Lippia sp.	Seed	Mature	7		20	3.45	3450	100
175	Communal Hearth	1	38	1	106	B	Euphorbiaceae	Euphorbia sp.	Seed	Mature	1		3	3.45	3450	100
175	Communal Hearth	1	38	1	106	B	Chenopodiaceae	Chenopodium sp.	Seed	Mature	4		12	3.45	3450	100
175	Communal Hearth	1	38	1	106	B	Erythoxylaceae	Erythoxylon coca	Seed	Mature	1		3	3.45	3450	100
175	Communal Hearth	1	38	1	106	B	Solanaceae	Nicotiana sp.	Seed	Mature	11		32	3.45	3450	100
175	Communal Hearth	1	38	1	106	B	Fabaceae	Prosopis sp.	Seed	Mature	9		26	3.45	3450	100
175	Communal Hearth	1	38	1	106	B	Malvaceae	cf. Malvaceae	Seed	Mature	1		3	3.45	3450	100
175	Communal Hearth	1	38	1	106	B	Cyperaceae	Scirpus sp.	Seed	Mature	1		3	3.45	3450	100
175	Communal Hearth	1	38	1	106	B		Shell 4mm	Shell		143		414	3.45	3450	100
175	Communal Hearth	1	38	1	106	B		Slag	Metal	Amorphous	1		3	3.45	3450	100
175	Communal Hearth	1	38	1	106	B	Fabaceae	Acacia sp.	Flower	Mature	3		9	3.45	3450	100
175	Communal Hearth	1	38	1	106	B		Ceramic Redware	Ceramic		2		6	3.45	3450	100
175	Communal Hearth	1	38	1	106	B		Unid.	Fruit	Stem	7		20	3.45	3450	100
175	Communal Hearth	1	38	1	106	B		2mm Charcoal	Stem	Branchwood		11.93	34.58	3.45	3450	100
175	Communal Hearth	1	38	1	106	B		4mm Charcoal	Stem	Branchwood		48.70	141.16	3.45	3450	100
175	Communal Hearth	1	38	1	106	B	Apocynaceae	Vallesia glabra	Stem	Mature		15.65	45.36	3.45	3450	100

175	Communal Hearth	1	38	1	106	B	Asteraceae	Tessaria integrifolia	Stem	Mature		8.45	24.49	3.45	3450	100
175	Communal Hearth	1	38	1	106	B	Fabaceae	Prosopis sp.	Stem	Branchwood		3.18	9.22	3.45	3450	100
175	Communal Hearth	1	38	1	106	B	Fabaceae	Acacia sp.	Stem	Branchwood		0.35	1.01	3.45	3450	100
175	Communal Hearth	1	38	1	106	B	Capparidaceae	Capparis ovalifolia	Stem	Branchwood		1.19	3.45	3.45	3450	100
175	Communal Hearth	1	38	1	106	B	Fabaceae	Fabaceae	Stem	Juvenile		4.56	13.22	3.45	3450	100
175	Communal Hearth	1	38	1	106	B		Llama sp.	Dung			0.28	0.81	3.45	3450	100
176	Communal Hearth	1	38	1	106	A	Poaceae	Zea mays	Seed	Mature	1		3	3.325	3325	100
176	Communal Hearth	1	38	1	106	A	Fabaceae	cf. Fabaceae	Seed	Mature	9		27	3.325	3325	100
176	Communal Hearth	1	38	1	106	A	Fabaceae	Acacia sp.	Seed	Mature	3		9	3.325	3325	100
176	Communal Hearth	1	38	1	106	A	Fabaceae	Prosopis sp.	Seed	Mature	14		42	3.325	3325	100
176	Communal Hearth	1	38	1	106	A		Wood Fungi	Spore		35		105	3.325	3325	100
176	Communal Hearth	1	38	1	106	A	Erythoxylaceae	Erythoxylon coca	Seed	Mature	1		3	3.325	3325	100
176	Communal Hearth	1	38	1	106	A	Solanaceae	cf. Solanaceae	Seed	Mature	1		3	3.325	3325	100
176	Communal Hearth	1	38	1	106	A		Unid.	Seed	Mature	1		3	3.325	3325	100
176	Communal Hearth	1	38	1	106	A		Lithic		Lithic	1		3	3.325	3325	100
176	Communal Hearth	1	38	1	106	A		Shell 4mm	Shell		7		21	3.325	3325	100
176	Communal Hearth	1	38	1	106	A		Mammal	Bone		3		9	3.325	3325	100
176	Communal Hearth	1	38	1	106	A		Unid.	Fruit	Stem	1		3	3.325	3325	100
176	Communal Hearth	1	38	1	106	A		2mm Charcoal	Stem	Branchwood		13.27	39.91	3.325	3325	100
176	Communal Hearth	1	38	1	106	A		4mm Charcoal	Stem	Branchwood		32.66	98.23	3.325	3325	100
176	Communal Hearth	1	38	1	106	A	Fabaceae	Prosopis sp.	Stem	Branchwood		28.57	85.92	3.325	3325	100
176	Communal Hearth	1	38	1	106	A	Capparidaceae	Capparis angulata	Stem	Branchwood		1.44	4.33	3.325	3325	100
176	Communal Hearth	1	38	1	106	A	Capparidaceae	Capparis ovalifolia	Stem	Branchwood		1.65	4.96	3.325	3325	100
176	Communal Hearth	1	38	1	106	A	Fabaceae	Acacia sp.	Stem	Branchwood		2.72	8.18	3.325	3325	100
176	Communal Hearth	1	38	1	106	A	Fabaceae	Fabaceae	Stem	(Desiccated) Juvenile		0.38	1.14	3.325	3325	100
176	Communal Hearth	1	38	1	106	A		Scoria	Metal	Amorphous		1.98	5.95	3.325	3325	100
177	Communal Hearth	1	38	1	106	C	Cucurbitaceae	Cyclanthera sp.	Seed	Mature	1		3	4	4000	100
177	Communal Hearth	1	38	1	106	C	Fabaceae	UKN#12-177	Seed	Mature	2		5	4	4000	100
177	Communal Hearth	1	38	1	106	C	Poaceae	Panicum sp.	Seed	Mature	1		3	4	4000	100
177	Communal Hearth	1	38	1	106	C	Malvaceae	Gossypium sp.	Seed	Mature	2		5	4	4000	100
177	Communal Hearth	1	38	1	106	C	Fabaceae	cf. Fabaceae	Seed	Mature	2		5	4	4000	100
177	Communal Hearth	1	38	1	106	C	Boraginaceae	Heliotropum sp.	Seed	Mature	1		3	4	4000	100
177	Communal Hearth	1	38	1	106	C	Euphorbiaceae	Euphorbia sp.	Seed	Mature	1		3	4	4000	100
177	Communal Hearth	1	38	1	106	C	Chenopodiaceae	Chenopodium sp.	Seed	Mature	1		3	4	4000	100
177	Communal Hearth	1	38	1	106	C	Solanaceae	Nicotiana sp.	Seed	Mature	1		3	4	4000	100
177	Communal Hearth	1	38	1	106	C	Fabaceae	Prosopis sp.	Seed	Mature	24		60	4	4000	100
177	Communal Hearth	1	38	1	106	C		Shell 4mm	Shell		23		58	4	4000	100
177	Communal Hearth	1	38	1	106	C		Unid.	Dung		6		15	4	4000	100
177	Communal Hearth	1	38	1	106	C		4mm Charcoal	Stem	Branchwood		30.46	76.15	4	4000	100
177	Communal Hearth	1	38	1	106	C		2mm Charcoal	Stem	Branchwood		18.00	45.00	4	4000	100
177	Communal Hearth	1	38	1	106	C	Asteraceae	Tessaria integrifolia	Stem	Mature		29.41	73.53	4	4000	100
177	Communal Hearth	1	38	1	106	C	Fabaceae	Prosopis sp.	Stem	Branchwood		2.26	5.65	4	4000	100
177	Communal Hearth	1	38	1	106	C	Fabaceae	Acacia sp.	Stem	Branchwood		7.05	17.63	4	4000	100

APPENDIX J

List of All Distinctions for Recovered Wood

Family	Determination	Item Description
Apocynaceae	*Vallesia glabra*	Mature
Asteraceae	*Tessaria integrifolia*	Desiccated
		Mature
Capparidaceae	*Capparis angulata*	Juvenile
		Mature
		Branchwood
	Capparis ovalifolia	Branchwood
Chenopodiaceae	Wood Type LL	Stem
Fabaceae	*Acacia* sp.	Juvenile
		Mature
		Branchwood
	Fabaceae	(Desiccated) Juvenile
		Juvenile
		Branchwood
	Prosopis sp.	(Desiccated) Chip
		Adze marks
		Juvenile
		Mature
		Branchwood
Poaceae	*Gynerium* sp.	Stem
	Wood Type FF	Stem
	Wood Type KK	Stem
	Zea mays	Stem
Undetermined	2mm Charcoal	Branchwood
	4mm Charcoal	Juvenile
		Branchwood
	Wood Type AA	Mature
		Branchwood
	Wood Type BB	Mature
	Wood Type CC	Branchwood
	Wood Type GG	Branchwood
	Wood Type JJ	Branchwood

APPENDIX K

Mound I, II, and Site-Wide Ubiquity and Diversity Scores

Determination	Family	Organ/ Type	Mound I Taxon Incidence	Mound I Ubiquity (n=38)	Mound I % Total Assemblage X Density (Density/Tot. Density Taxa Mound I)	Mound I % Total Assemblage X Total Taxa Present (Density Rank MI Assemblage)	Mound II Taxon Incidence	Mound II Ubiquity (n=18)	Mound II % Total Assemblage X Density (Density/Tot. Density Taxa Mound II)	Mound II % Total of Assemblage X # of Taxa Present (Density Rank MII Assemblage)	Site-Wide Taxon Incidence	Site-Wide Ubiquity (n=56)	Total % Site-Wide Assemblage X Density	Total % Site-Wide X # Taxa Present (Contribution to Diversity)
Zea mays	**Poaceae**	Seed	38	100.0%	0.9%	6.5%	18	100.0%	4.4%	5.4%	52	92.9%	1.5%	6.0%
2mm Charcoal	**Unid**	Stem	36	94.7%	14.4%	14.3%	18	100.0%	28.5%	14.1%	55	98.2%	20.4%	14.2%
Fabaceae	**Fabaceae**	Stem (Juv.)	17	44.7%	6.7%	6.8%	18	100.0%	5.7%	13.3%	35	62.5%	6.3%	9.1%
Zea mays	**Poaceae**	Fruit	16	42.1%	0.5%	3.1%	18	100.0%	5.3%	5.1%	34	60.7%	1.3%	3.9%
Lippia sp.	**Verbenaceae**	Seed	30	79.0%	75.7%	5.9%	17	94.4%	3.2%	4.8%	47	83.9%	62.9%	5.5%
Prosopis sp.	**Fabaceae**	Stem	38	100.0%	39.9%	15.5%	16	88.9%	21.3%	11.9%	55	98.2%	32.0%	14.2%
Prosopis sp.	**Fabaceae**	Seed	32	84.2%	3.4%	6.3%	15	83.3%	13.4%	4.3%	47	83.9%	5.1%	5.5%
4mm Charcoal	**Unid**	Stem	26	68.4%	23.0%	10.4%	15	83.3%	23.3%	11.1%	41	73.2%	23.2%	10.6%
Unid.	**Unid**	Seed	19	50.0%	0.8%	3.7%	13	72.2%	3.9%	3.7%	32	57.1%	1.3%	3.7%
Cavia sp.	**Caviinae**	Dung	13	34.2%	0.7%	2.6%	11	61.1%	8.6%	3.1%	24	42.9%	2.1%	2.8%
Acacia sp.	**Fabaceae**	Stem	25	65.8%	9.0%	10.0%	11	61.1%	8.1%	8.2%	36	64.3%	8.7%	9.3%
Acacia sp.	**Fabaceae**	Seed	21	55.3%	1.3%	4.1%	10	55.6%	1.6%	2.8%	31	55.4%	1.4%	3.6%
cf. Fabaceae	**Fabaceae**	Seed	26	68.4%	2.4%	5.1%	10	55.6%	1.7%	2.8%	36	64.3%	2.3%	4.2%
Gossypium sp.	**Malvaceae**	Seed	7	18.4%	0.8%	1.4%	10	55.6%	3.6%	2.8%	17	30.4%	1.3%	2.0%
Capparis ovalifolia	**Capparidaceae**	Stem	5	13.2%	0.2%	2.0%	9	50.0%	2.3%	6.7%	14	25.0%	1.1%	3.6%
Prosopis sp.	**Fabaceae**	Leaf	15	39.5%	0.8%	3.0%	9	50.0%	1.8%	2.6%	24	42.9%	0.9%	2.8%
Acacia sp.	**Fabaceae**	Fruit	2	5.3%	0.0%	0.4%	8	44.4%	1.4%	2.3%	10	17.9%	0.3%	1.2%
Fish	**Other**	Bone	7	18.4%	0.3%	1.4%	8	44.4%	1.0%	2.3%	15	26.8%	0.4%	1.7%
Shell 4mm	**Other**	Shell	21	55.3%	3.2%	4.1%	8	44.4%	4.8%	2.3%	29	51.8%	3.5%	3.4%
Capparis angulata	**Capparidaceae**	Stem	15	39.5%	2.0%	6.0%	7	38.9%	1.7%	5.2%	22	39.3%	1.9%	5.7%
Prosopis sp.	**Fabaceae**	Fruit	3	7.9%	0.1%	0.6%	7	38.9%	2.7%	2.0%	10	17.9%	0.5%	1.2%
Capsicum sp.	**Solanaceae**	Seed	5	13.2%	0.1%	1.0%	7	38.9%	0.8%	2.0%	12	21.4%	0.2%	1.4%
cf. Solanaceae	**Solanaceae**	Seed	5	13.2%	0.1%	1.0%	7	38.9%	1.1%	2.0%	12	21.4%	0.3%	1.4%
Nicotiana sp.	**Solanaceae**	Seed	7	18.4%	0.2%	1.4%	7	38.9%	0.5%	2.0%	14	25.0%	0.2%	1.6%
Amaranthus sp.	**Amaranthaceae**	Seed	5	13.2%	0.0%	1.0%	6	33.3%	1.8%	1.7%	11	19.6%	0.3%	1.3%
Chenopodium sp.	**Chenopodiaceae**	Seed	22	57.9%	0.9%	4.3%	6	33.3%	0.8%	1.7%	28	50.0%	0.9%	3.2%
Euphorbia sp.	**Euphorbiaceae**	Seed	12	31.6%	0.1%	2.4%	6	33.3%	1.6%	1.7%	18	32.1%	0.4%	2.1%
cf. Fabaceae	**Fabaceae**	Fruit	13	34.2%	0.2%	2.6%	6	33.3%	1.6%	1.7%	19	33.9%	0.5%	2.2%
Donax sp.	**Other**	Shell	12	31.6%	0.6%	2.4%	6	33.3%	2.8%	1.7%	18	32.1%	1.0%	2.1%
cf. Asteraceae	**Asteraceae**	Seed	2	5.3%	0.0%	0.4%	5	27.8%	0.2%	1.4%	7	12.5%	0.1%	0.8%
Tessaria integrifolia	**Asteraceae**	Stem	16	42.1%	1.4%	6.4%	5	27.8%	0.1%	3.7%	21	37.5%	0.8%	5.4%
Heliotropum sp.	**Boraginaceae**	Seed	21	55.3%	0.9%	4.1%	5	27.8%	0.3%	1.4%	26	46.4%	0.8%	3.0%
Lagenaria sp.	**Cucurbitaceae**	Seed		0.0%	0.0%	0.0%	5	27.8%	1.1%	1.4%	5	8.9%	0.2%	0.6%
Acacia sp.	**Fabaceae**	Flower	7	18.4%	0.2%	1.4%	5	27.8%	0.5%	1.4%	12	21.4%	0.3%	1.4%
Zea mays	**Poaceae**	Stem	4	10.5%	0.1%	1.6%	5	27.8%	0.0%	3.7%	9	16.1%	0.0%	2.3%
Physalis sp.	**Solanaceae**	Seed	1	2.6%	0.0%	0.2%	5	27.8%	0.4%	1.4%	6	10.7%	0.1%	0.7%
Lippia sp.	**Verbenaceae**	Fruit	13	34.2%	1.2%	2.6%	5	27.8%	1.2%	1.4%	18	32.1%	1.2%	2.1%
Lagenaria sp.	**Cucurbitaceae**	Fruit	2	5.3%	0.0%	0.4%	4	22.2%	0.3%	1.1%	6	10.7%	0.1%	0.7%
Phaseolus sp.	**Fabaceae**	Seed	2	5.3%	0.0%	0.4%	4	22.2%	0.4%	1.1%	6	10.7%	0.1%	0.7%
UKN#64-120.1	**Fabaceae**	Fruit	1	2.6%	0.1%	0.2%	4	22.2%	0.6%	1.1%	5	8.9%	0.2%	0.6%

Determination	Family	Organ/ Type	**Mound I Taxon Incidence**	**Mound I Ubiquity** (n=38)	**Mound I % Total Assemblage X Density** (Density/Tot. Density Taxa Mound I)	**Mound I % Total Assemblage X Total Taxa Present** (Density Rank MI Assemblage)	**Mound II Taxon Incidence**	**Mound II Ubiquity** (n=18)	**Mound II % Total Assemblage X Density** (Density/Tot. Density Taxa Mound II)	**Mound II % Total of Assemblage X # of Taxa Present** (Density Rank MII Assemblage)	**Site-Wide Taxon Incidence**	**Site-Wide Ubiquity** (n=56)	**Total % Site-Wide Assemblage X Density**	**Total % Site-Wide X # Taxa Present** (Contribution to Diversity)
Ceramic Redware	**Other**	Ceramic	6	15.8%	0.1%	1.2%	4	22.2%	0.3%	1.1%	10	17.9%	0.1%	1.2%
Ambrosia sp	**Asteraceae**	Seed		0.0%	0.0%	0.0%	3	16.7%	0.4%	0.9%	3	5.4%	0.1%	0.3%
Trianthema sp.	**Azioiaceae**	Seed	3	7.9%	0.0%	0.6%	3	16.7%	0.1%	0.9%	6	10.7%	0.0%	0.7%
Heliotropum sp.	**Boraginaceae**	Fruit		0.0%	0.0%	0.0%	3	16.7%	1.4%	0.9%	3	5.4%	0.2%	0.3%
Lepidium sp.	**Brassicaceae**	Seed	1	2.6%	0.0%	0.2%	3	16.7%	0.1%	0.9%	4	7.1%	0.0%	0.5%
Ipomoea sp.	**Convulvulaceae**	Seed		0.0%	0.0%	0.0%	3	16.7%	0.2%	0.9%	3	5.4%	0.0%	0.3%
Cucurbitaceae	**Cucurbitaceae**	Seed	1	2.6%	0.0%	0.2%	3	16.7%	0.3%	0.9%	4	7.1%	0.1%	0.5%
Cucurbita maxima	**Cucurbitaceae**	Seed		0.0%	0.0%	0.0%	3	16.7%	1.0%	0.9%	3	5.4%	0.2%	0.3%
Cucurbita sp.	**Cucurbitaceae**	Seed	3	7.9%	0.0%	0.6%	3	16.7%	0.6%	0.9%	6	10.7%	0.1%	0.7%
Acacia sp.	**Fabaceae**	Leaf	6	15.8%	0.1%	1.2%	3	16.7%	0.1%	0.9%	9	16.1%	0.1%	1.0%
Cu Alloy	**Other**	Metal	1	2.6%	0.0%	0.2%	3	16.7%	0.1%	0.9%	4	7.1%	0.0%	0.5%
Slag	**Other**	Metal	10	26.3%	0.5%	2.0%	3	16.7%	0.4%	0.9%	13	23.2%	0.5%	1.5%
Wood Fungi	**Other**	Spore	10	26.3%	2.0%	2.0%	3	16.7%	12.1%	0.9%	13	23.2%	3.8%	1.5%
Scoria	**Other**	Metal	2	5.3%	0.1%	0.8%	3	16.7%	0.7%	2.2%	5	8.9%	0.4%	1.3%
cf. Poaceae	**Poaceae**	Seed	2	5.3%	0.0%	0.4%	3	16.7%	0.3%	0.9%	5	8.9%	0.1%	0.6%
Gynerium sp.	**Poaceae**	Stem	2	5.3%	0.0%	0.8%	3	16.7%	1.6%	2.2%	5	8.9%	0.7%	1.3%
Unid.	**Unid**	Fruit	9	23.7%	0.3%	1.8%	3	16.7%	1.3%	0.9%	12	21.4%	0.5%	1.4%
Vallesia glabra	**Apocynaceae**	Stem	4	10.5%	0.4%	1.6%	2	11.1%	0.2%	1.5%	6	10.7%	0.3%	1.6%
cf. Cactaceae	**Cactaceae**	Seed		0.0%	0.0%	0.0%	2	11.1%	0.2%	0.6%	2	3.6%	0.0%	0.2%
Erythoxylon coca	**Erythoxylaceae**	Seed	8	21.1%	0.1%	1.6%	2	11.1%	0.1%	0.6%	10	17.9%	0.1%	1.2%
Neptunia sp.	**Fabaceae**	Seed		0.0%	0.0%	0.0%	2	11.1%	0.9%	0.6%	2	3.6%	0.2%	0.2%
UKN#43-121.1	**Fabaceae**	Fruit		0.0%	0.0%	0.0%	2	11.1%	0.1%	0.6%	2	3.6%	0.0%	0.2%
UKN#43-121.1	**Fabaceae**	Seed		0.0%	0.0%	0.0%	2	11.1%	0.4%	0.6%	2	3.6%	0.1%	0.2%
Portulaca sp.	**Portulacaceae**	Seed	3	7.9%	0.0%	0.6%	2	11.1%	0.4%	0.6%	5	8.9%	0.1%	0.6%
Unid.	**Unid**	Dung	4	10.5%	0.1%	0.8%	2	11.1%	1.0%	0.6%	6	10.7%	0.3%	0.7%
Bidens sp.	**Asteraceae**	Seed		0.0%	0.0%	0.0%	1	5.6%	2.7%	0.3%	1	1.8%	0.5%	0.1%
Taraxacum sp.	**Asteraceae**	Seed	1	2.6%	0.0%	0.2%	1	5.6%	0.0%	0.3%	2	3.6%	0.0%	0.2%
UKN#25-37	**Asteraceae**	Seed		0.0%	0.0%	0.0%	1	5.6%	0.0%	0.3%	1	1.8%	0.0%	0.1%
UKN#50-142	**Asteraceae**	Seed		0.0%	0.0%	0.0%	1	5.6%	0.0%	0.3%	1	1.8%	0.0%	0.1%
Bixa sp.	**Bixaceae**	Seed		0.0%	0.0%	0.0%	1	5.6%	0.0%	0.3%	1	1.8%	0.0%	0.1%
Cordia sp.	**Boraginaceae**	Fruit		0.0%	0.0%	0.0%	1	5.6%	0.0%	0.3%	1	1.8%	0.0%	0.1%
Cordia sp.	**Boraginaceae**	Seed		0.0%	0.0%	0.0%	1	5.6%	0.0%	0.3%	1	1.8%	0.0%	0.1%
Capparis sp.	**Capparidaceae**	Fruit		0.0%	0.0%	0.0%	1	5.6%	0.0%	0.3%	1	1.8%	0.0%	0.1%
UKN#66-166	**Capparidaceae**	Fruit		0.0%	0.0%	0.0%	1	5.6%	0.0%	0.3%	1	1.8%	0.0%	0.1%
Silene sp.	**Caryophyllaceae**	Seed	1	2.6%	0.0%	0.2%	1	5.6%	0.0%	0.3%	2	3.6%	0.0%	0.2%
Cavia sp.	**Caviinae**	Bone		0.0%	0.0%	0.0%	1	5.6%	0.0%	0.3%	1	1.8%	0.0%	0.1%
Suaeda sp.	**Chenopodiaceae**	Seed	3	7.9%	0.0%	0.6%	1	5.6%	0.0%	0.3%	4	7.1%	0.0%	0.5%
Arachis sp.	**Fabaceae**	Seed	2	5.3%	0.0%	0.4%	1	5.6%	0.1%	0.3%	3	5.4%	0.0%	0.3%
Parkinsonia sp.	**Fabaceae**	Seed		0.0%	0.0%	0.0%	1	5.6%	0.0%	0.3%	1	1.8%	0.0%	0.1%
Acacia sp.	**Fabaceae**	Stem		0.0%	0.0%	0.0%	1	5.6%	0.0%	0.7%	1	1.8%	0.0%	0.3%
Persea sp.	**Lauraceae**	Seed		0.0%	0.0%	0.0%	1	5.6%	0.0%	0.3%	1	1.8%	0.0%	0.1%
Malva sp.	**Malvaceae**	Seed	1	2.6%	0.0%	0.2%	1	5.6%	0.0%	0.3%	2	3.6%	0.0%	0.2%

Determination	Family	Organ/ Type	Mound I Taxon Incidence	Mound I Ubiquity (n=38)	Mound I % Total Assemblage X Density (Density/Tot. Density Taxa Mound I)	Mound I % Total Assemblage X Total Taxa Present (Density Rank MI Assemblage)	Mound II Taxon Incidence	Mound II Ubiquity (n=18)	Mound II % Total Assemblage X Density (Density/Tot. Density Taxa Mound II)	Mound II % Total of Assemblage X # of Taxa Present (Density Rank MII Assemblage)	Site-Wide Taxon Incidence	Site-Wide Ubiquity (n=56)	Total % Site-Wide Assemblage X Density	Total % Site-Wide X # Taxa Present (Contribution to Diversity)
cf. Myrtaceae	**Myrtaceae**	Fruit		0.0%	0.0%	0.0%	1	5.6%	0.0%	0.3%	1	1.8%	0.0%	0.1%
Psidium sp.	**Myrtaceae**	Seed	1	2.6%	0.0%	0.2%	1	5.6%	0.0%	0.3%	2	3.6%	0.0%	0.2%
UKN#41-121.1	**Nyctaginaceae**	Fruit		0.0%	0.0%	0.0%	1	5.6%	0.0%	0.3%	1	1.8%	0.0%	0.1%
Ceramic Blackware	**Other**	Ceramic	1	2.6%	0.0%	0.2%	1	5.6%	0.0%	0.3%	2	3.6%	0.0%	0.2%
Cosu Blanco	**Other**	Bone	1	2.6%	0.0%	0.2%	1	5.6%	0.0%	0.3%	2	3.6%	0.0%	0.2%
Lump	**Other**	Amorphous	1	2.6%	0.0%	0.2%	1	5.6%	0.0%	0.3%	2	3.6%	0.0%	0.2%
Mammal	**Other**	Bone	10	26.3%	0.3%	2.0%	1	5.6%	0.1%	0.3%	11	19.6%	0.3%	1.3%
Mica Bead	**Other**	Bead		0.0%	0.0%	0.0%	1	5.6%	0.0%	0.3%	1	1.8%	0.0%	0.1%
Wood Type CC	**Other**	Stem	2	5.3%	0.3%	0.8%	1	5.6%	0.0%	0.7%	3	5.4%	0.2%	0.8%
Wood Type GG	**Other**	Stem		0.0%	0.0%	0.0%	1	5.6%	0.6%	0.7%	1	1.8%	0.3%	0.3%
Piper sp.	**Piperaceae**	Flower		0.0%	0.0%	0.0%	1	5.6%	0.2%	0.3%	1	1.8%	0.0%	0.1%
Andropogon sp.	**Poaceae**	Seed	1	2.6%	0.0%	0.2%	1	5.6%	0.0%	0.3%	2	3.6%	0.0%	0.2%
Bromus sp.	**Poaceae**	Seed	1	2.6%	0.0%	0.2%	1	5.6%	0.1%	0.3%	2	3.6%	0.0%	0.2%
cf. Poaceae	**Poaceae**	Leaf	1	2.6%	0.0%	0.2%	1	5.6%	0.1%	0.3%	2	3.6%	0.0%	0.2%
cf. Poaceae	**Poaceae**	Root	1	2.6%	0.0%	0.2%	1	5.6%	0.0%	0.3%	2	3.6%	0.0%	0.2%
Panicum sp.	**Poaceae**	Seed	2	5.3%	0.0%	0.4%	1	5.6%	0.1%	0.3%	3	5.4%	0.0%	0.3%
Cenchrus sp.	**Poaceae**	Fruit		0.0%	0.0%	0.0%	1	5.6%	0.1%	0.3%	1	1.8%	0.0%	0.1%
Cenchrus sp.	**Poaceae**	Seed		0.0%	0.0%	0.0%	1	5.6%	1.0%	0.3%	1	1.8%	0.2%	0.1%
Gallium sp.	**Rubiaceae**	Seed		0.0%	0.0%	0.0%	1	5.6%	0.0%	0.3%	1	1.8%	0.0%	0.1%
Celtis sp.	**Ulmaceae**	Seed		0.0%	0.0%	0.0%	1	5.6%	0.0%	0.3%	1	1.8%	0.0%	0.1%
Unid.	**Unid**	Flower		0.0%	0.0%	0.0%	1	5.6%	0.0%	0.3%	1	1.8%	0.0%	0.1%
Unid.	**Unid**	Leaf	4	10.5%	0.1%	0.8%	1	5.6%	0.1%	0.3%	5	8.9%	0.1%	0.6%
Unid.	**Unid**	Nut		0.0%	0.0%	0.0%	1	5.6%	0.0%	0.3%	1	1.8%	0.0%	0.1%
Opuntia sp.	**Cactaceae**	Seed	1	2.6%	0.0%	0.2%		0.0%	0.0%	0.0%	1	1.8%	0.0%	0.1%
Capparis angulata	**Capparidaceae**	Flower	3	7.9%	0.0%	0.6%		0.0%	0.0%	0.0%	3	5.4%	0.0%	0.3%
Capparis crotonides	**Capparidaceae**	Seed	1	2.6%	0.1%	0.2%		0.0%	0.0%	0.0%	1	1.8%	0.1%	0.1%
Capparis ovalifolia	**Capparidaceae**	Fruit	1	2.6%	0.1%	0.2%		0.0%	0.0%	0.0%	1	1.8%	0.1%	0.1%
Capparis ovalifolia	**Capparidaceae**	Seed	1	2.6%	0.0%	0.2%		0.0%	0.0%	0.0%	1	1.8%	0.0%	0.1%
Capparis sp.	**Capparidaceae**	Flower	1	2.6%	0.0%	0.2%		0.0%	0.0%	0.0%	1	1.8%	0.0%	0.1%
Capparis sp.	**Capparidaceae**	Seed	3	7.9%	0.1%	0.6%		0.0%	0.0%	0.0%	3	5.4%	0.1%	0.3%
Wood Type LL	**Chenopodiaceae**	Stem	1	2.6%	0.0%	0.4%		0.0%	0.0%	0.0%	1	1.8%	0.0%	0.3%
Cyperus sp.	**Cyperaceae**	Seed	2	5.3%	0.0%	0.4%		0.0%	0.0%	0.0%	2	3.6%	0.0%	0.2%
Scirpus sp.	**Cyperaceae**	Seed	3	7.9%	0.1%	0.6%		0.0%	0.0%	0.0%	3	5.4%	0.0%	0.3%
Caesalpinia sp.	**Fabaceae**	Seed	1	2.6%	0.0%	0.2%		0.0%	0.0%	0.0%	1	1.8%	0.0%	0.1%
UKN#12-177	**Fabaceae**	Seed	2	5.3%	0.0%	0.4%		0.0%	0.0%	0.0%	2	3.6%	0.0%	0.2%
cf. Malvaceae	**Malvaceae**	Seed	2	5.3%	0.0%	0.4%		0.0%	0.0%	0.0%	2	3.6%	0.0%	0.2%
Sida sp.	**Malvaceae**	Seed	1	2.6%	0.0%	0.2%		0.0%	0.0%	0.0%	1	1.8%	0.0%	0.1%
UKN#44-126	**Malvaceae**	Seed	1	2.6%	0.0%	0.2%		0.0%	0.0%	0.0%	1	1.8%	0.0%	0.1%
Lithic	**Other**	(blank)	1	2.6%	0.0%	0.2%		0.0%	0.0%	0.0%	1	1.8%	0.0%	0.1%
Scoria	**Other**	Metal	1	2.6%	0.0%	0.2%		0.0%	0.0%	0.0%	1	1.8%	0.0%	0.1%
UKN#13-177	**Other**	Seed	1	2.6%	0.0%	0.2%		0.0%	0.0%	0.0%	1	1.8%	0.0%	0.1%
UKN#28-45	**Other**	Seed	1	2.6%	0.0%	0.2%		0.0%	0.0%	0.0%	1	1.8%	0.0%	0.1%

Determination	Family	Organ/ Type	**Mound I Taxon Incidence**	**Mound I Ubiquity** (n=38)	**Mound I % Total Assemblage X Density** (Density/Tot. Density Taxa Mound I)	**Mound I % Total Assemblage X Total Taxa Present** (Density Rank MI Assemblage)	**Mound II Taxon Incidence**	**Mound II Ubiquity** (n=18)	**Mound II % Total Assemblage X Density** (Density/Tot. Density Taxa Mound II)	**Mound II % Total of Assemblage X # of Taxa Present** (Density Rank MII Assemblage)	**Site-Wide Taxon Incidence**	**Site-Wide Ubiquity** (n=56)	**Total % Site-Wide Assemblage X Density**	**Total % Site-Wide X # Taxa Present** (Contribution to Diversity)
UKN#53-147	**Other**	Seed	1	2.6%	0.1%	0.2%		0.0%	0.0%	0.0%	1	1.8%	0.0%	0.1%
Burnt Clay	**Other**	Ceramic	3	7.9%	0.2%	1.2%		0.0%	0.0%	0.0%	3	5.4%	0.1%	0.8%
Ceramic Blackware	**Other**	Ceramic	3	7.9%	0.1%	1.2%		0.0%	0.0%	0.0%	3	5.4%	0.0%	0.8%
Ceramic Redware	**Other**	Ceramic	2	5.3%	0.6%	0.8%		0.0%	0.0%	0.0%	2	3.6%	0.3%	0.5%
Donax sp.	**Other**	Shell	1	2.6%	0.0%	0.4%		0.0%	0.0%	0.0%	1	1.8%	0.0%	0.3%
Mammal	**Other**	Dung	2	5.3%	0.0%	0.8%		0.0%	0.0%	0.0%	2	3.6%	0.0%	0.5%
Slag	**Other**	Metal	1	2.6%	0.1%	0.4%		0.0%	0.0%	0.0%	1	1.8%	0.0%	0.3%
Wood Type AA	**Other**	Stem	2	5.3%	0.1%	0.8%		0.0%	0.0%	0.0%	2	3.6%	0.0%	0.5%
Wood Type BB	**Other**	Stem	1	2.6%	0.0%	0.4%		0.0%	0.0%	0.0%	1	1.8%	0.0%	0.3%
Wood Type JJ	**Other**	Stem	1	2.6%	0.0%	0.4%		0.0%	0.0%	0.0%	1	1.8%	0.0%	0.3%
Oxalis sp.	**Oxalidaceae**	Root	1	2.6%	0.0%	0.2%		0.0%	0.0%	0.0%	1	1.8%	0.0%	0.1%
Oxalis sp.	**Oxalidaceae**	Seed	2	5.3%	0.0%	0.4%		0.0%	0.0%	0.0%	2	3.6%	0.0%	0.2%
Plantago sp.	**Plantaginaceae**	Seed	1	2.6%	0.0%	0.2%		0.0%	0.0%	0.0%	1	1.8%	0.0%	0.1%
Wood Type FF	**Poaceae**	Stem	1	2.6%	0.0%	0.4%		0.0%	0.0%	0.0%	1	1.8%	0.0%	0.3%
Wood Type KK	**Poaceae**	Stem	1	2.6%	0.0%	0.4%		0.0%	0.0%	0.0%	1	1.8%	0.0%	0.3%
cf. Portulacaceae	**Portulacaceae**	Seed	2	5.3%	0.0%	0.4%		0.0%	0.0%	0.0%	2	3.6%	0.0%	0.2%
Lippia sp.	**Verbenaceae**	Leaf	1	2.6%	0.0%	0.2%		0.0%	0.0%	0.0%	1	1.8%	0.0%	0.1%
Grand Total			251		100.0%	100.0%	135		100.0%	100.0%	386		100.0%	100.0%

REFERENCES CITED

Acevedo Mallque, Moisés and Yoji Kikata
1994 *Atlas of Peruvian Woods*. Universitaria Nacíonal de Agricultura La Molina, La Molina.

Adegbehin, J. O. and J. E. Omijeh
1994 A Survey of Fuel Wood Consumption and Alternative Sources of Domestic Energy in Kaduna State of Nigeria. *Discovery and Innovation* 6(2):166-75.

Agarwal, Bina
1986 *Cold Hearths and Barren Slopes: The Wood Fuel Crisis in the Third World*. The Riverdale Company, Inc. Publishers, Riverdale.

Allen, Catherine
2002 *The Hold Life Has: Coca and Cultural Identity in an Andean Community*. Smithsonian Institution Press, Washington D.C.

Allsopp, Dennis and Kenneth J. Seal
1986 *Introduction to Biodeterioration*. Edward Arnold Publishers, London.

Arnold, Dean E.
1985 *Ceramic Theory and Cultural Process*. Cambridge University Press, Cambridge.

1993 *Ecology and Ceramic Production in an Andean Community*. Cambridge University Press, New York.

1999 Advantages and Disadvantages of Vertical-Half Molding Technology: Implications for Production and Organization. In *Pottery and People: A Dynamic Interaction*, edited by James M. Skibo and Gary M. Feinman, pp. 59-80. University of Utah Press, Salt Lake City.

Arroyo Baldeon, Antenor J.
1997 *Consideraciones Ecologicas y Socioeconomicas de la Regeneración Natural del Algarrobo (Prosopis pallida) en el Valle de Olmos (Lambayeque-Peru)*. Facultad de Ciencias Forestales, Universidad Nacional Agraria La Molina.

Ashmore, Wendy A. and Wilk, Richard R.
1988 Household and Community in the Mesoamerican Past. In *Household and Community in the Mesoamerican Past*, edited by Wilk, Richard R. and Ashmore, Wendy A., pp. 1-27. University of New Mexico Press, Albuquerque.

Asouti, Eleni
2003 Woodland Vegetation and Fuel Exploitation at the Prehistoric Campsite of Pinarbasi, South-Central Anatolia, Turkey; The Evidence from the Wood Charcoal Macro-Remains. *Journal of Archaeological Science* 1-17.

Baas, Pieter and Regis B. Miller
1985 Functional and Ecological Wood Anatomy Some Introductory Comments. *IAWA Bulletin* 6(4): 281-2.

Baas, Pieter and Ronald E. Vetter
1989 Growth Rings in Tropical Woods. Printed as *IAWA Bulletin n.s.* 10(2):950174. International Association of Wood Anatomists, Leiden.

Baca, Victor F.
1953 El Algarrobo en Lambayeque. *La Vida Agricola* 30(357):657-661.

Baillie, Michael G. L.
1982 *Tree-Ring Dating and Archaeology*. University of Chicago Press, Chicago.

Balée, William and Clark L. Erickson, editors
2006 *Time and Complexity in Historical Ecology: Studies in the Neotropical Lowlands*. Columbia University Press, New York.

Barriga, Cesar, Ivan lares, and Efrain Malpartida
1993 *Manejo de Bosques y Pastos en la Costa Norte*. Universidad Nacíonal Agraria La Molina, Lima.

Basso, Kieth H.
1996 *Wisdom Sits in Places: Landscape and Language among the Western Apache*. University of New Mexico Press, Albuquerque.

Bawden, Garth
1982 Community Organization Reflected by the Household: A Study of Pre-Columbian Social Dynamics. *Journal of Field Archaeology* 9(2):165-181.

Beaumont, Peter
1993 *Drylands: Environmental Management and Development*. Routledge, London.

Bezúr, Aniko
2003 *Variability in Sicán Copper Alloy Artifacts*. PhD. Dissertation, Department of Materials Science and Engineering, University of Arizona.

Bloch, Maurice
1998 Why Trees, Too, Are Good to Think With: Towards an Anthropology of the Meaning of Life. In *The Social Life of Trees: Anthropological Perspectives on Tree Symbolism*, edited by Laura Rival, pp. 39-56. Berg, New York.

Block, Michaela and Michael Richter
2000 The Impact of Heavy Rainfalls on the Vegetation by El Niño 1997/98 in the Sechura

Desert/Northern Peru. *Phytocoenologia* 30, pp. 491-517.

Bogach, V. S.
1985 *Wood as Fuel: Energy for Developing Countries*. Praeger Scientific Press, New York.

Bohrer, Vorsila
1986 Guideposts in Ethnobotany. *Journal of Ethnobiology* 6(1):27-43.

Bourdieu, Pierre
1994 *Distinction: A Social Critique of the Judgment of Taste*. Routledge, New York.

Brack Egg, Antonio
1999 *Diccionario Enciclopedico de Plantas Utiles del Peru*. Centro Estudios Regionales Andinos "Bartolome de las Casas", Cuzco.

Brako, Lois and James L. Zarucchi
1993 *The Catalogue of Flowering Plants and Gymnosperms of Peru*. Missouri Botanical Garden, St. Louis.

Brown, Harrison
1970 Human Food Production as a Process in the Biosphere. In *The Biosphere*, pp. 95-103. W.H. Freeman Press and Company, San Francisco.

Brush, Stephen B.
1992 Ethnoecology, Biodiversity and Modernization in Andean Potato Agriculture. *Journal of Ethnobiology* 12(2):161-185.

Bullock, Stephen H.
1995 Plant Reproduction in Neotropical Dry Forests. In *Seasonally Dry Tropical Forests*, edited by Stephen H. Bullock, Harold A. Mooney, and Ernesto Medina, pp. 277-305. Cambridge University Press, Cambridge.

Bullock, Stephen H., Harold A. Mooney, and Ernesto Medina, editors
1995 *Seasonally Dry Tropical Forests*. Cambridge University Press, Cambridge.

Búrquez, Alberto and María de los Angeles Quintana
1994 Islands of Diversity: Ironwood Ecology and the Richness of Perennials in a Sonoran Desert Biological Reserve. In *Ironwood: An Ecological and Cultural Keystone of the Sonoran Desert*, edited by Gary P. Nabhan and John L. Carr, pp. 9-28. Occasional Papers in Conservation Biology, Occasional Paper No. 1, Conservation International, Washington D.C.

Camino, Lupe
1987 *Chicha de Maíz: Bebida y Vida del Pueblo Catacaos*. Centro de Investigación y Promoción del Campesinado, Piura.

Cavallaro, Raffael and Izumi Shimada
1988 Some Thoughts on Sicán Marked Brick Adobes and Labor Organization. *American Antiquity* 53(1):75-101.

Cayo, Ludwing
2003 *Report on the Use of On-Site Charcoal Production to Create Fuel for Traditional Ag-Pb Smelting, Porco, Bolivia*. Report on File with Dr. Mary Van Buren, Department of Anthropology, Colorado State University, Fort Collins, CO.

Centro de Investigación de Zonas Aridas and Universidad Nacíonal Agraria la Molina
1988 *Ecología y Uso del Algarrobo en el Peru*. UNALM, Lima.

Cleland, Kathryn M. and Izumi Shimada
1992 Sicán Bottles: Marking Time in the Peruvian Bronze Age--A Five-Part Typology and Seriation. Andean Past 3:193-235.

1998 Paleteada Potters: Technology, Production Sphere, and Sub-Culture in Ancient Peru. In *Andean Ceramics: Technology, Organization, and Approaches*, edited by Izumi Shimada, pp. 111-150. Supplement to MASCA Volume 15, 1998, Museum Applied Science Center for Archaeology, University of Pennsylvania Museum for Archaeology and Anthropology, Philadelphia.

Coles, Bryony and John Coles
1986 *Sweet Track to Glastonbury*. Thames and Hudson, London.

Cooper, John M.
1949 Fire Making. In *Handbook of South American Indians*, edited by Julian H. Steward, pp. 238-292. Vol. 5, U.S.G.P.O., Washington D.C.

Correa Álamo, Rosana
2000 Arquitectura Rural en la Costa Norte Túcume: continuidad y mestizaje. *Turismo y Patrimonia* 1(1):69-81.

Costin, Cathy L.
2000 The Use of Ethnoarchaeology for the Archaeological Study of Ceramic Production. *Journal of Archaeological Method and Theory* 7(4):377-403.

Costin, Cathy L. and Rita P. Wright
1998 *Craft and Social Identity*. American Anthropological Association, Washington D.C.

Craig, Alan K. and Izumi Shimada

1986 El Niño Flood Deposits at Batán Grande, Northern Peru. *Geoarchaeology* 1(1):29-38.

Creffield, J.W.
1996 *Wood Destroying Insects: Wood Borers and Termites*. CSIRO Publishing, Collingwood.

Crumley, Carole L., editor
1994 *Historical Ecology: Cultural Knowledge and Changing Landscapes*. SAR Press, Santa Fe.

Cuba Salerno, Amalia B., Agustin Silva Peralta, and Carlos Cornejo Flores
1998 *Bosques Secos y Desertificación*. INRENA, Lima.

Dawson, John and Rob Lucas
2005 *The Nature of Plants: Habitats, Challenges, and Adaptations*. Timber Press, Cambridge.

DeBoer, Warren
1990 Interaction, Imitation, and Communication as Expressed in Style: The Ucayali Experience. In *The Uses of Style in Archaeology*, edited by Margaret Conkey and Christine Hastorf, pp. 82-104. Cambridge University Press, Cambridge.

Delvaud, Claude C.
1984 *Las Regiones Costeñas del Peru Septentrional*. CIPCA/Pontifica Universidad Catolica del Peru, Lima.

DeVries, T. J., L. Ortlieb, A. Diaz, L. Wells, Cl. Hillaire-Marcel, J. S. Noller, D. H. Sandweiss, James B. Richardson III, Elizabeth J. Reitz, Harold B. Rollins, and Kirk A. Maasch
1997 Determining the Early History of El Niño. *Science* 276:965-7.

DeVries, Thomas J.
1987 A Review of Geological Evidence for Ancient El Niño Activity in Peru. *Journal of Geophysical Research* 92(C13):14471-9.

Earle, Timothy
1997 *How Chiefs Come to Power*. Stanford University Press, Stanford.

Elfadl, Mohamed A. and Olavi Luukkanen
2003 Effect of Pruning on Prosopis juliflora: Considerations for Tropical Dryland Agroforestry. *Journal of Arid Environments* 53:441-55.

Erickson, Clark L.
1999 Neo-Environmental Determinism and Agrarian 'Collapse' in Andean Prehistory. *Antiquity* 73:634-42.

FAO/UN
1983 *Wood Fuel Surveys*. United Nations FAO, Rome.

Ferreyra, Ramón
1987 *Estudio Sustemático de los Algarrobos de la Costa Norte del Peru*. Ministerio de Agricultura, Lima.

Folbre, Nancy
1986 Hearts and Spades: Paradigms of Household Economics. *World Development* 14(2):245-55.

Foster, John B.
2000 *Marx's Ecology: Materialism and More*. Monthly Review Press, New York.

Fowler, Catherine S.
1996 Historical Perspectives on Tmbisha Shoshone Land Management Practices, Death Valley, California. In *Case Studies in Environmental Archaeology*, edited by Elizabeth J. Reitz, Lee A. Newsom, and Sylvia J. Scudder, pp. 87-101. Plenum Press, New York.

Fralish, James S.
1997 Community Succession, Diversity, and Disturbance in the Central Hardwood Forest. In *Conservation in Highly Fragmented Landscapes*, edited by Mark W. Schwartz, pp. 234-66. Chapman and Hall, New York.

Fritts, H. C.
1976 *Tree Rings and Climate*. Academic Press, New York.

Gentry, Alwyn H.
1995 Diversity and Floristic Composition of Neotropical Dry Forests. In *Seasonally Dry Tropical Forests*, edited by Stephen H. Bullock, Harold A. Mooney, and Ernesto Medina, pp. 146-190. Cambridge University Press, Cambridge.

1996 *A Field Guide to the Families and Genera of Woody Plants of Northwest South America (Colombia, Ecuador and Peru)*. University of Chicago Press, Chicago.

Gero, Joan M. and Margaret W. Conkey, editors
1991 *Engendering Archaeology*. Blackwell Publishers Ltd, Cambridge.

Gillin, John
1947 *Moche: A Peruvian Coastal Community*. Smithsonian Institution Institute of Social Anthropology Greenwood Press Publishers, Westport.

Goldstein, David J. and Izumi Shimada
In Press Feeding the Fire: Food, Metal, and Ceramic Production in the Middle Sicán. In *From Subsistence to Social Strategies: New Directions in the Study of Daily Meals and Feasting Events*,

edited by Elizabeth Klarich. University of Colorado Press, Boulder.

Goldstein, David J. and Izumi Shimada
2007 Middle Sicán Multi-Craft Production: Resource Management and Labor Organization. In *Craft Production in Complex Societies: Multi-Crafting, Sequential Production*, and Producers, edited by Izumi Shimada, pp. 44-67. University of Utah Press, Salt Lake City.

Graber, David
2001 Chapter 3: Value as the Importance of Actions. In *Toward and Anthropological Theory of Value: The False Coin of Our Own Dreams*, by David Graber, pp. 49-90. Palgrave, New York.

Gummerman IV, George
In Press Big Hearths and Big Pots: Moche Feasting on the North Coast of Peru. In *From Subsistence to Social Strategies: New Directions in the Study of Daily Meals and Feasting Events*, edited by Elizabeth Klarich. University of Colorado Press, Boulder.

Halperin, Rhoda H.
1994 *Cultural Economies Past and Present.* University of Texas Press, Austin.

Harris, David R.
1989 An Evolutionary Continuum of People-Plant Interactions. In *Foraging and Farming: The Evolution of Plant Exploitation*, edited by David R. Harris and Gordon C. Hillman, pp. 11-26. Unwyn Hyman, London.

Hart, Gillian
1992 Imagined Unities: Constructions of 'The Household' in Economic Theory. In *Understanding Economic Process*, edited by Sutti Ortiz and Susan Jees, pp. 111-129. University Press of America, New York.

Hastorf, Christine A. and Melanie F. Wright
1998 Interpreting Wild Seeds from Archaeological Sites: A Dung Charring Experiment from the Andes. *Journal of Ethnobiology* 18(2):211-227.

Helms, Mary W.
1993 *Craft and the Kingly Ideal.* University of Texas Press, Austin.

Henman, Anthony
2005 *Mamacoca (Un Estudio Completo de la Coca).* Juan Gutenberg, Lima.

Herskovits, Melville
1952 *Economic Anthropology; A Study in Comparative Economics.* Knopf Publishing, New York.

Hocquenghem, Anne M.
1998 *Para Vencer la Muerte: Piura y Tumbes, Raíces en el Bosque Seco y en la Selva Alta-Horizontes en el Pacífico y en la Amazonia.* 109 Travaux de l'Institut Français d'Etudes Andines. Institut Français d'Etudes Andines, Lima.

Hocquenghem, Anne M. and Luc Ortlieb
1992 Eventos el Niño y Lluvias Anormales en la Costa del Peru: Siglos XVI-XIX. *Bulletin de l'Institut Français d'Etudes Andines* 21(1):197-278.

Holdridge, Leslie R.
1947 Determination of World Plant Formations from Simple Climatic Data. *Science* 105(2727):367-368.

Holm, Olaf
1966-1967 Money Axe from Ecuador. *Folk* 8-9:135-43.

Holmgren, Milena, Marten Scheffer, Exequiel Ezcurra, Julio R. Guitérrez, and Godefridus M. J. Mohren
2001 El Niño Effects on the Dynamics of Terrestrial Ecosystems. *Trends in Ecology and Evolution* 16(2):89-94.

Howe, Ellen G. and Ulrich Petersen
1993 Silver and Lead in the Late Prehistory of the Mantaro Valley, Peru. In *Archaeometry of Pre-Colombian Sites and Artifacts*, edited by D. A. Scott and Pieter Meyers, pp. 183-198. Getty Conservation Institute, Marina Del Rey.

Huertas, Luis
1999 *La Costa Peruana Vista a Través de Sechura: Espacio, Arte y Tecnología.* Prom Peru, Lima.

Ilic, Jugoslav
1991 *CSIRO Atlas of Hardwoods.* Crawford House Press for CSIRO, East Melbourne.

Ingold, Tim
2000 *The Perception of the Environment: Essays on Livelihood, Dwelling, and Skill.* Routledge, London.

INRENA and Proyecto Algarrobo
1993 *Mapa e Inventorio Forestal de los Bosques Secos de Lambayeque.* INRENA, Chiclayo.

InsideWood
2004 Published on the Internet. *http://insidewood/lib.ncsu.edu/search.* Various Dates of Accession.

Isbell, William H. and G. F. McEwen, editors
1991 *Huari Administrative Structure: Prehistoric Monumental Architecture and State Government.* Dumbarton Oaks, Washington D.C.

Johannessen, Sissel
1988 Plant Remains and Culture Change: Are Paleoethnobotanical Data Better Than We Think? In *Current Paleoethnobotany: Analytical Methods and Cultural Interpretations of Archaeological Plant Remains*, edited by Christine A. Hastorf and Virginia S. Popper, pp. 145-166. University of Chicago Press, Chicago.

Johannessen, Sissel and Christine A. Hastorf
1990 A History of Fuel Management (A.D. 500 to the Present) in the Mantaro Valley, Peru. *Journal of Ethnobiology* 10(1):61-90.

Jones, Martin
1988 The Arable Field: A Botanical Battleground. In *Archaeology and the Flora of the British Isles*, edited by M. Jones, pp. 86-92. Oxford Committee for Archaeology, Oxford.

Jones, Owain and Paul Cloke
2002 *Tree Cultures*. Berg, New York.

Kamp, Kathryn A.
2001 Children Working and Playing: A Southwestern Case Study in Learning Ceramics. *Journal of Anthropological Research*, 57(4):427-450.

Kaulicke, Peter
1993 Evidencia Paleoclimáticas en Asentamientos del Alto Piura Durante el Período Intermedio Temprano. *Bulletin de l'Institut Français d'Etudes Andines* 22(1):283-311.

Keefer, David K., Susan D. de France, Michael E. Moseley, James B. Richardson III, Dennis R. Satterlee, and Amy Day-Lewis
1998 Early Maritime Economy and El Niño Events at Quebrada Tacahuay, Peru. *Science* 281:1833-5.

Klaus, Haagen
2003 *Life and Death at Huaca Sialupe: The Mortuary Archaeology of a Middle Sicán Community, North Coast of Peru.* M.A. Thesis, Southern Illinois University, Carbondale.

Kolata, Alan L., Michael W. Binford, Mark Brenner, John W. Janusek, and Charles Ortloff
2000 Environmental Thresholds and the Empirical Reality of State Collapse: A Response to Erickson (1999). *Antiquity* 74 :424-6.

Kroeber, Alfred L.
1930 *Archaeological Explorations in Peru, Part II: The Northern Coast. 2(2).* Field Museum of Natural History, Anthropology Memoirs, Chicago.

Lechtman, Heather and Arthur Steinberg
1979 The History of Technology: An Anthropological Point of View. In *The History and Philosophy of Technology*, edited by G Bugliarello and D. B. Doner, pp. 135-60. The University of Illinois Press, Champaign.

Lemonnier, Pierre
1986 The Study of Material Culture Today: Toward an Anthropology of Technical Systems. *Journal of Anthropological Archaeology* 5:147-86.

1992 *Elements for an Anthropology of Technology.* 88. Anthropological Papers of the Museum of Anthropology, University of Michigan, Ann Arbor.

Lennstrom, Heidi A. and Christine S. Hastorf
1995 Interpretation in Context: Sampling and Analysis in Paleoethnobotany. *American Antiquity* 60(4):701-21.

Lentz, David L.
1999 Plant Resources of the Ancient Maya: The Paleoethnobotanical Evidence. In *Reconstructing Ancient Maya Diet*, edited by Christine D. White, pp. 3-18. University of Utah Press, Salt Lake City.

Leroi-Gourhan, André
1964 *Le Geste et la Parole: 1. Technique et Langage*. Albin Michel, Paris.

de Lavalle, José Antonio
1989 *Lambayeque.* Banco de Credito del Peru, Lima.

Lopinot, Neal H. and William I. Woods
1993 Wood Overexploitation and the Collapse of Cahokia. In *Foraging and Farming in the Eastern Woodlands*, edited by C. Margaret Scarry, pp. 206-31. University Press of Florida, Gainsville.

Luna-Suárez, S., M. L. Luna-Guido, J. T. Frias-Hernández, V. Olalde-Portugal, and L. Dendooven
1998 Soil Processes as Affected by Replacement of Natural Mesquite Ecosystem with Maize Crop. *Biology of Fertilized Soils* 27:274-8.

Lyon, Patricia
1991 Andean Art and Its Cultural Implications. In *Kodai Andes Bijutsu* (Ancient Art of the Andean World), edited by Shozo Masuda and Izumi Shimada, pp. 27-45. Iwanami Publishers, Tokyo.

Maas, J. Manuel
1995 Conversion of Tropical Dry Forest to Pasture and Agriculture. In *Seasonally Dry Tropical Forests*, edited by Stephen H. Bullock, Harold A. Mooney, and Ernesto Medina, pp. 399-422. Cambridge University Press, Cambridge.

Mabres Torrello, Antonio, Ronald Woodman, and Rosa Zeta

1993 Algunos Apuntes Históricos Adiconales sobre la Cronolgía de El Niño. *Bulletin de l'Institut Français d'Etudes Andines* 22(1):395-406.

Machare, J and L. Ortlieb
1993 Registros del Fenomeno El Nino en el Peru. *Bulletin de l'Institut Français d'Études Andines* 22(1):35-52 .

Maquet, Jacques C.
1979 *Aesthetic Anthropology*. Undena, Malibu.

Masuda, Shozo
1985 Algae Collectors and Lomas. In Andean Ecology and Civilization, edited by Shozo Masuda , Izumi Shimada, and Craig Morris, pp. 233-250. University of Tokyo Press, Tokyo.

Matson, F. R.
1966 Power and Fuel Resources in the Ancient Near East. *Advancement of Science* 23(10):146-53.

1989 Ceramics: The Hub of Ancient Craft. In *Cross-Craft and Cross-Cultural Interactions in Ceramics*, edited by P. E. Mc Govern, M. D. Notis, and W. D. Kingery, pp. 13-29. American Ceramics Society, Westerville, Ohio.

Medina, Ernesto
1995 Diversity of Life Forms of Higher Plants in Neotropical Dry Forests. In *Seasonally Dry Tropical Forests*, edited by Stephen H. Bullock, Harold A. Mooney, and Ernesto Medina, pp. 221-242. Cambridge University Press, Cambridge.

Merkel, John, Izumi Shimada, C. P. Swann, and R. Doonan
1994 Investigation of Prehistoric Copper Production at Batán Grande, Peru: Interpretation of the Analytical Data for Ore Samples. In *Archaeometry of Pre-Columbian Sites and Artifacts*, edited by David A. Scott and P. Meyers, pp. 199-227. The Getty Conservation Institute, Marina Del Rey.

Merkel, John F.
1983 Summary of Experimental Results for Late Bronze Age Copper Smelting and Refining. *MASCA Journal* 2(6):173-8.

Miller, Naomi F.
1985 Paleoethnobotanical Evidence for Deforestation in Ancient Iran: A Case Study of Urban Malyan. *Journal of Ethnobiology* 5(1):1-19.

Miller, Naomi F. and Tristine L. Smart
1984 Intentional Burning of Dung as Fuel: A Mechanism for the Incorporation of Charred Seeds into the Archaeological Record. *Journal of Ethnobiology* 4(1):15-28.

Minnis, Paul E.
1985 *Social Adaptation to Food Stress: A Prehistoric Southwest Example*. University of Chicago Press, Chicago.

Moore, Henrietta L.
1992 Households and Gender Relations: The Modeling of the Economy. In *Understanding Economic Process*, edited by Sutti Ortiz and Susan Jees, pp. 131-148. University Press of America, New York.

Moore, Jerry D.
1989 Pre-Hispanic Beer in Coastal Peru: Technology and Social Context of Prehistoric Production. *American Anthropologist* 91:682-95.

Moorehart, Christopher, David Lentz, and Keith Prufer
2003 Wood of the Gods: The Ritual Use of Pine (Pinus spp.) by the Ancient Lowland Maya. *Latin American Antiquity* 16(3):255-274.

Morris, Craig
1982 The Infrastructure of Inka Control in the Peruvian Central Highlands. In *The Inca and Aztec States 1400-1800: Anthropology and History*, edited by George A. Collier, Renato I. Rosaldo, and John D. Wirth, pp. 153-71. Academic Press, New York.

Moseley, Michael E.
1975 Prehistoric Principles of Labor Organization in the Moche Valley, Peru. *American Antiquity* 40:191-6.

1978 An Empirical Approach to Prehistoric Agrarian Collapse: The Case of the Moche Valley, Peru. In *Social and Technological Management in Dry Lands*, edited by Nancie L. Gonzalez, pp. 9-43. Westview Press, Boulder.

1990 Structure and History in the Dynastic Lore of Chimor. In *The Northern Dynasties: Kingship and Statecraft in Chimor*, edited by Michael E. Moseley and A. Cordy-Collins, pp. 1-39. Dumbarton Oaks, Washington D.C.

Moseley, Michael E., Robert A. Feldman, and Charles R. Ortloff
1981 Living with Crises: Human Perception of Process and Time. In *Biotic Crises in Biological and Evolutionary Times*, edited by M. Nitecki, pp. 231-67. Academic Press, New York.

Moutarde, Fanny
2006 *L'Évolution du Couvert Ligneux et de son Exploitation par l'Homme dans la Vallée du Lurin (Côte Centrale du Pérou), de l'Horizon Ancien (900-100 av. J.-C.) á l'Horizon Tardif (1460-1532 apr. J.-C.)- Approche Anthracologique*. Ph. D.

Dissertation, Faculty of Archaeology, University of Paris I, Panthéon-Sorbonne.

Narvaez Vargas, Alfredo
2001 *Dioses, Encantos, y Gentiles: Introducción al Estudio Tradición Oral Lambayecana.* Impressiones del Castillo, S.A., Chiclayo.

Nash, Stephen E.
1999 *Time, Trees, and Prehistory: Tree-Ring Dating and the Development of North American Archaeology, 1914-1950.* University of Utah Press, Salt Lake City.

Netherly, Patricia J.
1977 *Local Level Lords on the North Coast of Peru.* Ph.D. Dissertation, Department of Anthropology, Cornell University.
1984 The Management of Late Andean Irrigation Systems on the North Coast of Peru. American Antiquity 49(2):227-254.

Newsom, Lee A.
1993 Plants and People: Cultural, Biological, and Ecological Responses to Wood Exploitation. In *Foraging and Farming in the Eastern Woodlands,* edited by C. Margaret Scarry, pp. 115-137. University of Florida Press, Gainsville.

Nordt, Lee, Frances Hayashida, Tom Hallmark, and Corey Crawford
2004 Late Prehistoric Soil Fertility, Irrigation Management, and Agricultural Production in Northwest Coastal Peru. *Geoarchaeology* 19(1):21-46.

Ocampo, Miguel
1991 *Metodologia para la Determinación de la Edad y las Tasas de Crecimiento para las Especias Forestales de los Bosques Secos de la Costa Norte del Peru.* Proyecto Algarrobo con Fundación para el Desarrollo Nacional, Chiclayo.

Ortloff, Charles R. and Alan L. Kolata
1993 Climate and Collapse: Agro-Ecological Perspectives on the Decline of the Tiwanaku State. *Journal of Archaeological Science* 20:195-221.

Ortloff, Charles R., Michael E. Moseley, and Robert A. Feldman
1983 The Chicama-Moche Intervalley Canal: Social Explanations and Physical Paradigms. *American Antiquity* 48(2):375-89.

Panshin, A. J. and Carl de Zeeuw
1980 *Textbook of Wood Technology.* McGraw-Hill, Inc., New York.

Pasiecznik, N.
2002 *Exploiting the Value of Prosopis for Dryland Forestry and Agroforestry Systems.* HDRA Brief, Coventry.

Pasiecznik, N. M., P. Felker, P. J. C. Harris, L. N. Harsh, G. Cruz, J. C. Tewari, K. Cadoret, and L. J. Maldonado
2001 *The Prosopis juliflora- Prosopis pallida Complex: A Monograph.* HDRA, Coventry.

Pastor Rozas, Marcos
1993 *Diagnostico Actual de Los Bosques de Batan Grande.* Internal Report for the Foundation for National Development, Proyecto Algarrobo, Chiclayo, Lambayeque, Peru.

Pearsall, Deborah M.
1983 Evaluating the Stability of Subsistence Strategies by Use of Paleoethnobotanical Data. *Journal of Ethnobiology* 3(2):121-37.

2001 *Paleoethnobotany: A Handbook of Procedures.* Academic Press, New York.

Polanyi, Karl
1944 *The Great Transformation.* Holt, Rheinhart and Winston, New York.

Popper, Virginia S.
1988 Quantitative Measurements in Paleoethnobotany. In *Current Paleoethnobotany,* edited by Christine A. Hastorf and Virginia S. Popper, pp. 53-71. University of Chicago Press, Chicago.

Proyecto Algarrobo
1997 *Inventorio Forestal de los Bosques Secos de Batán Grande.* Universidad Nacíonal Agraria la Molina, Lima.

Pyne, S. J.
1998 Forged in Fire: History, Land, and Anthropogenic Fire. In *Advances in Historical Ecology,* edited by W. Balee, pp. 64-103. Colombia University Press, New York.

Ramírez, Susan E.
1996 *The World Turned Upside Down: Cross Cultural Contact and Conflict in Sixteenth- Century Peru.* Stanford University Press, Stanford .

Raven, Peter H., Ray F. Every, and Susan E. Eichhorn
1999 *Biology of Plants.* W.H. Freeman Publishers, New York.

Raymond, J. S., Augusto Oyuela-Caycedo, and Patrick H. Carmichael
1998 The Earliest Ceramic Technologies of the Northern Andes: A Comparative Analysis. In *Andean Ceramics: Technology, Organization, and*

Approaches, edited by Izumi Shimada, pp. 153-172. Supplement to MASCA Volume 15, 1998, Museum Applied Science Center for Archaeology, University of Pennsylvania Museum for Archaeology and Anthropology, Philadelphia.

Record, Samuel J.
1942-8 Keys to Tropical American Woods. *Topical Woods*. Various Issues.

Record, Samuel J. and Robert W. Hess
1972 *Timbers of the New World.* Arno Press, New York.

Reed, Richard K.
1995 *Prophets of Agroforestry: Guaraní Communities and Commercial Gathering.* University of Texas Press, Austin.

Reiss, W. and Stübel, A.
1880-1987 *Peruvian Antiquities: The Necropolis of Ancon in Peru; A Series of Illustrations of the Civilization and Industry of the Empire of the Incas.* Translated by A.B. Keane, New York, 8 parts.

Risco, Jorge J.
1998 Tecnologías Locales Mejoradas en el Uso de los Rescuros del Bosque Seco en un Contexto Familiar: Cocinas Mejoradas y Algarroberas. In *Bosques Secos y Desertificacion*, edited by Amalia B. Cuba Salerno, Agustin Silva Peralta, and Carlos Cornejo Flores, pp. 129-138. INRENA, Lima.

Rival, Laura, editor
1998 *The Social Life of Trees*. Berg, Oxford.

Rollins, Harold B., James B. Richardson III, and Daniel H. Sandweiss
1986 The Birth of El Niño: Geoarchaeological Evidence and Implications. *Geoarchaeology* 1(1):3-15.

Rospigliosi, Christina
2007 *Ceramic Production and State Control: A View from a Hinterland Middle Sicán Ceramic Workshop.* Paper presented at 35th Annual Meeting of the Midwestern Conference of Andean Archaeology, Anthropology, and Ethnology, Southern Illinois University Carbondale.

Rostworowski de Diez Canseco, Maria
1981 *Recursos Naturales Renovables y Pesca Siglos XVI y XVII.* Instituto de Estudios Peruanos, Lima.

1989 *Costa Peruana Prehispánica.* Instituto de Estudios Peruanos, Lima.

Russell, Glenn S., Banks L. Leonard, and Jesus Briceño
1998 The Cerro Mayal Workshop: Addressing Issues of Craft Specialization in Moche Society. In *Andean Ceramics: Technology, Organization, and Approaches*, edited by Izumi Shimada, pp. 63-90. Supplement to MASCA Volume 15, 1998, Museum Applied Science Center for Archaeology, University of Pennsylvania Museum for Archaeology and Anthropology, Philadelphia.

Sagastegui Alva, Abundio
1973 *Manual de las Malezas de la Costa Norperuana.* Universidad Nacional de Trujillo, Trujillo.

Sahlins, Marshall D.
1972 *Stone Age Economics.* Aldine, Chicago.

Santos-Granero, Fernando
1998 Writing History into the Landscape: Space, Myth, and Ritual in Contemporary Amazonia. *American Ethnologist* 25(2):128-148.

Schaedel, Richard P.
1987 2000 Años de la Continuidad Cultural de los Muchik en la Costa Norte del Peru. *Ibero-Americankisches Archiv* 13(1):117-28.

1988 *La Etnografia Muchik en las Fotografias de H. Bruning: 1886-1925.* Ediciones COFIDE, Lima.

1992 Paleotechnology and Flood Control on the Peruvian North Coast. In *Paleo ENSO Records, International Symposium: Extended Abstracts*, edited by L. Ortlieb and J. Macharé, pp. 285-6. CONCYTEC and ORSTOM, Lima.

Schmidt, Peter R.
1994 Historical Ecology and Landscape Transformation in Eastern Equatorial Africa. In *Historical Ecology: Cultural Knowledge and Changing Landscapes*, edited by Carole L. Crumley, pp. 99-126. SAR Press, Santa Fe.

1997 *Iron Technology in East Africa: Symbolism, Science, and Archaeology.* Indiana University Press, Bloomington.

Schwartz, Mark W.
1997 *Conservation in Highly Fragmented Landscapes.* Chapman and Hall, New York.

Schweingruber, Fritz H.
1996 Tree Rings and Environment: Dendroecology. Paul Haupt Publishers, Berne.

Scott, James C.
1998 *Seeing Like a State.* Yale University Press, New Haven.

Service, Elman

1962 *Primitive Social Organization: An Evolutionary Perspective.* Random House, New York.

Sherbondy, Jeanette E.
1986 *Mallki: Ancestros y Cultivo de Árboles en los Andes.* FAO-Holanda, Lima.

Shimada, Izumi
1981 The Batán Grande-La Leche Archaeological Project. *Journal of Field Archaeology* 8:405-446.

1985 Perception, Procurement, and Management of Resources: Archaeological Perspectives. In *Andean Ecology and Civilization*, edited by Shozo Masuda , Izumi Shimada, and Craig Morris, pp. 357-399. University of Tokyo Press, Tokyo.

1986 Batán Grande and Cosmological Unity in the Prehistoric Central Andes. In Andean Archaeology: Papers in Memory of Clifford Evans, edited by R. Matos, S. Turpin, and H. Eling, pp. 163-88. Institute of Archaeology, University of California, Los Angeles.

1990 Cultural Continuities and Discontinuities on the Northern Coast of Peru, Middle-Late Horizons. In The Northern Dynasties: Kingship and Statecraft in Chimor, edited by Michael E. Moseley and A. Cordy-Collins, pp. 297-391. Dumbarton Oaks, Washington D.C.

1994a Pampa Grande and the Mochica Culture. University of Texas Press, Austin.

1994b Pre-Hispanic Metallurgy and Mining in the Andes: Advances and Future Tasks. In In Quest of Mineral Wealth: Aboriginal and Colonial Mining and Metallurgy, edited by Alan K. Craig and Robert C. West, pp. 37-73. Louisiana State University Press, Baton Rouge.

1995 *Cultura Sicán: Dios, Riqueza y Poder en la Costa Norte del Peru.* Fundacion del Banco Continental para el Fomento del la Educación y la Cultura, Edubanco, Lima.

1997a *Organizational Responses to Severe Drought: Case Studies from the North Coast of Peru.* Paper presented at th 63rd Annual Meeting of the Society for American Archaeology, Seattle, Washington, March 23-26.

1997b Organizational Significance of Marked Bricks and Associated Construction Features on the North Peruvian Coast. In *Archaeologica Peruana 2: Prehispanic Architecture and Civilization in the Andes*, edited by Elizabeth Bonnier and Henning Bischoff, pp. 62-89. Sociedad Arqueológica Peruano-Alemana, Mannheim.

1997c The Variability and Evolution of Prehispanic Kilns on the Peruvian Coast. In *Prehistory and History of Ceramic Kilns*, edited by Prudence M. Rice and W. David Kingery, pp. 103-127. American Ceramic Society, Columbus.

2000 *Report of the 1999 Field Season of the Sicán Archaeological Project: Excavations at Huaca Sialupe, Lower La Leche Valley.* Manuscript on File with National Institute of Culture, Lima, Peru.

2007 Introduction. In *Craft Production in Complex Societies: Multi-Crafting, Sequential Production, and Producers*, edited by Izumi Shimada, pp. 1-21. University of Utah Press, Salt Lake City.

Shimada, Izumi, Ken B. Anderson, Herbert Haas, and Jean H. Langenheim
1997 Amber from 1000-year Old Prehispanic Tombs in Ancient Peru. In *Materials Issues in Art and Archaeology V*, edited by Pamela B. Vandiver, James R. Druzik, John F. Merkel, and John Stewart. Vol. 462, Materials Research Society, Pittsburgh.

Shimada, Izumi, Victor Chang, Ursel Wagner, Rupert Gebhart, Hector Neff, Michael Glascock, and David Killick
1998 Formative Ceramic Kilns and Production in Batán Grande, North Coast of Peru. *In Andean Ceramics: Technology, Organization, and Approaches*, edited by Izumi Shimada, pp. 23-62. Supplement to MASCA Volume 15, 1998, Museum Applied Science Center for Archaeology, University of Pennsylvania Museum for Archaeology and Anthropology, Philadelphia.

Shimada, Izumi, Stephen Epstein, and Alan K. Craig
1982 Batán Grande: A Prehistoric Metallurgical Center in Peru. *Science* 216:952-9.

Shimada, Izumi, David J. Goldstein, Ursel Wagner, and Aniko Bezúr
In press Pre-Hispanic Sicán Furnaces and Metalworking: Toward a Holistic Understanding. *Bulletin de l'Institut Français d'Etudes Andines.*

Shimada, Izumi, David J. Goldstein, Werner Haüsler, José Sosa, and Ursel Wagner
2003 Early Pottery Making in Northern Coastal Peru: Part II: Field Firing Experiments. *Hyperfine Interactions* 150(1-4):91-105.

Shimada, Izumi, Adon Gordus, Jo A. Griffin, and John F. Merkel
1999 Sicán Alloying, Working, and Use of Precious Metals: An Interdisciplinary Perspective. In *Metals in Antiquity*, edited by Suzanne M. M. Young, A. M. Pollard, Paul Budd, and Robert A. Ixer. Vol. 792, BAR International Series, Oxford.

Shimada, Izumi and John F. Merkel
1991 Copper Alloy Metallurgy in Ancient Peru. *Scientific American* 265(1):80-6.

Shimada, Izumi and Jorge Antonio Montenegro
2002 *Second Season (2001) of Excavations in the Middle Sicán Workshop of Huaca Sialupe, Lower La Leche Valley*. Manuscript on File with National Institute of Culture, Lima, Peru.

Shimada, Izumi, Ken-Ichi Shimoda, Julie Farnham, Robert Corruccini, and Hirokatsu Watananbe
2004 An Integrated Analysis of Pre-Hispanic Mortuary Practices: A Middle Sicán Case Study. *Current Anthropology* 45(3):369-402.

Shimada, Izumi and Ursel Wagner
2001 Peruvian Black Pottery Production and Metalworking: A Middle Sican Craft Workshop at Huaca Sialupe. *MASCA Journal* 26(1):25-30.

2007 A Holistic Approach to Pre-Hispanic Craft Production. In *Archaeological Anthropology: Perspectives on Method and Theory*, edited by James M. Skibo, Michael W. Graves, and Miriam T. Stark, pp. 163-197. University of Arizona Press, Tucson.

Shimada, Melody and Izumi Shimada
1985 Prehistoric Llama Breeding and herding on the North Coast of Peru. *American Antiquity* 50(1):3-26.

Sillar, B.
2000 Dung by Preference: The Choice of Fuel as an Example of How Andean Pottery Production is Embedded within Wider Technical, Social, and Economic Practices. *Archaeometry* 42(1):43-60.

Smart, Tristine L. and Ellen S. Hoffman
1988 Environmental Interpretation of Archaeological Charcoal. In *Current Paleoethnobotany: Analytical Methods and Cultural Interpretations of Archaeological Plant Remains*, edited by Christine A. Hastorf and Virginia S. Popper, pp. 167-205. University of Chicago Press, Chicago.

Smith, Bruce D.
2007 The Ultimate Ecosystem Engineers. *Science* 315:1797-8.

Smith, Cyril S.
1965 Materials and the Development of Civilization and Science. *Science* 148:908-917.

Stahle, David W.
1996 The Hydroclimatic Application of Tree-Ring Chronologies. In *Tree Rings, Environment, and Humanity*, edited by J. S. Dean, D. M. Meko, and T. W. Swetnam, pp. 119-126. Radiocarbon.

Taylor, Sara
2002 *Artisan Autonomy in the Middle Sicán State: Variability in Mold-Made Ceramic Production.* M.A. Thesis, Southern Illinois University, Department of Anthropology.

Thompson, G. B.
1994 Wood Charcoals from Tropical Sites: A Contribution to Methodology and Interpretation. In *Tropical Archaeobotany*, edited by Jon G. Hather, pp. 9-33. Routledge, New York.

Tosi Jr., Joseph A.
1960 *Zonas de Vida Natural en el Peru: Memoria Explicativa sobre el Mapa Ecologico del Peru.* Instituto Interamericano de Ciencias Agricolas de OEA.

Towle, Margaret A.
1961 *The Ethnobotany of Pre-Colombian Peru.* Aldine, Chicago.

Tschauner, H., M. Vetters-Tschauner, J. Dulanto, M. Saco, and C. Wester
1994 Un Taller Alfarero Chimu en el Valle de Lambayeque. In *Technologia y organizacion de la Producción de Ceramica Prehistorica*, edited by Izumi Shimada, pp. 349-393. Ponitifica Universidad Catolica del Peru, Fondo Editorial, Lima.

Tucker, Robert C.
1978 *The Marx and Engels Reader*. Second ed. W.W. Norton.

Uceda, Santiago and Jose Armas
1998 An Urban Pottery Workshop at the Site of Moche, North Coast of Peru. In *Andean Ceramics: Technology, Organization, and Approaches*, edited by Izumi Shimada, pp. 91-110. Supplement to MASCA Volume 15, 1998, Museum Applied Science Center for Archaeology, University of Pennsylvania Museum for Archaeology and Anthropology, Philadelphia.

Upton, Martin
1996 *The Economics of Tropical Farming Systems.* Cambridge University Press, Cambridge.

Vreeland, James M.
1986 Una Perspectiva Antropologica de la Paleotecnologia en el Desarrollo Agrario del Norte de Peru. *America Indigena* 46(2):275-318.

Wagner, Ursel, Rupert Gebhart, Enver Murad, Josef Riederer, Izumi Shimada, Cornelius Ulbert, and Fritz E. Wagner
1998 Production of Formative Ceramics: Assessment by Physical Methods. In *Andean Ceramics: Technology, Organization, and Approaches*, edited by Izumi Shimada, pp. 173-198.

Supplement to MASCA Volume 15, 1998, Museum Applied Science Center for Archaeology, University of Pennsylvania Museum for Archaeology and Anthropology, Philadelphia.

Wagner, Ursel, Izumi Shimada, David J. Goldstein, and Werner Haüsler
2001 Sicán Kilns and Furnaces: Field Firings and Archaeometric Studies. *Hyperfine Interaction* C(1-4).

Weberbauer, Augustus
1945 *El Mundo Vegetal de los Andes Peruanos: Estudio Fitogeográfico*. Ministerio de Agricultura, Lima.

Weir, Glendon H., Robert A. Benfer, and John G. Jones
1988 Preceramic to Early Formative Subsistence on the Central Coast. In *Economic Prehistory of the Central Andes*, edited by Elizabeth S. Wing and Jane C. Wheeler, pp. 56-92. Vol. 427, BAR International Series, Oxford.

Weir, Glendon H. and Herbert H. Eling Jr.
1986 Pollen Evidence for Economic Plant Utilization in Prehistoric Agricultural Fields of the Jequetepeque Valley, Northern Peru. In *Andean Archaeology: Papers in Memory of Clifford Evans*, edited by R. Matos, S. Turpin, and H. Eling, pp. 150-162. Institute of Archaeology, University of California, Los Angeles.

Weismantel, Mary J.
1998 *Food, Gender, and Poverty in the Ecuadorian Andes*. Waveland Press, Walnut Creek.

Wells, Lisa E. and Jay S. Noller
1999 Holocene Co-evolution of the Physical Landscape and Human Settlement in Northern Coastal Peru. *Geoarchaeology* 14(8):755-89.

Wertime, Theodore and Steven F. Wertime
1982 *Early Pyrotechnology: The Evolution of the First Fire Using Industries*. Smithsonian Institution Press, Washington D.C.

Wertime, Theodore A.
1964 Man's First Encounters with Metallurgy. *Science* 164:1257-1267.

Wheeler, Elisabeth A. and Pieter Baas
1998 Wood Identification-A Review. *IAWA Journal* 19(3):241-264.

Wheeler, Elizabeth A., Pieter Baas, and Peter E. Gasson
1989 *IAWA List of Microscopic Features for Hardwood Identification*. International Association of Wood Anatomists, Leiden.

Whitmore, Timothy C.
1998 *Introduction to Tropical Rain Forests*. Oxford University Press, New York.

Winterhalder, Bruce, Robert Larsen, and R. B. Thomas
1974 Dung as an Essential Fuel Resource in a Highland Peruvian Community. *Human Ecology* 2(2):89-104.

Wolf, Eric R.
1997 *Europe and the People Without History*. University of California Press, Los Angeles.

Zevallos Pollito, Percy A.
1986 *Caracterización Dendrológoca de 30 Especies Forestales de Lambayeque*. Facultad de Ciencias Forestales, Universidad Nacional Agraria La Molina.

ACKNOWLEDGEMENTS

There are many inspirational characters behind the development of my career and the development of this project. Foremost is Dr. Lee Newsom who allowed me to convince her to take me on. Through folly and disappointment, hard work and demonstration, I have tried to build this work based on her ideas, and most importantly those of her mentors William Stern and Elizabeth Wing. This dissertation is a humble attempt at synthesizing the social and ecological to reconstruct part of the past that many take for granted and very few, myself included, understand in its true complexity. For me the directions of Dr. Newsom and her mentors present a visionary way of interpreting the past in light of the planet that we have all shared in past and present.

Second is Dr. Izumi Shimada, who opened his Peruvian world to me without reservation, permitting me to hit the ground running with excellent support and, moreover, friends in a country where I once did not speak the language. I was given not only a project, but also a dataset that I could freely take charge of and construct my own interpretations as I developed methods new to both of us while participating in his project. Most importantly, I have learned from Dr. Shimada how to 'be' in Peru professionally, making sure that the legacy of Peruvian, as in by and for Peruvians, research in archaeology and anthropology is never neglected in the work that we do as guests in their country.

Third is Jorge Montenegro, a friend, colleague, and inspiration. I have learned from him what it is to truly be dedicated to the ethics and moral of a discipline, especially when it means that one must live as an alien for more than a decade. He has offered me his friendship, family, and, on more than one occasion his bed, all are exchanges of kindness that I can never fully reciprocate. I do, however, make an attempt to do so, with this dissertation, by demonstrating that the approach and methods employed here and in future research offer much to Peruvian archaeology. Without his encouragement and insistence, I would never have started working in Peru, met my wife, or found a calling. Hence, the completion of this research from certain angles has more to do with his family and ideas, than my own.

Fourth, I thank Don Julio Navarro and his family. Don Julio opened a whole new dimension, literally, for me that included not only seeing and knowing plants, but also talking to them. The entire reference set for the analysis and the interpretive dataset would never have achieved the level of depth without his participation in this project. Don Julio brought my work to a level of relevance in the modern world that is due to his direct participation and benediction of the work that I present here. His friendship demonstrates to me that friends, not informants, make great fieldwork and life worth living.

Caught up within these four people are four worlds of my academic life that are made whole with their aid and participation in this project. Additionally, there are others to be named and have played a significant supporting role in my intellectual development. Among this cast of characters are an august group of informal editors and counselors: David Driver, Claudia Krugavoy, William Whitehead, Patrick R. Willams, D. Ann Treiu, Jon Hageman, Jason Gonzales, William N. Duncan, Carolyn M. Goldstein, Phebe Goldstein, Jon Haber, Charles Elmes, Tina Ray, Ursel Wagner, Chip Perkins, and John Duffy.

Additionally, there is an amazing world of people who have contributed to parts of my intellectual growth throughout this project in Peru, the United States, and elsewhere that I name here: Emily Stovel, Fanny Moutarde, Luis Huaman, Lizette Muñoz, Natalia Guzman, Victor Curay, Aline Magnoni, Mike Haller, Yenny Ihue, Familia Quispe-Valencia, Megan E. Reed and Family, Elizabeth Brumfiel, Christine Hastorf, Mike Dillon, Ann Underhill, Don Rice, Steven Kraft, Jon Hill, Susan Ford, Don Ugent, Melody Shimada, Carlos Elera, Ismael Llauce, Juan Carlos Santoyo, Mary Weismantel, Mary Van Buren, Peter Coleman, Ralph Murray, Susan de France, Sofia Chacaltana, Regis Miller, Tina Lerner, Carlos Reynel, A. Dasa, Jose Sosa, Aniko Bezúr, Sarah Taylor, Haagen Klaus, Julie Farnham, Gene Gutenberg, Liz Klarich, Judith K. Brown, Alexandre Chevallier, Matthew Sayre, Frances Hayashida, Luis Huaman, Anibal Calderon, Elaine Edelman and Family, Ashley Leasure, and Felice Gianluca Sperone, Charlie Cobb, Terri Price, Susan Lowe, and Susan Davis..

Financial support for the various stages of work was generously contributed by Dr. and Mrs. Sidney Goldstein and the sponsors for Dr. Izumi Shimada's research, the H. John Heinz III Fund for Latin American Studies, Committee for Research and Exploration of the National Geographic Society, and the Wenner-Gren Foundation for Anthropological Research. The Graduate School of Southern Illinois University at Carbondale and National Science Foundation Dissertation Improvement Grant #0211364 sponsored my research and write up. All research undertaken in Peru was done under permit to either Dr. Izumi Shimada through the National Institute of Culture, Ministry of Education, Peru, or myself through the National Institute for Natural Resources, Ministry of

Agriculture. The National Sicán Museum, Ferreñafe, Peru, the Contisuyo Museum, Moquegua, Peru, the Field Museum of Natural History and Northeastern Illinois University, both located in Chicago, Illinois, and The South Carolina Institute of Archaeology and Anthropology, Columbia, South Carolina, all offered logistical support throughout the analysis phase of this project.

Finally, without my wife Robin, my best friend, consultant, inspiration, and dearest confidant, this research would never have ended.

While I owe a great debt to my mentors and friends, all errors, oversight, missing or incomplete information is a reflection of my own negligence.

My work is dedicated to my grandfather David, my grandmother Helen, my father Sidney and my wife Robin, as it takes three generations of support to make science happen.

www.ingramcontent.com/pod-product-compliance
Lightning Source LLC
La Vergne TN
LVHW071209210926
848482LV00015B/759